MODELING IN MATERIALS PROCESSING

Mathematical modeling and computer simulation have been widely embraced in industry as useful tools for improving materials processing. Although courses in materials processing have covered modeling, they have traditionally been devoted to one particular class of materials, that is, polymers, metals, or ceramics. This text offers a new approach, presenting an integrated treatment of metallic and nonmetallic materials. The authors show that a common base of knowledge – specifically, the fundamentals of heat transfer and fluid mechanics – provides a unifying theme for these seemingly disparate areas. Emphasis is placed on understanding basic physical phenomena and knowing how to include them in a model. Thus, chapters explain how to decide which physical phenomena are important in specific applications, and how to develop analytical models. A unique feature is the use of scaling analysis as a rational way to simplify the general governing equations for each individual process. The book also treats selected numerical methods, showing the relationship among the physical system, analytical solution, and the numerical scheme. A wealth of practical, realistic examples are provided, as well as homework exercises. Students, and practicing engineers who must deal with a wide variety of materials and processing problems, will benefit from the unified treatment presented in this book.

Jonathan A. Dantzig is Professor of Mechanical Engineering, Department of Mechanical and Industrial Engineering, University of Illinois at Urbana-Champaign. His research focuses on materials processing – especially solidification and casting processes, finite element methods, heat transfer, and fluid dynamics.

Charles L. Tucker III is W. Grafton and Lillian B. Wilkins Professor, Department of Mechanical and Industrial Engineering, University of Illinois at Urbana-Champaign. His research interests include the processing of polymers and composite materials, the modeling and simulation of manufacturing processes, and the use of numerical methods.

MODELING IN MATERIALS PROCESSING

JONATHAN A. DANTZIG
University of Illinois

CHARLES L. TUCKER III
University of Illinois

PUBLISHED BY THE PRESS SYNDICATE OF THE UNIVERSITY OF CAMBRIDGE
The Pitt Building, Trumpington Street, Cambridge, United Kingdom

CAMBRIDGE UNIVERSITY PRESS
The Edinburgh Building, Cambridge CB2 2RU, UK
40 West 20th Street, New York, NY 10011-4211, USA
10 Stamford Road, Oakleigh, VIC 3166, Australia
Ruiz de Alarcón 13, 28014 Madrid, Spain
Dock House, The Waterfront, Cape Town 8001, South Africa

http://www.cambridge.org

© Jonathan A. Dantzig and Charles L. Tucker III 2001

This book is in copyright. Subject to statutory exception
and to the provisions of relevant collective licensing agreements,
no reproduction of any part may take place without
the written permission of Cambridge University Press.

First published 2001

Printed in the United States of America

Typefaces Times Ten 10/13 pt. and Helvetica Neue Condensed   *System* LATEX 2_ε [TB]

A catalog record for this book is available from the British Library.

Library of Congress Cataloging in Publication Data
Modeling in materials processing / Jonathan A. Dantzig, Charles L. Tucker, III.
    p. cm.
    Includes bibliographical references and index.
    ISBN 0-521-77063-7 – ISBN 0-521-77923-5 (pb.)
    1. Manufacturing processes–Mathematical models.   I. Dantzig, J. A.   II. Tucker, Charles L.

TS183 .M612   2001
670.42'01'5118 – dc21   00-054673

ISBN 0 521 77063 7 hardback
ISBN 0 521 77923 5 paperback

Contents

Preface		*page* ix
Principal Nomenclature		xi
1	**Introduction**	1
	1.1 What Is a Model?	1
	1.2 A Simple Pendulum	2
	1.3 One-Dimensional Traffic Flow	6
	1.4 Summary	19
	Bibliography	19
	Exercises	20
2	**Governing Equations**	24
	2.1 Preliminaries	24
	2.2 Mass Balance	29
	2.3 Momentum Balance	33
	2.4 Energy Balance	45
	2.5 Summary	52
	Bibliography	53
	Exercises	53
	Appendix	58
3	**Scaling and Model Simplification**	60
	3.1 Introduction	60
	3.2 Basic Scaling Analysis	62
	3.3 Small Parameters and Boundary Layers	69
	3.4 Classical Dimensionless Groups	76
	3.5 Nondimensionalization for Numerical Solutions (Advanced)	78
	3.6 Summary	81
	Bibliography	81
	Exercises	81
4	**Heat Conduction and Materials Processing**	87
	4.1 Steady Heat Conduction in Solids	90
	4.2 Transient Heat Conduction	93
	4.3 Conduction with Phase Change	106

	4.4	Summary	120
		Bibliography	121
		Exercises	121
		Appendix	130

5 Isothermal Newtonian Fluid Flow — 132
- 5.1 Newtonian Flow in a Thin Channel — 132
- 5.2 Other Slow Newtonian Flows — 143
- 5.3 Free Surfaces and Moving Boundaries — 149
- 5.4 Flows with Significant Inertia — 161
- 5.5 Summary — 177
- Bibliography — 178
- Exercises — 178

6 Non-Newtonian Fluid Flow — 190
- 6.1 Non-Newtonian Behavior — 190
- 6.2 The Power Law Model — 192
- 6.3 Power Law Solutions for Other Simple Geometries — 200
- 6.4 Principles of Non-Newtonian Constitutive Equations — 202
- 6.5 More Non-Newtonian Constitutive Equations — 209
- 6.6 The Generalized Hele–Shaw Approximation — 217
- 6.7 Summary — 228
- Bibliography — 228
- Exercises — 229

7 Heat Transfer with Fluid Flow — 239
- 7.1 Uncoupled Advection — 239
- 7.2 Temperature-Dependent Viscosity and Viscous Dissipation — 250
- 7.3 Buoyancy-Driven Flow — 259
- 7.4 Summary — 275
- Bibliography — 275
- Exercises — 276

8 Mass Transfer and Solidification Microstructures — 282
- 8.1 Governing Equations for Diffusion — 282
- 8.2 Solid-State Diffusion — 285
- 8.3 Solidification Microstructure Development — 295
- 8.4 Summary — 319
- Bibliography — 320
- Exercises — 321

A Mathematical Background — 327
- A.1 Scalars, Vectors, and Tensors: Definitions and Notation — 327
- A.2 Vector and Tensor Algebra — 331
- A.3 Differential Operations in Rectangular Coordinates — 335
- A.4 Vectors and Tensors in Cylindrical and Spherical Coordinates — 337
- A.5 The Divergence Theorem — 339
- A.6 Curvature of Curves and Surfaces — 339
- A.7 The Gaussian Error Function — 343
- Bibliography — 345
- Exercises — 345

B Balance and Kinematic Equations — 348
- B.1 Continuity Equation: General Form — 348
- B.2 Continuity Equation: Constant ρ — 348

B.3	Rate-of-Deformation Tensor	349
B.4	Vorticity Tensor	350
B.5	General Equation of Motion	350
B.6	Navier–Stokes Equation: Constant ρ and μ	352
B.7	Heat Flux Vector: Isotropic Material	353
B.8	Energy Balance: General Form	354
B.9	Energy Balance: Constant ρ, k, and μ	355
	Bibliography	356

Index   357

Preface

After some years of teaching separate courses on metal solidification and polymer processing, we realized that the two subjects shared a substantial base of common material. All the models started with the same basic equations and were built by using the same general procedure. We began to teach a single course on materials processing, and we found that our unified treatment gave students a better overall perspective on modeling. We also discovered that we needed a new book, as existing texts were almost all devoted exclusively to polymers, or to metals, or to ceramics. In this book, we treat metal and polymer processing problems together, building around the transport equations as a unifying theme.

We were also dissatisfied with ad hoc model development, in which terms were arbitrarily dropped from the governing equations, or simplifications were made without a clear explanation. Simplifying the general governing equations is a critical step in modeling, but it is a skill, not an art. In this text we introduce scaling analysis as a systematic way to reduce the governing equations for any particular problem. Scaling provides a way for both novices and experts to simplify a model, while ensuring that all of the important phenomena are included.

After deriving the governing equations in their general form and introducing scaling analysis, we examine physical phenomena such as heat conduction and fluid flow. We work out many problems that include only a few of these phenomena – problems that can be solved analytically. One might call these "canonical problems." They allow the reader to study each phenomenon in isolation, and then to explore how that phenomenon interacts with others. Real processes frequently involve multiple physical phenomena, and the ability to isolate a single phenomenon and understand its role is one of the great benefits of modeling. Canonical problems help students place different phenomena in perspective, and give them the ability to anticipate which phenomena will be important in any particular process. We once overheard a student describe our materials processing course as "the place where you finally understand what they taught you in heat transfer and fluid mechanics." We hope so.

We present examples for many different materials and processes, including polymer extrusion and injection molding, as well as metal casting and microstructure development. In each example we begin with the governing equations, and we use scaling to arrive at the final set of equations to be solved. This systematic approach makes problems for many different materials accessible.

Of course, most practical materials processing models require a numerical solution, and any accomplished modeler knows a great deal about numerical methods. We chose not to say much about numerical methods, preferring to give solid coverage to the governing equations and physical phenomena. However, our canonical problems provide excellent test cases for numerical solution methods, and we use them to demonstrate some of the pitfalls of numerical modeling. By showing examples in which numerical schemes may be inaccurate or unstable, we help the reader become a more intelligent user of modeling software.

There is a lot to learn here, and many of the exercises at the end of each chapter go beyond the examples in the chapter. These exercises are written in a way that guides the student through the problem, step by step. This style emphasizes the overall pattern of problem solving, and it allows students to do more complex problems than they could otherwise attempt. A full set of solutions is available to instructors who adopt the book as a course text. Please contact the authors for the Solutions Manual.

We started writing this book to distill the important lessons from our own experience, one of us in polymer processing and the other in metal solidification. We eventually found that knowing more about modeling of all types of materials made us better at modeling the materials and processes we were so familiar with. We hope you will have the same experience.

Jonathan A. Dantzig
Charles L. Tucker III
Urbana, Illinois
November 1, 2000

Principal Nomenclature

Symbol	Quantity Represented
A	area
a	parameter in exponential model for temperature-dependent viscosity
$\mathbf{b}$	vector of body force per unit mass
C_A	mass fraction of species A
C_ℓ	mass fraction of solute in liquid
C_s	mass fraction of solute in solid
$\hat{C}_A$	concentration (mass per unit volume) of species A
c_0, c_1, etc.	constants of integration
c_p	specific heat at constant pressure
c_v	specific heat at constant volume
D	mass diffusivity of solute; diameter
D_{AB}	intrinsic mass diffusivity of species A in B
$\tilde{D}$	interdiffusion coefficient of species A in B
$\mathbf{D}$	rate-of-deformation tensor $= (\mathbf{L}+\mathbf{L}^T)/2$
E	internal energy of a system
E_η, E_D	activation energies for viscosity, diffusion
$\mathbf{e}_1, \mathbf{e}_x$, etc.	unit vectors
$\mathbf{F}$	deformation gradient tensor, $F_{ij} = \partial X_i/\partial x_j$
f_ℓ	mass fraction of liquid
f_s	mass fraction of solid
g	acceleration that is due to gravity $= 9.82$ m/s^2
G	temperature gradient; magnitude of pressure gradient
h	heat transfer coefficient; height above a reference level; half-gap height
H	gap height; specific enthalpy
$\mathbf{I}$	unit tensor (identity tensor); the ij component is δ_{ij}
I_D, II_D, III_D	scalar invariants of the tensor $\mathbf{D}$
i	$\sqrt{-1}$
$\mathbf{J}_A$	mass fraction flux for species A, equal to $\hat{\mathbf{J}}_A/\rho$
$\hat{\mathbf{J}}_A$	mass flux vector for species A (mol/area $\times$ time)
J	Jacobian $= \det \mathbf{F}$

J_0, J_1	Bessel functions of the first kind, of order zero and one
k	thermal conductivity (tensor)
k	thermal conductivity (scalar)
k_0	partition coefficient
L	length
L_d	primary dendrite arm length
L_f	latent heat of fusion
L_p	latent heat of pressure change
L_v	latent heat of volume change
L	velocity gradient tensor, $L_{ij} = \partial v_i/\partial x_j$
ℓ	filled length of a mold
m	material constant in the power law model
m_ℓ, m_S	slopes of the liquidus and solidus lines
M	mass of a system; morphological number
n	power law index
n	unit vector normal to a surface
P	power input to a system
p	pressure
$\hat{p}$	modified pressure $= p + \rho_0 g h$
q	heat flux vector
Q	volume flow rate; power of a point heat source; heat input to a system
Q	rotation matrix
R	radius; gas constant $= 8.31$ J/mol K
$\dot{R}$	specific heat generation rate
R_A	generation rate of species A
$\mathcal{R}$	thermal resistance
r, θ, z	cylindrical coordinates
r, θ, ϕ	spherical coordinates
S	bounding surface; specific entropy; flow conductance
s	inverse of power law index $= 1/n$
T	temperature; torque
T_m	melting temperature
T_L	liquidus temperature
T_S	solidus temperature
t	time
t	surface traction vector
$\hat{\mathbf{t}}$	unit vector tangent to a curve
v	velocity vector
V	volume; average velocity
$\hat{V}$	specific volume
W	width
W	vorticity tensor $= (\mathbf{L} - \mathbf{L}^T)/2$
X	material coordinate vector
x	position vector
x, y, z	Cartesian coordinates; also x_1, x_2, x_3

PRINCIPAL NOMENCLATURE

α	thermal diffusivity $= k/(\rho c_p)$
β	volumetric thermal expansion coefficient
Γ	surface tension (interfacial tension)
$\dot{\gamma}$	scalar strain rate $= (2\mathbf{D}:\mathbf{D})^{1/2}$
δ	solidified layer thickness; boundary layer thickness
δ_{ij}	unit tensor in indicial notation (the Kronecker delta)
ε	specific internal energy
ϵ_{ijk}	permutation symbol
κ	curvature of a curve; ratio of an outer radius to inner radius
κ_m, κ_G	mean and Gaussian curvatures of a surface
Λ	magnitude of the pressure gradient
λ	dilatational viscosity, compressible Newtonian fluid
λ_2	secondary dendrite arm spacing
μ	viscosity, Newtonian fluid
ν	kinematic viscosity $= \mu/\rho$
η	viscosity, non-Newtonian fluid
π	3.14159...
ρ	density
ρ_0	density at reference temperature and pressure
θ	dimensionless temperature; angular coordinate
σ	total stress tensor
$\boldsymbol{\tau}$	extra stress tensor
τ	scalar magnitude of $\boldsymbol{\tau}$
τ_Y	yield stress
Ω	angular velocity
ω	vorticity vector $= \nabla \times \mathbf{v}$
ξ	scaled length within a boundary layer
ζ	similarity variable

CHAPTER ONE

Introduction

1.1 WHAT IS A MODEL?

In recent years, modeling has been embraced by the materials processing community as a tool for understanding and improving manufacturing processes. Models are often implemented in computer programs, but there are important differences between a model and the computer code that implements it. A model is a set of equations used to represent a physical process. Finite element or finite difference methods, and the computer programs that implement them, are techniques to *solve* the equations of the model, but they are not the model itself. Our main emphasis will be on creating models – on reducing a physical process to a set of equations – especially models whose solution accurately describes the behavior of the process. Occasionally we also will explore numerical solution methods, often to point out where unenlightened use can lead you astray.

Whenever we create a model, we make assumptions about what phenomena are important to the behavior of the physical process. This is both good and bad. Assumptions help define the mathematical model and make it amenable to analysis. However, incorrect assumptions and erroneous information become part of the model and may well distort the results. Sometimes assumptions greatly simplify the model and permit an easy solution, but they may also cause important physical phenomena to be misrepresented or overlooked. Limiting the number of assumptions helps to avoid this problem but may make the model overly complex. Then the solution becomes difficult, and important information may be obscured. Building a good model requires making careful and informed decisions about the assumptions.

In this book we describe a systematic approach to modeling. We start from the physical system and proceed through mathematical formulation to solution of the model equations. We pay particular attention to the issue of choosing what assumptions to make. Our goal is to ensure that the reduction of the physical system to a mathematical model is done correctly. We present the task of modeling as a series of steps.

1. Define the scope and goals of the model.
2. Make a conceptual sketch and define basic quantities.
3. Develop a mathematical description of the conceptual sketch.
4. Write fundamental equations that govern the primary variables.

5. Introduce constitutive relations between the primary variables.
6. Reduce the governing equations by making assumptions.
7. Scale the variables and governing equations.
8. Solve the remaining equations to infer the behavior of the system.

In the following sections we present two simple examples to illustrate this stepwise approach. The first example is a physical system whose mathematical model should be quite familiar to the reader: a simple pendulum. The second example, a model of traffic flow, examines a familiar physical situation, but it develops a mathematical representation that will be new to most readers. These two problems provide concrete examples of the overall approach, but they are simple enough to be easily followed. In later chapters we apply this same stepwise approach to problems in materials processing.

1.2 A SIMPLE PENDULUM

Consider the simple pendulum, constructed by connecting a mass to one end of a string and fixing the string at its opposite end. We will now construct a mathematical model of this system, following the procedure outlined above.

Step 1: Define the scope and goals of the model. The physical system is, of course, the pendulum. Defining the goals is not as easy as it might seem, because this requires us to decide what aspects of the system behavior are most important – at least to us. Let us assume that we are going to use this pendulum as a timepiece, so we are interested in computing its period of oscillation. Other aspects of the motion are secondary.

Step 2: Make a conceptual sketch and define basic quantities. A sketch of the pendulum is shown in Fig. 1.1. The figure defines some quantities that we will use to analyze the system. The suspended mass is m, and gravitational acceleration g is oriented vertically down the page. The distance from the fixed end of the string to the center of the mass is r, and we introduce the variable θ to represent the angular position of the pendulum at any time t. Note that m and g are *parameters* of the

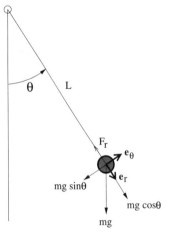

Figure 1.1: Conceptual sketch of a simple pendulum, showing forces and primitive variables; $\mathbf{e}_r$ and $\mathbf{e}_\theta$ are the unit vectors in the r and θ directions.

1.2 A SIMPLE PENDULUM

problems (i.e., they are constant for any specific pendulum), whereas r and θ are *variables*, because they may change as a function of time t; t itself is an independent variable.

Step 3: Develop a mathematical description of the conceptual sketch. As shown in Fig. 1.1, a force due to gravity with magnitude mg acts on the mass. It is convenient to resolve this force into radial and tangential components, as shown in the figure. The tension in the string also exerts a force, which has only a radial component F_r. We also will need the acceleration of the mass in terms of r, θ, and their derivatives. If both r and θ change with time, then the radial component of acceleration is $\ddot{r} - r\dot{\theta}^2$, and the tangential component is $r\ddot{\theta} + 2\dot{r}\dot{\theta}$; the dots indicate ordinary time derivatives.

Step 4: Write fundamental equations that govern the primary variables. We know that the motion of the mass will be governed by Newton's law, $\mathbf{f} = m\mathbf{a}$. We resolve the forces and accelerations into their r and θ components, and we get two equations:

$$m(\ddot{r} - r\dot{\theta}^2) = mg\cos\theta - F_r$$
$$m(r\ddot{\theta} + 2\dot{r}\dot{\theta}) = -mg\sin\theta \tag{1.1}$$

We also need initial conditions on r, θ, and their first derivatives. For the moment we will just write the conditions for θ and $\dot{\theta}$, which we choose as

$$\theta(t=0) = \theta_0$$
$$\dot{\theta}(t=0) = 0 \tag{1.2}$$

The initial angle θ_0 is an additional parameter in the problem.

Step 5: Introduce constitutive relations between the primary variables. The governing equations are sufficient in this problem, and this step is not necessary. We will take it up in the next example.

Step 6: Reduce the governing equations by making assumptions. There are many reasons for making assumptions, each having its own inherent advantages and dangers. We will indicate some of these as we make various assumptions.

In fact, we have already made two important assumptions implicitly. Let us now state them explicitly.

- The mass of the string is negligible and the mass is concentrated at a point. The analysis is simplified because now we need only consider a point mass, and we can neglect such factors as the moment of inertia of the mass and string, and their angular acceleration. (A different model that includes these factors, called the *physical pendulum*, is well known.) Bear in mind, however, that we have introduced attributes into the model that may not necessarily match those of the real system.
- Friction has been neglected. Thus, our model represents a perpetual motion machine. Friction does play a role in determining the period of the pendulum, a phenomenon explored in one of the exercises.

We will now further assume that the string has constant length (independent of the tension F_r), so that $r = L$ for all time. Note that L is now another parameter of the problem. In the second of Eqs. (1.1), we replace r by the constant L and set $\dot{r} = 0$, which allows us to compute $\theta(t)$. The first of Eqs. (1.1) reduces to an equation to determine F_r once θ and $\dot{\theta}$ are known. The assumption of constant r reduces the dimensionality of the problem, greatly simplifying the solution. However, it further restricts the range of validity of our model.

Finally, we will limit our attention to small motions of the pendulum, that is, to small values of the angle θ. When using words like "small," one must define them carefully in the context of the present problem. Here we define small as that range of angles in which $\sin\theta \simeq \theta$. The error associated with this approximation is less than 1% for angles up to 14°. This assumption also greatly simplifies the problem, because the remaining equation is now linear:

$$L\ddot{\theta} = -g\theta \tag{1.3}$$

We have finally obtained an equation we can easily solve. Notice how simple our model of the pendulum has become in comparison to the real system.

Step 7: Scale the variables and governing equations. To *scale* a variable, one divides the variable by its characteristic value, expressed in terms of the parameters of the problem. This produces a new, dimensionless variable whose order of magnitude is one. For example, the angle θ has a characteristic value θ_0. Using this, we define a dimensionless variable ϕ as

$$\phi = \frac{\theta}{\theta_0} \tag{1.4}$$

Note that θ, while dimensionless, was not scaled to be of order one. The new variable ϕ is order one because, in the absence of external forces, $\phi \in [-1, 1]$.

The reasons for scaling and the methods for doing it will be discussed in Chapter 3. For the time being let us simply accept that it is convenient if all of the variables in the governing equation have been scaled to eliminate their dimensions and to make them of order one.

We also need to scale the time variable, though it is not obvious how to get a time scale from the problem parameters. We can deduce the time scale by going ahead and defining a scaled time variable τ as

$$\tau = \frac{t}{t_c} \tag{1.5}$$

where t_c is a characteristic time whose value we do not yet know. That is, we choose to measure time in units that are somehow related to our model system, but we still need to find how t_c depends on the known parameters of the problem.

Next we scale the governing equation by recasting it in terms of the scaled variables. Applying the chain rule for differentiation shows that

$$\frac{d}{dt} = \frac{d}{d\tau}\frac{d\tau}{dt} = \frac{1}{t_c}\frac{d}{d\tau} \tag{1.6}$$

Similarly, the second derivative is

$$\frac{d^2}{dt^2} = \frac{1}{t_c^2}\frac{d^2}{d\tau^2} \tag{1.7}$$

1.2 A SIMPLE PENDULUM

Substitution of this expression into Eq. (1.3) transforms the governing equation to

$$\left(\frac{L}{g}\right)\left(\frac{1}{t_c^2}\right)\frac{d^2\phi}{d\tau^2} = -\phi \tag{1.8}$$

Now because ϕ is of order one, the right-hand side of Eq. (1.8) is of order one, and thus the left-hand side of Eq. (1.8) side must be of order one as well. Because τ is of order one, we expect that $d^2\phi/d\tau^2$ on the left-hand side also will be of order one. The only way this can be true is for the collection of parameters on the left-hand side of Eq. (1.8) to equal one. From this we deduce that the characteristic time is

$$t_c = \sqrt{L/g} \tag{1.9}$$

This gives the characteristic value t_c in terms of known parameters L and g. Now all of our scaled variables are clearly defined, and the governing equation, Eq. (1.8), simplifies to

$$\frac{d^2\phi}{d\tau^2} = -\phi \tag{1.10}$$

Note that scaling has given us an estimate for the period for the pendulum, without ever solving a differential equation. This is one of the many benefits of scaling analysis.

The boundary conditions must also be scaled. Following the same procedure, we find that Eqs. (1.2) become

$$\phi(\tau = 0) = 1$$
$$\frac{d\phi}{d\tau}(\tau = 0) = 0 \tag{1.11}$$

Step 8: Solve the remaining equations to infer the behavior of the system. The solution of Eq. (1.10) is given by

$$\phi(\tau) = c_1 \sin \tau + c_2 \cos \tau \tag{1.12}$$

The two constants c_1 and c_2 are evaluated from the initial conditions, Eqs. (1.11). This gives the solution, in terms of the scaled dimensionless variables, to be

$$\phi = \cos \tau \tag{1.13}$$

If desired, one can substitute Eqs. (1.4) and (1.5) into Eq. (1.13) to express the solution in dimensional form.

$$\frac{\theta}{\theta_0} = \cos\left(\frac{t}{\sqrt{L/g}}\right) \tag{1.14}$$

Inspecting these equations shows that our *model* of the simple pendulum undergoes a simple harmonic motion of period $\tau = 2\pi$, or $t = 2\pi\sqrt{L/g}$. This is the result we were looking for.

At this point one might go back over the analysis, consider the validity of the various assumptions, and explore the sensitivity of the results to those assumptions. Usually this requires constructing and solving a new model without the assumptions, and comparing the results to the original model. The exercises at the end of

1.3 ONE-DIMENSIONAL TRAFFIC FLOW

Our second example deals with traffic flow. Highway engineers are often called upon to design roads leading to a planned site, such as a new shopping center or stadium. They need to ensure that the roads have sufficient capacity to carry the anticipated traffic to the site, without undue backups and delays. The models for such problems can be quite sophisticated. In this section we consider a very simple form to illustrate some of the concepts involved.

1.3.1 MODEL FORMULATION

Step 1: Define the scope and goals of the model. We will consider vehicles traveling in one direction on a one-lane highway, with no entrances or exits. We seek to answer questions that a highway planner might ask. How much traffic can the road carry? How long does it take to travel between two points on the road, and how does that travel time depend on the amount of traffic? If the road gets backed up with traffic, how long does it take the road to clear? Note that we are interested in the global characteristics of the highway, rather than the movement of individual vehicles.

Step 2: Make a conceptual sketch and define basic quantities. The sketch in Fig. 1.2 shows a piece of the roadway with several vehicles. The x coordinate measures distance along the road, and each vehicle is identified by its position x_i and its length L_i. Each vehicle also has an associated speed $v_i = \dot{x}_i$ and an acceleration $a_i = \ddot{x}_i$.

Step 3: Develop a mathematical description of the conceptual sketch. There are several different approaches to this task. In what might be called a "molecular dynamics" approach, we could develop a data structure for each vehicle, consisting of, say, $[x_i, v_i, a_i, L_i]$, then seek microscopic models for relationships between the vehicles. For example, we might choose a model in which each vehicle's motions is governed by itself and its nearest neighbors. Such a model would be represented mathematically as

$$[x_i, v_i, a_i, L_i] = f([x_j, v_j, a_j, L_j]; j = i-1, i, i+1) \tag{1.15}$$

We might deduce the function f from experimental observations of many drivers under various conditions. Given such a function, initial conditions, and suitable data-management capacity, one could completely describe the evolution of the system.

Unfortunately, this would not be a very useful model. It would contain an enormous amount of detail, but it would not easily address our original questions about

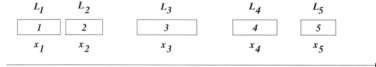

Figure 1.2: Conceptual sketch for one-dimensional traffic flow.

1.3 ONE-DIMENSIONAL TRAFFIC FLOW

global trends. Does a highway engineer really want to know how closely vehicle 7 follows vehicle 8 when they are separated by 30 m and traveling at 32.7 miles per hour; or whether vehicle 3 will drive more slowly because vehicle 4 is longer than an average car?

An altogether different approach is to model the traffic as a continuum. We step back from the individual vehicles and consider phenomena that occur *on average*. To this end, we define two functions: vehicle density $\rho(x, t)$, having units of vehicles per unit distance, and velocity $v(x, t)$, with units of distance per unit time. We assume these functions to be continuous in x and t. That is, $\rho(x, t)$ represents the *average* number of vehicles per unit distance traveling on the road in the neighborhood of position x at time t, and $v(x, t)$ is the average velocity of those vehicles. We also note that the vehicle flow rate is $Q \equiv \rho v$. $Q(x, t)$ equals the number of vehicles per unit time passing the position x. From the original problem statement, this is the main quantity we are interested in.

To compute $\rho(x, t)$ from a snapshot of the vehicles on the road, one might choose an averaging aperture around the point x, count the vehicles within this aperture, and divide by the aperture's length. The size of this averaging aperture affects the behavior of the measured function ρ. If the aperture is too small, say of the order of a vehicle length, then ρ will be a very noisy function and quantities such as $\partial \rho / \partial x$ will not have much meaning. Making the aperture larger will smooth out these meaningless fluctuations, but if the aperture is too large then real fluctuations in traffic will be lost in the averaging process. This phenomenon is explored in the exercises at the end of the chapter. Note that the averaging aperture affects only the interpretation of experimental data. We can formulate the continuum model and analyze it mathematically without ever saying exactly how big the averaging aperture is.

This continuum model does exclude some phenomena. Obviously anything that occurs on a length scale smaller than the averaging aperture will be invisible to the model. Less obvious, but equally important, is that all information about the behavior of individual vehicles is suppressed. This is an extremely important, basic point: all models have shortcomings that derive from their original assumptions.

Step 4: Write fundamental equations that govern the primary variables. To derive a fundamental governing equation, we use the principle that cars are neither created nor destroyed (at least not in the practice of normal driving). We choose a fixed control volume V, which extends from x to $x + \Delta x$ as shown in Fig. 1.3. We may write a balance equation for the vehicles in the control volume as

$$\frac{\text{vehicles in}}{\text{time}} - \frac{\text{vehicles out}}{\text{time}} = \frac{\text{vehicles accumulated}}{\text{time}} \quad (1.16)$$

The rate of vehicles entering at the left is $Q(x)$, the rate of vehicles leaving at the right is $Q(x + \Delta x)$, and the number of vehicles inside the control volume is $\rho \Delta x$.

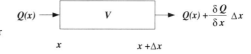

Figure 1.3: A control volume near position x for the one-dimensional traffic flow problem.

Thus, the balance equation for the control volume is

$$\underbrace{Q(x)}_{\text{vehicles in}} - \underbrace{Q(x+\Delta x)}_{\text{vehicles out}} = \underbrace{\frac{\partial}{\partial t}(\rho \Delta x)}_{\text{vehicles accumulated}} \qquad (1.17)$$

We rearrange terms and divide by Δx (which does not change with time) to find

$$\frac{\partial \rho}{\partial t} + \frac{Q(x+\Delta x) - Q(x)}{\Delta x} = 0 \qquad (1.18)$$

As $\Delta x \to 0$, the second term becomes $\partial Q/\partial x$, and the balance equation is

$$\frac{\partial \rho}{\partial t} + \frac{\partial Q}{\partial x} = 0 \qquad (1.19)$$

or, in terms of ρ and v,

$$\frac{\partial \rho}{\partial t} + \frac{\partial (\rho v)}{\partial x} = 0 \qquad (1.20)$$

The reader may recognize this as a one-dimensional form of the mass balance equation.

Step 5: Introduce constitutive relations between the primary variables. Unlike the previous example, we are not ready to solve Eq. (1.20), because we have one equation but two unknowns, ρ and v. An additional equation is needed, in this case a *constitutive* equation. In materials processing models a constitutive equation describes the behavior of a specific material or class of materials. This is different from a fundamental physical law such as the conservation of mass, of which Eq. (1.20) is a special case, which applies to any situation. In this problem we need a constitutive equation that relates v and ρ.

Constitutive equations are usually concocted by combining a few simple assumptions with experimental observations. In this example we will start by assuming that velocity is a function of density, $v = v(\rho)$. We can get a pretty good idea of what this function should look like from the following "rules."

1. Fear of flying rule: If there is nobody else on the road, vehicles will travel at some maximum speed. Hence, $v(\rho = 0) = v_{\max}$.
2. Get out of my way rule: Increasing the density always decreases the velocity, so that $(dv/d\rho) \le 0$ over the entire range of ρ.
3. Gridlock rule: At some maximum density the velocity will drop to zero, $v(\rho = \rho_{\max}) = 0$, and the road turns into a parking lot.

Clearly there are many functional forms that satisfy these rules. To keep the subsequent analysis simple, we choose a linear relationship:

$$v = v_{\max}\left(1 - \frac{\rho}{\rho_{\max}}\right) \qquad (1.21)$$

In practice, the parameters $v_{\max}$ and $\rho_{\max}$ would be determined from experiments. More experimental data might also suggest something other than a linear form. However, we will see that even this simple constitutive equation provides some useful insights about traffic flow.

1.3 ONE-DIMENSIONAL TRAFFIC FLOW

Returning to the overall model, because $v = v(\rho)$ we rewrite Eq. (1.20) as

$$\frac{\partial \rho}{\partial t} + \left(\frac{\partial(\rho v)}{\partial \rho}\right) \frac{\partial \rho}{\partial x} = 0 \qquad (1.22)$$

Substitution of constitutive relation (1.21) yields

$$\frac{\partial \rho}{\partial t} + v_{\max}\left(1 - \frac{2\rho}{\rho_{\max}}\right) \frac{\partial \rho}{\partial x} = 0 \qquad (1.23)$$

Now we have a single primary equation in the variable ρ. This first-order differential equation requires an initial condition for $\rho(x, t = 0)$, as well as a boundary condition on the end of the domain where traffic enters. When Eq. (1.23) has been solved for $\rho(x, t)$, we can substitute the results back into Eq. (1.21) to find $v(x, t)$.

Step 6: Reduce the governing equations by making assumptions. We already took a major step in this direction when we assumed that the traffic flow was one dimensional. We also assumed that no vehicles would enter or leave the highway at intermediate points. In subsequent chapters we will be more careful about assumptions, first writing very general governing equations and then reducing them by making various assumptions.

Before developing the model further, let us make some observations from what we have so far. Figure 1.4 shows a graph of the flow rate Q as a function of traffic density. Recall that increasing ρ always decreases the velocity v. We see in Fig. 1.4 that, on one hand, when the density is low, an increase in density leads to an increase in flow rate, because the increased number of vehicles traveling outweighs the decrease in speed. On the other hand, when the density is high, the reduction in speed overwhelms the increased number of vehicles that are traveling, and adding more vehicles actually decreases the flow rate. The maximum flow rate $Q_{\max}$, also called the *capacity* of the road, occurs at some density less than $\rho_{\max}$. This explains why some freeways have stoplights at every entrance. The lights are controlled to keep the traffic density on the freeway at or below the density that produces $Q_{\max}$.

Because we chose a linear constitutive relation between v and ρ, Fig. 1.4 is a parabola with its maximum value at $\rho_{\max}/2$. However, any constitutive relation satisfying the three rules above will display similar behavior.

Step 7: Scale the variables and governing equations. We first make the variables dimensionless. Let the total length of the highway we are modeling be L. This

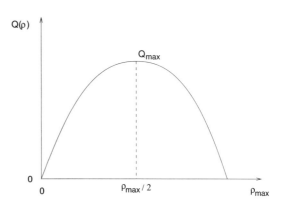

Figure 1.4: Traffic flow rate Q as a function of traffic density ρ, for the linear constitutive relation in Eq. (1.21).

provides a natural scaling for the x coordinate. We also scale ρ on ρ_{max} and v on v_{max}. As in the last section, we introduce t_c as a characteristic time, whose value will be determined once we scale the governing equation. Using asterisks to denote the scaled dimensionless variables, we have

$$x^* = \frac{x}{L}$$

$$t^* = \frac{t}{t_c}$$

$$\rho^* = \frac{\rho}{\rho_{max}}$$

$$v^* = \frac{v}{v_{max}} \quad (1.24)$$

Note that all of the dimensionless variables are now of order one. Substituting these relations into Eq. (1.23) gives

$$\frac{\partial \rho^*}{\partial t^*} + \left(\frac{v_{max} t_c}{L}\right)(1 - 2\rho^*)\frac{\partial \rho^*}{\partial x^*} = 0 \quad (1.25)$$

It is now apparent that the characteristic time t_c should be chosen as (L/v_{max}), which is the time required to travel the full length of the highway at the maximum velocity. We are then left with the dimensionless form

$$\frac{\partial \rho^*}{\partial t^*} + (1 - 2\rho^*)\frac{\partial \rho^*}{\partial x^*} = 0 \quad (1.26)$$

The constitutive equation, initial condition, and boundary condition can also be recast in dimensionless form.

Equation (1.26) is a *wave equation*, with (dimensionless) propagation speed $c = 1 - 2\rho^*$. Because $\rho^* \in [0, 1]$, the wave speed is positive for small densities, negative for large densities, and exactly zero at $\rho^* = 1/2$. This observation explains several common phenomena that occur on highways. When a highway is carrying light to moderate traffic, the cars tend to form into packs. If you are driving at the average speed, then you tend to catch up to a pack from the back, slowly work your way forward, and then leave the pack from the front. The pack represents a forward-moving compression wave, with a propagation speed c that is positive at this low density. This wave speed is slower than the average driving speed, so individual cars move faster than the pack.

Now suppose that traffic is heavy and vehicles slow for some reason, say to gawk at an accident. This initiates a compression wave, but at high traffic density the compression wave propagates backward because c is negative. This gradually pushes the beginning of the slow-down zone backward (upstream) of the accident, causing the familiar traffic jam. You join the traffic jam well upstream of the accident, and you creep forward slowly. With a negative wave speed there is no effect on the traffic downstream of the accident, so once you finally pass the accident the traffic density decreases immediately and you speed up.

Finally, when the highway is carrying its maximum capacity Q_{max}, the wave speed is exactly zero. Any small perturbation in density, caused for instance by a driver tapping his brakes, will remain stationary. This explains the curious phenomenon

1.3 ONE-DIMENSIONAL TRAFFIC FLOW

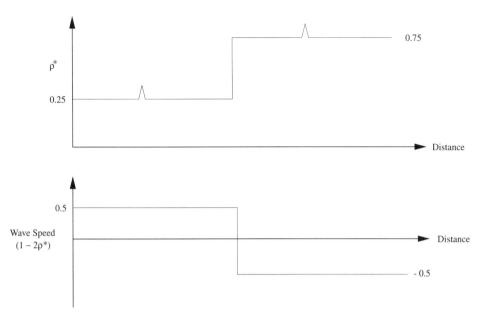

Figure 1.5: An initial traffic density distribution (dimensionless) with its corresponding wave speed distribution. For this initial condition, a shock develops where the two disturbances meet.

seen on busy highways, in which traffic slows briefly and then speeds up again with no apparent cause.

Step 8: Solve the remaining equations to infer the behavior of the system. Solution of Eq. (1.26) is a challenging task, because the density-dependent wave speed makes the equation nonlinear. One might linearize the equation by selecting a reference density ρ_0^* and using it to fix the wave speed. This makes the problem more tractable, but considerably less interesting. The solution in this case (see the exercises) is that the initial condition propagates with the wave speed $(1 - 2\rho_0^*)$.

One of the more interesting aspects of Eq. (1.26) can be understood by considering a case in which the initial density is a step function of position, as shown in Fig. 1.5. The wave speed will then be positive in the first section and negative in the second, as shown in the figure. If we introduce a small density variation in the first section, it propagates forward in time because its wave speed is positive. However, a small density variation on the right will propagate backward because its wave speed is negative. When the two disturbances reach the boundary where the initial density changes, a stationary shock wave forms. Further discussion of this phenomenon is beyond the scope of this book, but may be found, for example, in Whitham (1974).

1.3.2 NUMERICAL SOLUTIONS FOR TRAFFIC FLOW (ADVANCED)

A few sections in this book are more appropriate for more advanced readers, and we mark them accordingly. Most of these sections deal with a numerical implementation of the models, and they may be omitted in an introductory course with no loss of continuity.

The fact that the wave speed in Eq. (1.26) is a function of ρ^* makes the equation nonlinear. Thus, even this simple problem may have to be solved numerically to determine the evolution of the traffic pattern from a given initial condition. We now examine numerical solutions for this example, to display some important characteristics that will appear again in later chapters.

We introduce a finite difference representation of the solution, with nodes along the x axis that are separated by equal increments of Δx^*. Similarly, we will develop the solution for equal time increments Δt^*. The density value at node j and time step n will be denoted by $\rho_j^n = \rho^*(j\Delta x^*, n\Delta t^*)$. We can then write finite difference approximations for the derivatives at $x^* = j\Delta x^*$, $t^* = n\Delta t^*$ as

$$\frac{\partial \rho^*}{\partial t^*} \simeq \frac{\rho_j^{n+1} - \rho_j^n}{\Delta t^*}$$

$$\frac{\partial \rho^*}{\partial x^*} \simeq \frac{\rho_{j+1}^n - \rho_{j-1}^n}{2\Delta x^*} \qquad (1.27)$$

This scheme is called the forward time-centered space (FTCS) scheme. When substituted into the governing equation, Eq. (1.26), it produces the following simple formula for time integration:

$$\rho_j^{n+1} = \rho_j^n - \frac{\Delta t^*}{2\Delta x^*}(1 - 2\rho_j^n)(\rho_{j+1}^n - \rho_{j-1}^n) \qquad (1.28)$$

We also need to represent boundary conditions at both ends of the domain. (The boundary condition at the exit is required for numerical reasons that will be apparent in a moment.) One often develops finite difference boundary conditions by writing the difference equation at nodes 1 and J (the last node). This introduces fictitious nodes 0 and $J+1$, whose values are determined by the boundary conditions. We will use the following boundary conditions, which allow the interior information to propagate out of the domain but do not introduce any new information into the domain:

$$\rho_{J+1}^{n+1} = \rho_J^n$$

$$\rho_0^{n+1} = \rho_1^n \qquad (1.29)$$

The computer code shown in Fig. 1.6 demonstrates how simple it is to program and solve this system of equations, and to integrate the governing equation from an initial condition. The program begins with a pulse at $x^* = 0.5$, and we expect the pulse to propagate with no distortion either forward or backward, depending on the density and the wave speed.

The FTCS scheme is simple, easy to understand, and easy to program. The only thing wrong with this scheme is that it is always unstable! For any choice of Δt and Δx we find that the solution diverges, as illustrated in Fig. 1.7. Before investigating the reason for this problem, let us introduce a "fix," known as Lax's method. In this method, the term ρ_j^n on the right-hand side of Eq. (1.28) is replaced by an average value at the previous time step,

$$\rho_j^n \approx \frac{1}{2}(\rho_{j+1}^n + \rho_{j-1}^n) \qquad (1.30)$$

1.3 ONE-DIMENSIONAL TRAFFIC FLOW

```fortran
      program traffic
      parameter (nmax=41)
      dimension r_old(nmax), r_new(nmax)
c
      write(*,'(a,$)')'Enter initial density [0,1] -> '
      read(*,*)r_0
c
c  Set density to r_0 everywhere, adding a spike at x = 0.5
c
      dx = 1.0/(nmax-1)
      call init(r_old,nmax,dx,r_0)
c
c  Choose dt to satisfy the CFL stability condition
c
      if(r_0.ne.0.5)then
        dt = abs(dx/(1.0-2.*r_0))
      else
        dt = dx
      endif
c
c  Integrate up to time = 1
c
      nsteps = 1.0/dt
      do 100 n=1,nsteps
        time = dt*(n-1)
        do 50 i=2,nmax-1
c  For FTCS scheme, r_avg = r_old. Modify for Lax's method
          r_avg = r_old(i)
          r_new(i) = r_avg - 0.5*(dt/dx)*(1.0-2.0*r_old(i))*
     +                              (r_old(i+1) - r_old(i-1))
50      continue
        r_new(1) = r_old(1)
        r_new(nmax) = r_old(nmax)
        call update(r_new,r_old,nmax,n,time,dt,dx,nsteps)
100   continue
      stop
      end
```

Figure 1.6: FORTRAN program for the FTCS scheme. Subroutine `init` initializes the density, and subroutine `update` prints out results and shifts the new solution vector into the old one.

The result of this change is illustrated in Figs. 1.8 and 1.9, where we see that the pulse does indeed propagate in the correct direction, and with the correct speed. However, we also see that the pulse may gradually distort and spread, depending on the time step size and grid spacing. This effect is known as *dispersion*.

This example provides a graphic illustration of both the strength and the potential pitfalls of numerical solution methods. A code to solve an analytically intractable

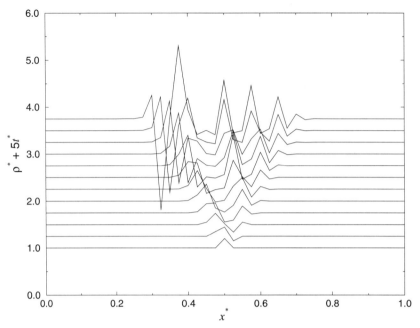

Figure 1.7: Computed propagation of a pulse beginning at $x^* = 0.5$ on an otherwise uniform density of 0.75, using the FTCS algorithm. The ordinate is $\rho^* + 5t^*$, so that several time steps can be shown superimposed. The instability of the algorithm is apparent.

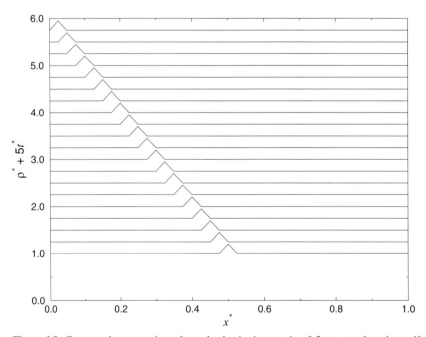

Figure 1.8: Computed propagation of a pulse beginning at $x^* = 0.5$ on an otherwise uniform density of 0.75, using Lax's algorithm with $|c|\Delta t^*/\Delta x^* = 1$. Note the undistorted propagation of the pulse in the $-x^*$ direction.

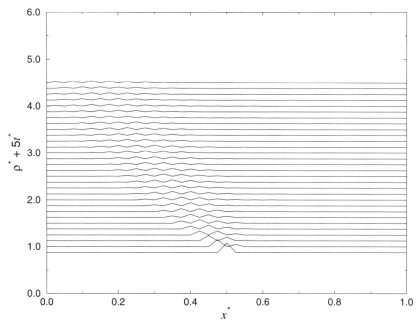

Figure 1.9: Computed propagation of a pulse beginning at $x^* = 0.5$ on an otherwise uniform density of 0.75, using Lax's algorithm, with $|c|\Delta t^*/\Delta x^* = 0.5$. Note the oscillatory dispersion superimposed on the backward propagation.

problem was fairly easy to write. Unfortunately, it was also easy to write a code that failed to solve the problem correctly. Why doesn't the simple FTCS scheme work? What justification is there for using Lax's method? Why does it improve stability? Where does the oscillatory dispersion come from? More importantly to the model user, when the results look plausible, does that mean they are correct? It is very instructive to pursue some of these questions in the context of this simple example, in which the analysis is relatively straightforward.

To simplify the consideration of stability, suppose that the wave speed is constant; that is, we rewrite Eq. (1.26) as

$$\frac{\partial \rho^*}{\partial t^*} = -c \frac{\partial \rho^*}{\partial x^*} \tag{1.31}$$

where c is the wave speed. The difference equation then becomes

$$\rho_j^{n+1} = \rho_j^n - \frac{c\Delta t^*}{2\Delta x^*}\left(\rho_{j+1}^n - \rho_{j-1}^n\right) \tag{1.32}$$

Constant wave speed still provides a meaningful model, because it will tell us how the numerical solution will behave for small perturbations around an average density. If a method is unstable for small perturbations, we expect it to be unstable for large perturbations too.

If the equations for all the nodes are assembled into matrix form, we obtain

$$\{\rho^{n+1}\} = [B]\{\rho^n\} \tag{1.33}$$

where $\{\rho^{n+1}\}$ and $\{\rho^n\}$ represent the vectors of nodal values of ρ^* at the new and old time steps, respectively, and $[B]$ is a matrix of constant coefficients. The form of $[B]$

can be deduced from Eq. (1.32) for $j \neq 1, J$ as

$$B_{jk} = \begin{cases} -\frac{c\Delta t^*}{2\Delta x^*}, & k = j - 1 \\ 1, & k = j \\ +\frac{c\Delta t^*}{2\Delta x^*}, & k = j + 1 \\ 0, & \text{otherwise} \end{cases} \quad (1.34)$$

At the endpoints the boundary conditions modify $[B]$, but the matrix remains tridiagonal.

If we begin with the initial condition denoted by $\{\rho^0\}$, then application of Eq. (1.33) yields

$$\{\rho^1\} = [B]\{\rho^0\} \quad (1.35)$$

and repeated application for n time steps yields

$$\{\rho^n\} = \underbrace{[B][B]\cdots[B]}_{n \text{ times}}\{\rho^0\} = [B]^n\{\rho^0\} \quad (1.36)$$

where $[B]^n$ indicates that the matrix $[B]$ is raised to the power n.

A numerical scheme is said to be *stable* if an error in the solution stays bounded with succeeding time steps. To examine the stability of this algorithm, let us suppose that the initial vector contained some error, that is,

$$\{\bar{\rho}^0\} = \{\rho^0\} + \{\varepsilon^0\} \quad (1.37)$$

where $\{\rho^0\}$ is again the initial solution vector and $\{\varepsilon^0\}$ is a small error. After n iterations using this starting vector, we would obtain a solution of

$$\{\bar{\rho}^n\} = [B]^n\{\rho^0\} + [B]^n\{\varepsilon^0\} \quad (1.38)$$

The error at time step n is the difference between the correct numerical solution $\{\rho^n\}$ and the actual solution $\{\bar{\rho}^n\}$. This difference is given by

$$\{\varepsilon^n\} = -[B]^n\{\varepsilon^0\} \quad (1.39)$$

We can see clearly from Eq. (1.39) that if the error grows, even infinitesimally, at a time step, then the error will grow without bound by repeated application of the algorithm at subsequent time steps. Such an algorithm is said to be unstable. There are two principal methods for examining algorithms for evidence of such instability. The first is called a von Neumann or Fourier analysis, in which one examines the amplification of errors of individual frequency components. This method is the easiest to apply, and we shall examine it first.

When the density is known only as values at discrete points, as in a numerical solution, then not all possible functional forms for ρ can be represented. The distributions that may appear can be written as a truncated Fourier series

$$\varepsilon \sim \sum_{j=0}^{J} b_j(t^*) e^{ik_j x^*} \quad (1.40)$$

1.3 ONE-DIMENSIONAL TRAFFIC FLOW

where the $b_j(t^*)$ are time-dependent coefficients, J is the number of grid points, $i = \sqrt{-1}$, and $k_j = j\pi/L$ and L is the length of the domain. Substitution of the series solution into the differential equation reveals that $b_j(t^*)$ is proportional to e^{at^*}. A sufficient condition for the errors to not grow in time is that the real part of a, $\text{Re}(a)$, be less than or equal to zero: $\text{Re}(a) \leq 0$.

It is sufficient to examine the behavior of a single component b_j, and infer that stability for each coefficient implies stability for the entire series. To that end we consider a single term, and we examine its behavior as follows. Substituting $e^{at^*}e^{ik_jx^*}$ into the difference equation for ε_j, Eq. (1.32), yields

$$e^{a(t^*+\Delta t^*)}e^{ik_jx^*} = e^{at^*}e^{ik_jx^*} + \frac{c\Delta t^*}{2\Delta x^*}\left(e^{at^*}e^{ik_j(x^*+\Delta x^*)} + e^{at^*}e^{-ik_j(x^*-\Delta x^*)}\right) \quad (1.41)$$

Notice that the left-hand side of Eq. (1.41) can be written as $e^{a\Delta t^*}(e^{at^*}e^{ik_jx^*})$, and thus the change in the original signal at time t^* at time $t^* + \Delta t^*$ is embodied in the term $e^{a\Delta t^*}$, which is called the *amplification factor*. Dividing through by $e^{at^*}e^{ik_jx^*}$ yields

$$e^{a\Delta t^*} = 1 + \frac{c\Delta t^*}{2\Delta x^*}\left(e^{ik_j\Delta x^*} - e^{-ik_j\Delta x^*}\right) \quad (1.42)$$

Application of a trigonometric identity allows us to rewrite this expression as

$$e^{a\Delta t^*} = 1 - \frac{c\Delta t^*}{2\Delta x^*}(2i \sin k_j \Delta x^*) \quad (1.43)$$

We introduce the shorthand notation $\nu = c\Delta t^*/\Delta x^*$, called the *Courant number*, and $\beta = k_j \Delta x^*$, and we write

$$e^{a\Delta t^*} = 1 - i\nu \sin \beta \quad (1.44)$$

Finally, let us rewrite

$$e^{a\Delta t^*} = |e^{a\Delta t^*}|e^{i\phi} \quad (1.45)$$

where

$$\phi = \tan^{-1}(-\nu \sin \beta) \quad (1.46)$$

and

$$|e^{a\Delta t^*}| = 1 + \nu^2 \sin^2 \beta \quad (1.47)$$

We can see from this form that the amplification factor applies a change in amplitude of the error, given by its magnitude, and a phase shift ϕ. Stability of the algorithm requires that the magnitude of the amplification factor be at most one. Clearly, for this algorithm, there is no nonzero value of ν that satisfies this condition, and the algorithm is said to be unconditionally unstable.

It is left as an exercise to apply the same procedure to Lax's form of the difference equation, that is,

$$\rho_j^{n+1} = \frac{1}{2}\left(\rho_j^{n+1} + \rho_j^{n-1}\right) - \frac{c\Delta t^*}{2\Delta x^*}\left(\rho_{j+1}^n - \rho_{j-1}^n\right) \quad (1.48)$$

to show that the amplification factor for this case becomes

$$e^{a\Delta t^*} = \cos\beta + i\nu\sin\beta \tag{1.49}$$

and

$$|e^{a\Delta t^*}| = (\cos^2\beta + \nu^2\sin^2\beta)^{1/2} \tag{1.50}$$

This leads to the *conditional stability criterion*

$$\nu = \frac{c\Delta t^*}{\Delta x^*} \leq 1 \tag{1.51}$$

known as the *Courant–Friedrichs–Lewy stability condition*, also called the Courant condition or the CFL limit. The CFL limit can be interpreted physically as the ratio of the true wave speed c to the speed $(\Delta x^*/\Delta t^*)$ at which information propagates in the numerical scheme. When this number exceeds unity, the numerical scheme cannot propagate information as fast as the physical system requires, and instability results. For a stable solution the time step and grid spacing must be chosen so that the waves propagate one grid step or less in each time step.

We may further observe that if $\nu < 1$, then the magnitude of the amplification factor $(\cos^2\beta + \nu^2\sin^2\beta)^{1/2}$ is less than one, and the error will decrease geometrically with time. Unfortunately, the same geometric decrease will apply to the true solution, as well, and the algorithm is said to be *dissipative*. For $\nu < 1$, the amplification factor also depends on the wave number, and some frequency components are dissipated more rapidly than others.

Note also that if we choose Δt^* and Δx^* such that the equality holds in Eq. (1.51), then the magnitude of the amplification factor is exactly unity for all β, and no dissipation will occur. This gives the result shown in Fig. 1.8, which represents a bit of cheating on the part of the authors. For this very special choice of time step and grid size, there is no dissipation. However, a slightly larger time step would be unstable, and a slightly smaller step would show dissipation.

We may make a further observation about Lax's method, which is actually characteristic of all numerical schemes. If we rearrange terms in Eq. (1.48), we may obtain

$$\frac{\rho_j^{n+1} - \rho_j^n}{\Delta t^*} = -c\left(\frac{\rho_{j+1}^n - \rho_{j-1}^n}{2\Delta x^*}\right) + \frac{(\Delta x^*)^2}{2\Delta t^*}\left(\frac{\rho_{j+1}^n - 2\rho_j^n + \rho_{j-1}^n}{(\Delta x^*)^2}\right) \tag{1.52}$$

We now ask ourselves: What differential equation would have this form as its FTCS finite difference representation? The answer is

$$\frac{\partial\rho^*}{\partial t^*} = -c\frac{\partial\rho^*}{\partial x^*} + \frac{(\Delta x^*)^2}{2\Delta t^*}\frac{\partial^2\rho^*}{\partial x^{*2}} \tag{1.53}$$

This equation contains a second derivative that was not present in Eq. (1.26), arising from our choice of discretization. The additional term in this case corresponds to a diffusion process. Thus, we see that Lax's algorithm may be viewed as introducing an *artificial* diffusivity, whose magnitude is $(\Delta x^*)^2/(2\Delta t^*)$. This term produces the dissipation seen in Fig. 1.9. However, it also produced the stability we so much appreciated.

A shortcoming of the von Neumann stability analysis is that it does not consider boundary conditions, and these may affect stability. An analysis that considers the boundary conditions would examine the eigenvalues λ_j, $j = 1, \ldots, J$ of the matrix $[B]$ and show that, if $|\lambda_j| \leq 1$ for all j, then the method is stable. This approach can be difficult to implement in practice, however, where finding the eigenvalues of large matrices can be very costly. In some cases, for example those in which the boundary conditions are periodic, the two methods produce the same stability criteria.

Is there a way to get stability without distorting the solution? Yes. Using a fine grid (small Δx^*) keeps the artificial diffusivity small. But as the grid size is reduced, one must also reduce the time step to maintain stability, as shown in Eq. (1.51). The best solutions come from a very fine grid and a very small time step, where the two factors are carefully balanced. We should note that there are better algorithms for this problem than Lax's method (e.g., Tannehill et al., 1997). These algorithms are stable and less dissipative. Most of these are implicit algorithms, which require somewhat more complex computation. In some problems this issue never arises. However, most real-life models present the analyst with the same trade-off between stability and accuracy and solution time and complexity. Only by understanding when this will occur can one find useful and reliable solutions.

1.4 SUMMARY

A model is a set of equations that represents a physical system. We have outlined a general procedure for constructing mathematical models. This procedure will be just as applicable for problems in materials processing as it was for the pendulum or for traffic flow. The elements of any model include general physical laws, constitutive equations, and boundary or initial conditions.

Invariably one must make assumptions when constructing a model, and each assumption contributes its own benefits and liabilities to the model. The authors are firm believers in making all the assumptions of a model as explicit and clear as possible. In this chapter, scaling the variables and the governing equations provided an easy way to obtain estimates for some parts of the solution. In later chapters we will see that scaling can also be used to evaluate the accuracy of various assumptions and to guide model simplification. This is the reason we place such strong emphasis on scaling.

Numerical methods provide an important way to solve model equations, particularly when the equations are nonlinear or the solution is multidimensional. However, numerical solutions also have their own pitfalls. It is an unfortunate fact of numerical life that the simplest algorithms are rarely the best. In many problems a numerical solution must be pursued with care, with a good deal of understanding about the method, in order to achieve the twin goals of stability and accuracy.

BIBLIOGRAPHY

C. L. Dym and E. S. Ivey. *Principles of Mathematical Modeling*. Academic Press, New York, 1980.

R. Habermann. *Mathematical Models*. Prentice-Hall, Englewood Cliffs, NJ, 1977.

J. C. Tannehill, D. A. Anderson, and R. H. Pletcher. *Computational Fluid Mechanics and Heat Transfer*. Taylor and Francis, Washington, DC, 2nd edition, 1997.

G. B. Whitham. *Linear and Nonlinear Waves*. Wiley, New York, 1974.

EXERCISES

1 Pendulum with large oscillations. Consider the behavior of the pendulum for large-amplitude oscillations. Use the nonlinear form of Eq. (1.3), that is,

$$\ddot{\theta} + \sin\theta = 0$$

(a) Multiply the equation by $\dot{\theta}$, and show that the result is $d(\cdot)/dt = 0$.

(b) Integrate the equation and evaluate the constant from the initial condition.

(c) Rearrange the result into an expression of the form $\dot{\theta} = (\cdot)$, and then integrate from $t = 0$ to $t = T/4$. What are the corresponding integration limits on θ? Show that the result can be cast in terms of a complete elliptic integral of the first kind:

$$K(m) = \int_0^{\pi/2} \frac{d\psi}{\sqrt{1 - m\sin^2\psi}}$$

where $m = \sin^2(\theta_0/2)$. Hint: Define a new variable ψ from $\sin(\theta/2) = \sin(\theta_0/2)\sin\psi$.

(d) After this equation is solved, the computed period is approximated by

$$T = 2\pi\left(1 + \left(\frac{1}{2}\right)^2 \sin^2\frac{\theta_0}{2} + \left(\frac{1 \times 3}{2 \times 4}\right)^2 \sin^4\frac{\theta_0}{2} + \cdots\right)$$

Truncating this expression after the first term gives the period predicted by the linear model. Find the value of θ_0 where the error in the linear-model period is 1%. Explain why this value is different from the 14° error bound discussed in regard to Eq. (1.3).

2 Pendulum with friction. Add friction to the analysis of the pendulum, considering only the linear case. Frequently the friction force is assumed to be proportional to the velocity. Analyze this case by revising Eq. (1.3) to read

$$mL\ddot{\theta} = -CL\dot{\theta} - mg\theta$$

(a) Redo the scaling analysis performed in the text for this new equation. The first and last terms in the equation will again have no coefficients, but there will be a parameter multiplying the $\dot{\theta}$ term. Call the parameter F. What are its units?

(b) Assume that F is small compared to one. Solve the differential equation, then compute and plot the period of the pendulum for increasing values of F.

(c) Is there a critical value of F where the behavior of the pendulum changes from oscillatory to some other form?

3 Pendulum with a flexible string. Relax the assumption that the pendulum string was inextensible. Take the following steps.

(a) Introduce a constitutive equation relating the string tension F_r to its length r. Let the unstressed length be L, the cross-sectional area A, and assume the string is linearly elastic with Young's modulus E.

EXERCISES

(b) Develop the governing equations for this model and give an appropriate set of initial conditions. Simplify the equations by assuming that θ is small.

(c) Scale the variables and the governing equations. Your equations will contain some dimensionless combinations of problem parameters. If one of these combinations is small, the extension of the string can be neglected. Identify this dimensionless group.

Important: Do not solve the governing equations.

4 *Passing in the traffic model.* In the one-dimensional traffic flow model presented in this chapter, can one car pass another? If so, how would you detect that in the model results? Explain your answer.

5 *Vehicle density data.* Table 1.1 gives data for the number of cars within each successive 0.02-mile interval over 4 miles of road. Enter these data into a spreadsheet or plotting program. Plot the data as vehicle density (cars per mile). Now choose an averaging aperture of two intervals (0.04 miles) and plot the new density function. Repeat for averaging apertures of 10 and 50 intervals. Which averaging apertures give a meaningful density function?

6 *Traffic constitutive behavior.* The traffic flow model used Eq. (1.21) as a constitutive model, giving traffic velocity v as a function of traffic density ρ.

Table 1.1 Vehicle Density Data for Problem 5

1	1	1	2	2	2	2	2	1	1
1	1	2	2	2	2	2	2	2	1
1	1	1	1	2	2	1	2	1	1
1	1	1	1	2	2	2	2	1	0
1	1	2	2	2	2	2	2	1	1
1	1	1	2	2	2	2	1	1	1
1	1	2	2	2	1	2	2	1	1
1	1	2	2	2	3	2	2	1	1
2	1	1	2	2	2	2	2	1	1
2	1	1	2	2	3	2	1	1	1
1	1	2	2	2	2	2	2	1	2
1	1	2	2	2	2	2	1	1	2
1	1	2	1	2	2	2	1	1	1
1	2	2	2	2	2	2	1	1	1
0	1	2	2	2	2	2	1	1	1
1	2	2	2	2	2	2	1	0	1
1	1	2	2	2	2	1	2	1	1
1	1	2	2	2	2	2	2	1	0
1	1	2	2	2	2	2	1	1	1
0	1	2	2	2	1	2	1	1	1

Values are the number of cars within successive 0.02-mile intervals. Read down the first column, then down the second column, etc.

(a) Based on your own knowledge and experience, provide reasonable numbers for v_{max} and ρ_{max}. Convenient units are miles per hour and cars per mile. Using these numbers, calculate the capacity of a one-lane road.

(b) Construct an improved constitutive model for $v(\rho)$ that includes the following phenomena. When traffic density is low, cars travel at an average speed that is independent of density. This speed is determined by a legal speed limit. However, once the density increases above a certain value, the average speed goes down. Again, provide reasonable numbers for the constants in your constitutive model.

(c) For your two models, plot the curves of flow rate Q as a function of density (like Fig. 1.4). When v was a linear function of ρ, we found that Q was maximum at some intermediate density between zero and ρ_{max}, and that the wave speed was positive below this density and negative above it. Will the different shapes of your new curve alter these behaviors of the traffic model?

7 *Traffic model with an intersection.* Extend the one-dimensional traffic model to include an intersecting road. Think of the main road as a limited-access highway, and the intersecting road as an on ramp, bringing traffic from a second highway onto the first. Complete Steps 1 through 7 of the modeling procedure. Be sure to state which equations represent physical laws, which are constitutive assumptions, and which are boundary conditions. Do not attempt to solve the model equations. However, based on your understanding of the original model, you should be able to identify conditions under which traffic will back up along the on ramp.

8 *The wave equation.* Determine a general solution to the wave equation,

$$\frac{\partial \rho}{\partial t} + c \frac{\partial \rho}{\partial x} = 0$$

when the wave speed c is constant. Hint: Let $\xi = x - ct$, and use the chain rule to show that this equation reduces to $d\rho/d\xi = 0$.

9 *Continuously stirred tank.* A very different physical system, which appears frequently in chemical plants, is the continuously stirred tank. The tank has volume V, and it is always filled with a fluid. Fluid enters an inlet port of the tank continuously at a volume flow rate Q, and it leaves the outlet port at the same rate. The entering fluid contains a dissolved substance (solute) with concentration $C_i(t)$ (units of moles per liter), and the concentration of solute for fluid leaving the tank is $C_o(t)$.

The tank is vigorously stirred, so much so that we are willing to assume that any solute entering the tank is immediately mixed uniformly with the current contents of the tank. This is a very optimistic assumption about mixing, but it provides a simple model that tells us a good deal about the best case. One purpose of such a tank is to smooth out fluctuations in the solute concentration. That is, we want the variations in $C_o(t)$ with time to be much smaller than the variations in $C_i(t)$.

Formulate a model for this system that would allow you to calculate $C_o(t)$ given $C_i(t)$. Take the following steps.

(a) Restate the scope and goals of the model.

(b) Sketch the system and list the basic quantities.

(c) Develop a mathematical description of the sketch. Here you should write an expression for the amount of solute inside the tank (in moles), the rate at which solute is entering the tank at the inlet (in moles per second), and the rate at which solute is leaving the tank.

(d) Write fundamental equations that govern the primary variables. In this case you want to make a mass balance on the solute in the tank.

(e) Introduce constitutive equations. The constitutive information here is about mixing in the tank: If we assume that the tank is always well mixed (the concentration is the same everywhere in the tank) then the concentration of fluid leaving the tank equals the concentration for fluid in the tank. (You may have used this information already.)

(f) There are no more assumptions to be made here, so you can proceed to scale the variables and governing equations. Use a reference concentration C_{ref} to scale the concentrations. You should be able to find a characteristic time scale in terms of the parameters of the problem.

(g) Solve your governing equation for the case in which the tank initially has concentration C_{ref} and, for all $t > 0$, the inlet concentration is $C_i = 0$. The result will tell you how long it takes to flush the solute out of the tank.

10 **Lax's method (advanced).** Modify the code given in Fig. 1.6 to represent Lax's formulation, and show that a disturbance on an otherwise uniform density field propagates undistorted for $\rho_0 = 0.25$ and $\rho_0 = 0.75$.

11 **Stability analysis for Lax's method (advanced).** Follow the procedure for the von Neumann stability analysis to determine the stability conditions for Lax's method applied to the one-dimensional wave equation.

(a) Start with the explicit form for Lax's method,

$$\rho_j^{n+1} = \frac{1}{2}\left(\rho_j^{n+1} + \rho_j^{n-1}\right) - \frac{c\Delta t^*}{2\Delta x^*}\left(\rho_{j+1}^n - \rho_{j-1}^n\right)$$

and substitute $\rho \sim e^{at^*}e^{ik_j x^*}$; then reduce to a form with the amplification factor alone on the left-hand side.

(b) Find the magnitude of the amplification factor and determine the condition for stability.

CHAPTER TWO

Governing Equations

Materials processing models often deal with the transport of mass, momentum, and energy. In this chapter we derive governing equations for the conservation of these quantities. We focus on differential forms, which are used in continuum models. We also include some integral forms, which help in the physical interpretation of the equations. The forms we derive are quite general, and you may wonder if the complexity this adds is worthwhile. Remember that we are seeking equations that have a *very* wide range of validity, so that, when we simplify them for a particular model, we can be sure that no term has been left out.

Although the focus of this chapter is on balance equations, we also introduce a few classical constitutive equations: the Newtonian fluid and Fourier's law of heat conduction. However, we also retain forms of the balance equations that can be used with other constitutive equations.

Insofar as possible, we avoid reference to particular coordinate systems, by using general vector and tensor notation. Readers who are unfamiliar with this style may want to review Appendix A. All of the governing equations derived in this chapter are written out in terms of their components for rectangular, cylindrical, and spherical coordinates in Appendix B. These component forms are the actual starting point for most models.

2.1 PRELIMINARIES

2.1.1 MATERIAL AND SPATIAL DESCRIPTIONS

In this chapter we want to develop equations for quantities that associate a value with each point in the material, quantities that include velocity, temperature, pressure, and stress. We also want to deal with materials that are moving and deforming. This gives us two choices for the mathematical descriptions of these quantities. With the use of temperature as an example, the two choices are:

> *a spatial description*, which gives the value of temperature at each point in space as a function of time, and
> *a material description*, which gives the value of temperature at each material point as a function of time.

2.1 PRELIMINARIES

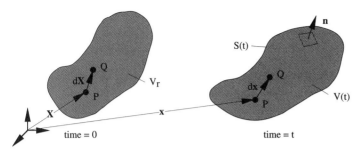

Figure 2.1: An arbitrary material volume at two different times; P and Q are material points, **X** is the material coordinate of point P, and **x** is the spatial coordinate of P at time t. The triad of axes represents the reference origin, fixed in space.

To write this mathematically, we use **x** as the *position vector*. That is, each value of **x** denotes a unique, fixed position in space, regardless of what material might occupy that point at a given time. Here **x** is the coordinate for a spatial description, which we could write as

$$T = g_s(\mathbf{x}, t) \tag{2.1}$$

with g_s representing some function. We can also choose some instant (say $t = 0$) as the reference time, and conveniently label any point in the material by using its position (e.g., its x_1, x_2, and x_3 coordinates) at that time. Use **X** to denote this vector quantity, which we call the *material coordinate*. Every point in the material has its own value of **X**, and it keeps that value of **X** through all subsequent motion. Here **X** is the coordinate for a material description, which could be written as

$$T = g_m(\mathbf{X}, t) \tag{2.2}$$

Note that g_m is a different function from g_s, because it takes different arguments, even though it describes the same information.

Figure 2.1 illustrates the difference between material and spatial coordinates. A material volume is shown at two different times, time = 0 (which we choose here as the reference time), and some other time t. The material point P has coordinates **X** at the reference time, and it keeps this material coordinate even at the later time, when it has some other position **x**.

We can translate between the material and spatial descriptions if we know how the material is moving. To describe the motion of a material we can specify the location of each material particle at any time. That is, we give the position **x** as a function of material coordinate **X** and time t,

$$\mathbf{x} = \mathbf{x}(\mathbf{X}, t) \tag{2.3}$$

If the material does not fracture or fragment then this function will be continuous, and one customarily assumes that it is differentiable. We also assume that the function can be inverted, to give

$$\mathbf{X} = \mathbf{X}(\mathbf{x}, t) \tag{2.4}$$

That is, one could ascertain which material point **X** occupies the position **x** at time t. The functions in Eqs. (2.3) and (2.4) are called *displacement functions*.

As a simple example of all this, consider a vehicle traveling along a highway during the winter. It is cold outside, so the heater is on inside the vehicle, keeping the occupants warm. A point inside the vehicle, for example on the dashboard, is a material point, as it always consists of the same material. The temperature at that point, made by an observer sitting in the front seat, does not change with time. In contrast, an observer standing by the side of the road, measuring the temperature at a particular point in space, would have a different view. When there is no vehicle at that point, the observer would measure the ambient temperature, which is much lower than the temperature of our material point. When the vehicle crosses his path of observation, the temperature would change with time, first increasing to the temperature of the interior, then falling back to the ambient as the car passes. These different results are precisely what we mean when we refer to the Lagrangian (inside the car) and Eulerian (fixed at a point on the road) representations.

Now let us do something similar by using equations. Suppose we have a material that is translating with a uniform velocity v_0 in the x_1 direction. If we choose $t = 0$ to define the reference configuration, then the displacement functions corresponding to Eq. (2.3) are

$$x_1 = X_1 + v_0 t \tag{2.5}$$

$$x_2 = X_2 \tag{2.6}$$

$$x_3 = X_3 \tag{2.7}$$

The inverse transformation would simply take $v_0 t$ to the other side of Eq. (2.5). Now suppose that the material has a temperature gradient in the x_1 direction, such that the material description of temperature corresponding to Eq. (2.2) is

$$T = 20 + 10 X_1 \tag{2.8}$$

Substituting Eq. (2.5) to eliminate X_1 gives us the spatial description corresponding to Eq. (2.1) as

$$T = 20 + 10 x_1 - 10 v_0 t \tag{2.9}$$

In this example the temperature in the material description Eq. (2.8) is steady, whereas the temperature in the spatial description of Eq. (2.9) varies with time.

Material and spatial descriptions are both used in modeling materials processing problems. When the material is a solid and its deformation is not too large, a material description is convenient. Most finite element codes for solid mechanics problems use a mesh that is fixed in the material so the mesh deforms as the material deforms. This is a material description. Material descriptions are also called *Lagrangian* descriptions.

When the material is a fluid it may undergo extremely large deformations, and a spatial description is usually more convenient. Most computational fluid dynamics software uses a mesh or grid that is fixed in space, and the fluid "flows through" the grid. This is a spatial description. Spatial descriptions are also called *Eulerian* descriptions.

In this chapter we will develop Eulerian-type governing equations that can be applied at fixed points in space, or to fixed control volumes. However, the basic

2.1 PRELIMINARIES

conservation principles for mass, momentum, and energy apply to fixed bodies of material. Thus, we will need some mathematical tools for moving back and forth between the material and spatial viewpoints. These tools are the material derivative and a change of configuration.

2.1.2 VELOCITY AND THE MATERIAL DERIVATIVE

One use of the displacement functions is to find the velocity of the material. Velocity is the rate of change of position ($\mathbf{x}$), for a fixed material particle. More formally, the velocity vector $\mathbf{v}$ is defined as

$$\mathbf{v} \equiv \left.\frac{\partial \mathbf{x}}{\partial t}\right|_{\mathbf{X}} \tag{2.10}$$

where $\partial \mathbf{x}/\partial t$ means the derivative of the *function* on the right-hand side of Eq. (2.3).

Applying this to the example in the preceding section, we take the derivative of Eq. (2.5) to find

$$v_1 = \left.\frac{\partial x_1}{\partial t}\right|_{X_1} = v_0 \tag{2.11}$$

and the other components of $\mathbf{v}$ are zero.

Equation (2.10) contains a particular kind of time derivative, in which the material point is held constant. This is called the *material derivative*. We will see this type of derivative frequently, and we will denote it as D/Dt. Note that the material derivative "follows" a material point, no matter where it may be in space. In contrast, we will normally use the partial derivative $\partial/\partial t$ to mean a time derivative at fixed spatial position $\mathbf{x}$, regardless of what material points occupy that position at different times.

If we have a material description of some quantity, such as temperature, then we can take the material derivative by simply differentiating with respect to time. For example, differentiating Eq. (2.8), we find that $DT/Dt = 0$. This means that the temperature at any material point does not change with time.

When we have a spatial description, it is easy to take the partial time derivative, and from Eq. (2.9) we find $\partial T/\partial t = -10v_0$. However, taking the material derivative of a spatial description requires a bit more effort. Let us develop this in terms of some general spatial function $\psi(\mathbf{x}, t)$.

We start by applying the chain rule

$$\frac{D\psi}{Dt} = \left.\frac{\partial \psi}{\partial t}\right|_{\mathbf{x}} + \left.\frac{\partial \psi}{\partial x_1}\frac{\partial x_1}{\partial t}\right|_{\mathbf{X}} + \left.\frac{\partial \psi}{\partial x_2}\frac{\partial x_2}{\partial t}\right|_{\mathbf{X}} + \left.\frac{\partial \psi}{\partial x_3}\frac{\partial x_3}{\partial t}\right|_{\mathbf{X}} \tag{2.12}$$

where we have expanded the derivative in terms of the Cartesian components of $\mathbf{x}$. In a spatial description the partial time derivative of ψ means the derivative at fixed *position* $\mathbf{x}$. A material particle may change position over time, so we also need the additional terms provided by the chain rule.

We can simplify Eq. (2.12) by remembering that the derivative of x_i with respect to t is simply the component v_i of the velocity vector, as in Eq. (2.10). This allows us to write the material derivative as

$$\frac{D\psi}{Dt} = \frac{\partial \psi}{\partial t} + \sum_{i=1}^{3} v_i \frac{\partial \psi}{\partial x_i}$$
$$= \frac{\partial \psi}{\partial t} + v_i \frac{\partial \psi}{\partial x_i} \qquad (2.13)$$
$$= \frac{\partial \psi}{\partial t} + \mathbf{v} \cdot \nabla \psi$$

In the second line we have introduced the summation convention, where a repeated index in a product indicates summation over that index. Equation (2.13) is the tool we were looking for: it tells us how to obtain the material derivative of a quantity when we have a spatial description.

Returning to the example in Eq. (2.9), we apply Eq. (2.13) to compute the material derivative. Because $v_2 = v_3 = 0$ and v_1 is given by Eq. (2.11), we find

$$\frac{DT}{Dt} = \frac{\partial T}{\partial t} + v_1 \left(\frac{\partial T}{\partial x_1} \right) \qquad (2.14)$$
$$= -10 v_0 + v_0(10) \qquad (2.15)$$
$$= 0 \qquad (2.16)$$

Of course we get the same answer as we obtained from the material description, but this time we only used the spatial field of temperature and the velocity distribution to find it.

In this book we will work mostly with spatial formulations, and the material derivative D/Dt will serve as a shorthand expression that can be expanded by using any of the right-hand sides of Eq. (2.13). However, one should always remember its physical interpretation, which is the time derivative following a material point. Note that the last form of Eq. (2.13) is not tied to any coordinate system, and it can be expanded for cylindrical, spherical, and other coordinate systems. The expressions for DT/Dt in rectangular, cylindrical, and spherical coordinates are written out in Appendix B.9.

A special case that arises occasionally is the time derivative seen by an observer who moves with some arbitrary velocity $\mathbf{V}$. Call this derivative $d\psi/dt$. The same type of chain rule expansion as Eq. (2.12) can be developed, but now the spatial derivatives $\partial x_1/\partial t$ and so on represent the observer's motion. The result is simply Eq. (2.13), with $\mathbf{v}$ replaced by $\mathbf{V}$:

$$\frac{d\psi}{dt} = \frac{\partial \psi}{\partial t} + V_i \frac{\partial \psi}{\partial x_i} \qquad (2.17)$$

2.1.3 CHANGE OF CONFIGURATION

Another concept that will be helpful is the *configuration* of the material. We specify a particular configuration by giving the location of each material particle. If a material body is moving or deforming, then its configuration changes as a function of time.

The displacement function in Eq. (2.3) is one way of describing these configurations. We already chose a reference time, to define the material coordinates **X**. The configuration at that time is called the *reference configuration*. Suppose we choose some arbitrary part of the material, as shown in Fig. 2.1. At time t this material occupies the volume $V(t)$ and is enclosed by the surface $S(t)$. Let V_r denote the volume that the same material occupies in the reference configuration, $V_r = V(t_{\text{ref}})$. Note that V varies in time, but V_r does not.

Now choose any time t other than the reference time, and examine the difference between the two configurations. This is conveniently done in terms of the relative positions of two nearby material points, say P and Q in Fig. 2.1. The vector $d\mathbf{X}$ connects P to Q in the reference configuration, whereas at time t this vector becomes $d\mathbf{x}$. We can find out how a material vector such as $d\mathbf{X}$ changes, and thus how the material deforms in the neighborhood of point P, by taking the derivatives of the displacement functions. These derivatives are the components of the *deformation gradient* tensor **F**. In Cartesian coordinates, the components of this tensor are

$$F_{ij} = \frac{\partial x_i}{\partial X_j} \tag{2.18}$$

Using **F**, we can calculate anything else we wish to know about the deformation. For instance, if we consider an infinitesimal material volume dV_r in the reference configuration, the magnitude of this volume dV in the current configuration is

$$dV = J dV_r \tag{2.19}$$

where

$$J = \det \mathbf{F} \tag{2.20}$$

That is, J is the Jacobian of the coordinate transformation between **x** and **X**. See the Appendix at the end of this chapter for a derivation of Eq. (2.19). This result allows us to transform any integral over the material volume $V(t)$ in its current configuration to an integral over the volume V_r in the reference configuration:

$$\int_{V(t)} \cdots dV = \int_{V_r} \cdots J dV_r \tag{2.21}$$

This is the second tool we were looking for. Note also that we are free to choose any reference time that is convenient.

2.2 MASS BALANCE

We are now ready to derive the balance equations. We start with conservation of mass.

2.2.1 DIFFERENTIAL FORM

Mass is continuously distributed throughout the material volume, and we refer to the mass per unit volume as the density, $\rho(\mathbf{x}, t)$. We postulate that the total mass M of the system is constant; that is, mass is neither created nor consumed. (It is possible for mass to be converted from one species to another, which we consider in

Chapter 8.) This is the principle of conservation of mass, a statement that can be written mathematically as

$$\frac{dM}{dt} = \frac{d}{dt}\int_{V(t)} \rho dV = 0 \qquad (2.22)$$

Because the domain of integration $V(t)$ changes over time, we must be careful in taking the derivative of this integral. It is convenient to first transform the integral to the reference configuration by using Eq. (2.21):

$$\int_{V(t)} \rho dV = \int_{V_r} \rho J dV_r \qquad (2.23)$$

We substitute this relation into Eq. (2.22), and, now that the domain of integration is constant, we can move the differentiation inside the integral:

$$\int_{V_r} \left(\frac{D\rho}{Dt} J + \rho \frac{DJ}{Dt}\right) dV_r = 0 \qquad (2.24)$$

In taking the derivative inside the integral we have written $d\rho/dt$ and dJ/dt as material derivatives. This is because they are now taken at a volume increment dV_r, which is fixed in the material.

In the Appendix at the end of this chapter we show that the material derivative of J is given by

$$\frac{DJ}{Dt} = J \nabla \cdot \mathbf{v} \qquad (2.25)$$

Using this fact and expanding the material derivative of ρ by means of Eq. (2.13) yields

$$\int_{V_r} \left(\frac{\partial \rho}{\partial t} + \mathbf{v} \cdot \nabla \rho + \rho \nabla \cdot \mathbf{v}\right) J dV_r = 0 \qquad (2.26)$$

We may now return this expression to the instantaneous material volume,

$$\int_{V(t)} \left(\frac{\partial \rho}{\partial t} + \nabla \cdot (\rho \mathbf{v})\right) dV = 0 \qquad (2.27)$$

where we have also consolidated the last two terms. Because this equation holds for any choice of the volume $V(t)$, the integrand must vanish, so that

$$\frac{\partial \rho}{\partial t} + \nabla \cdot (\rho \mathbf{v}) = 0 \qquad (2.28)$$

This is the mass balance equation we were looking for. It is also called the *continuity equation*. An alternate form, using the material derivative, is

$$\frac{D\rho}{Dt} + \rho (\nabla \cdot \mathbf{v}) = 0 \qquad (2.29)$$

Equation (2.28) is sometimes called the *conservative form* of the mass balance, because it contains a partial derivative with respect to time, and the divergence of some quantity, in this case $\rho \mathbf{v}$. The latter quantity is usually called the *flux* associated with the conserved quantity; in this case it would be called the *mass flux*. The reason for this interpretation becomes clear in the integral form.

2.2 MASS BALANCE

2.2.2 INTEGRAL FORM

To see why Eq. (2.28) is called the mass balance equation, return to the integral form of Eq. (2.27). Separating the two terms, we obtain

$$\int_{V(t)} \frac{\partial \rho}{\partial t} \, dV = -\int_{V(t)} \nabla \cdot (\rho \mathbf{v}) \, dV \tag{2.30}$$

Applying the divergence theorem, Eq. A.59, to the right-hand side yields

$$\int_{V(t)} \frac{\partial \rho}{\partial t} \, dV = -\int_{S(t)} \rho \mathbf{v} \cdot \mathbf{n} \, dS \tag{2.31}$$

where $S(t)$ is the external surface of the volume $V(t)$, and $\mathbf{n}$ is the outwardly directed unit normal vector at the surface.

Because all the differentiation takes place inside the integral, the fact that V and S change with time is unimportant. This allows us to apply this equation to an arbitrary volume, including one that is fixed in space. Let V represent such a *control volume*, and let S be its surface. Then, because the volume V is fixed, we can interchange the order of integration and differentiation on the left-hand side, and we write

$$\frac{\partial}{\partial t} \int_V \rho \, dV = -\int_S \rho \mathbf{v} \cdot \mathbf{n} \, dS \tag{2.32}$$

Equation (2.32) is a convenient integral form of the mass balance equation. The left-hand side represents the rate of mass accumulation in the volume, whereas the right-hand side represents the net influx of mass through the surface. If mass is to be conserved then these two quantities must balance; hence the name of the equation.

2.2.3 CONSTANT DENSITY

An important special case is a fluid whose density is constant. In that event we know that

$$\frac{\partial \rho}{\partial t} = 0$$
$$\nabla \rho = 0 \tag{2.33}$$

and Eq. (2.29) reduces to

$$\nabla \cdot \mathbf{v} = 0 \tag{2.34}$$

or, in indicial notation,

$$\frac{\partial v_i}{\partial x_i} = 0 \tag{2.35}$$

This reduction greatly simplifies the solution of the governing equations. It is sometimes stated incorrectly that Eq. (2.34) applies when the material is incompressible. However, incompressibility means only that the density is independent of the pressure. If the density depends on temperature, for example, and the temperature is not uniform in the volume, then Eq. (2.34) does not hold, and one must use the general form of Eq. (2.28).

2.2.4 SPECIFIC VOLUME

In later sections it will be convenient to speak of "specific" quantities, defined as the quantity per unit mass. For example, the *specific volume* $\hat{V}$ is the volume per unit mass. Evidently, $\hat{V} = (1/\rho)$. The material derivative of $\hat{V}$ is

$$\frac{D\hat{V}}{Dt} = -\frac{1}{\rho^2}\frac{D\rho}{Dt} = -\hat{V}^2\frac{D\rho}{Dt} \tag{2.36}$$

Referring back to Eq. (2.29) and substituting, we obtain

$$\rho\frac{D\hat{V}}{Dt} = \nabla \cdot \mathbf{v} \tag{2.37}$$

Thus, we identify the divergence of the velocity field $\nabla \cdot \mathbf{v}$ as the local rate of volume change per unit volume.

2.2.5 THE REYNOLDS TRANSPORT THEOREM

The type of integral we manipulated in the mass balance equation will appear in other derivations, and it is convenient at this point to generalize this treatment. Consider again some generic quantity ψ, and compute the derivative of the integral of $\rho\psi$ over the material volume $V(t)$. This is

$$\frac{d}{dt}\int_{V(t)} \rho\psi\, dV = \frac{d}{dt}\int_{V_r} \rho\psi J\, dV_r \tag{2.38}$$

where once again we have transformed the integral to the reference configuration. Moving the derivative inside the integral, where it becomes a material derivative, we obtain

$$\frac{d}{dt}\int_{V_r} \rho\psi J\, dV_r = \int_{V_r} \left(\frac{D\rho}{Dt}\psi J + \rho\frac{D\psi}{Dt}J + \rho\psi\frac{DJ}{Dt}\right) dV_r \tag{2.39}$$

However, Eq. (2.24) shows that the sum of the first and third terms on the right-hand side is exactly zero. [Remember that V_r is arbitrary, so the integrand in Eq. (2.24) must vanish.] This leaves

$$\frac{d}{dt}\int_{V_r} \rho\psi J\, dV_r = \int_{V_r} \rho\frac{D\psi}{Dt} J\, dV_r \tag{2.40}$$

Finally, returning to the current configuration $V(t)$ gives

$$\frac{d}{dt}\int_{V(t)} \rho\psi\, dV = \int_{V(t)} \rho\frac{D\psi}{Dt}\, dV \tag{2.41}$$

Equation (2.41), known as the *Reynolds transport theorem*, will be used in the derivations in the next several sections.

EXAMPLE 2.2.1 Transverse Velocity in Fiber Spinning

There are few problems that can be solved by using the continuity equation alone. However, every flow field must satisfy the continuity equation, and we can sometimes use this to simplify a problem. Here is an example. In the analysis, we will make lots of assumptions. In the next chapter, we will discuss in detail how

2.3 MOMENTUM BALANCE

Figure 2.2: Geometry and nomenclature for axisymmetric fiber spinning.

and when to make such assumptions. For now, we ask that you take it on faith that the assumptions we make are reasonable.

In the manufacture of synthetic textile fibers, a thin, cylindrical thread of polymer is stretched along its length. It is natural to adopt a cylindrical coordinate system for this problem, with r as the radial coordinate and the z coordinate being directed along the axis of the cylinder (see Fig. 2.2). We assume that the velocity field is axisymmetric so that $v_\theta = 0$ and $\partial/\partial\theta = 0$. Because the thread is thin, it is reasonable to assume that the z-direction velocity does not vary over the cross section, and we will assume that locally this velocity is given by

$$v_z = V_0 + \dot{\varepsilon} z \tag{2.42}$$

where V_0 and $\dot{\varepsilon}$ are constants. Here $\dot{\varepsilon}$ represents the rate of stretching in the axial direction. We now wish to find the radial velocity distribution $v_r(r, z)$ by using this information.

We further assume that the density of the material is constant. Then, from Appendix B.2, we select the continuity equation for constant density, in cylindrical coordinates:

$$\frac{1}{r}\frac{\partial}{\partial r}(rv_r) + \frac{1}{r}\frac{\partial v_\theta}{\partial \theta} + \frac{\partial v_z}{\partial z} = 0 \tag{2.43}$$

With the assumption of axisymmetry, the second term is zero, and we can differentiate Eq. (2.42) with respect to z and substitute the result for the third term. This leaves

$$\frac{1}{r}\frac{\partial}{\partial r}(rv_r) = -\dot{\varepsilon} \tag{2.44}$$

Multiplying by r and then integrating with respect to r gives

$$rv_r = -\frac{\dot{\varepsilon} r^2}{2} + f(z) \tag{2.45}$$

The function $f(z)$ must be determined by a boundary condition. If the fiber remains centered on its axis, then $v_r = 0$ at $r = 0$, and this requires $f(z) = 0$ for all z. Using this, we find the radial velocity distribution to be

$$v_r = -\frac{\dot{\varepsilon} r}{2} \tag{2.46}$$

That is, if the fiber is stretching axially at a rate $\dot{\varepsilon}$, then it must be contracting radially according to Eq. (2.46) to satisfy continuity.

2.3 MOMENTUM BALANCE

2.3.1 SURFACE FORCES, BODY FORCES, AND STRESS

In many problems we will consider the response of a material to imposed forces. These come in two distinct varieties, surface forces and body forces. *Surface forces* are transmitted through mechanical contact at the surface of the material. The surface force acting on any point of a surface is described by the *surface traction vector*, denoted by $\mathbf{t}(\mathbf{x}, t)$. This is defined in terms of the force $d\mathbf{f}$ acting on an area dS as

$$\mathbf{t} = \lim_{dS \to 0} \frac{d\mathbf{f}}{dS} \tag{2.47}$$

Note that $\mathbf{t}$ has units of force over area.

Forces may also be applied to a material without physical contact, for example by the action of gravity or electromagnetic fields. We refer to these forces as *body forces*. The vector $\mathbf{b}(\mathbf{x}, t)$ is the body force per unit mass acting at the point $\mathbf{x}$. More formally, if a body force $d\mathbf{f}$ acts on a volume dV, then

$$\mathbf{b} = \frac{1}{\rho} \lim_{dV \to 0} \frac{d\mathbf{f}}{dV} \tag{2.48}$$

The vector $\mathbf{b}$ has dimensions of force over mass. When gravity is the only body force, then the magnitude of $\mathbf{b}$ equals g, the acceleration that is due to gravity.

Returning to surface forces, if we choose some internal surface within a body as our surface, then the surface traction vector $\mathbf{t}$ describes the force transmitted across that surface. However, this is not a complete description because the value of $\mathbf{t}$ will depend on the orientation of the surface. To obtain a complete description we use the Euler–Cauchy stress principle, which states that there exists a second-order tensor σ such that

$$\mathbf{t} = \mathbf{n} \cdot \sigma \tag{2.49}$$

Here $\mathbf{n}$ is a unit vector normal to the surface; σ is called the *total stress tensor*, or the Cauchy stress. With this definition, the stress component σ_{ij} is the force in the j direction acting on an area normal to the i direction. However, conservation of angular momentum requires that the stress tensor be symmetric, $\sigma_{ij} = \sigma_{ji}$, and we seldom need to be careful about the order of the indices. (We have made an implicit assumption here that the material contains no internal couples. This is valid for most conventional materials. However, there are some materials, such as liquid crystals, in which internal couples do exist. The stress tensor for such materials is not symmetric.)

Equation (2.49) has two important uses. First, it shows that the total stress $\sigma(\mathbf{x}, t)$ completely describes the *internal* forces transmitted through the material at point $\mathbf{x}$. Second, when $\mathbf{x}$ lies on an *external* surface of the body, Eq. (2.49) relates the stress at the surface to the externally applied traction. Thus, Eq. (2.49) is used whenever a surface traction boundary condition is applied.

When a body experiences only a hydrostatic stress corresponding to some pressure p, then the total stress tensor is

$$\sigma = \begin{bmatrix} -p & 0 & 0 \\ 0 & -p & 0 \\ 0 & 0 & -p \end{bmatrix} = -p\mathbf{I} \tag{2.50}$$

Here **I** is the identity tensor. The pressure p has the same units as stress, force per unit area, and the identity tensor is isotropic, so the same traction is applied to any area regardless of its orientation. The negative sign appears because a positive pressure corresponds to a negative (compressive) stress.

In many cases it is useful to separate the total stress tensor into two parts, such that

$$\boldsymbol{\sigma} = -p\mathbf{I} + \boldsymbol{\tau} \tag{2.51}$$

The first term is a hydrostatic stress corresponding to the pressure p, and $\boldsymbol{\tau}$ is called the *extra stress tensor*. Written out in Cartesian components, this is

$$\begin{bmatrix} \sigma_{xx} & \sigma_{xy} & \sigma_{xz} \\ \sigma_{yx} & \sigma_{yy} & \sigma_{yz} \\ \sigma_{zx} & \sigma_{zy} & \sigma_{zz} \end{bmatrix} = \begin{bmatrix} -p & 0 & 0 \\ 0 & -p & 0 \\ 0 & 0 & -p \end{bmatrix} + \begin{bmatrix} \tau_{xx} & \tau_{xy} & \tau_{xz} \\ \tau_{yx} & \tau_{yy} & \tau_{yz} \\ \tau_{zx} & \tau_{zy} & \tau_{zz} \end{bmatrix} \tag{2.52}$$

The shear stress components of $\boldsymbol{\sigma}$ are the same as the shear stress components of $\boldsymbol{\tau}$, whereas the normal stress (diagonal) components of $\boldsymbol{\sigma}$ are divided between $\boldsymbol{\tau}$ and $-p\mathbf{I}$. Note that the symmetry of $\boldsymbol{\sigma}$ implies that $\boldsymbol{\tau}$ is symmetric. When modeling incompressible materials, we will divide the stress according to Eq. (2.51) and take $\boldsymbol{\tau}$ to be the stress associated with fluid deformation.

The choice of p is somewhat arbitrary, as we could add a constant to p and subtract the same constant from the diagonal entries in $\boldsymbol{\tau}$ (e.g., Astarita and Marrucci, 1974, pp. 35–36). There is one special choice of $p = -\text{tr}(\boldsymbol{\sigma})/3$, which gives the property $\text{tr}(\boldsymbol{\tau}) = 0$. In this case, $\boldsymbol{\tau}$ is called the *deviatoric stress*, and we denote it as $\boldsymbol{\sigma}'$. The deviatoric stress is frequently used in solid mechanics, in which the theory of plastic deformation for many materials holds that the deformation is independent of hydrostatic stress. In most fluid dynamics problems, the choice of p is unimportant, and we will not use the deviatoric stress there.

2.3.2 GENERAL LINEAR MOMENTUM BALANCE

Now return to the material volume $V(t)$ shown in Fig. 2.1. Newton's second law of motion states that the time rate of change of the linear momentum of the body equals the sum of all external forces acting on it, that is,

$$\frac{d}{dt}\int_{V(t)} \rho \mathbf{v}\, dV = \int_{S(t)} \mathbf{t}\, dS + \int_{V(t)} \rho \mathbf{b}\, dV \tag{2.53}$$

The right-hand-side terms represent the surface and body forces, respectively. Using the Reynolds transport theorem, Eq. (2.41), on the left-hand side yields

$$\int_{V(t)} \rho \frac{D\mathbf{v}}{Dt}\, dV = \int_{S(t)} \mathbf{t}\, dS + \int_{V(t)} \rho \mathbf{b}\, dV \tag{2.54}$$

For the surface force term we use Eq. (2.49) to replace **t** with $\mathbf{n} \cdot \boldsymbol{\sigma}$, and then we apply the divergence theorem, Eq. (A.59), to transform the surface integral into a volume integral. This gives

$$\int_{V(t)} \rho \frac{D\mathbf{v}}{Dt}\, dV = \int_{V(t)} \nabla \cdot \boldsymbol{\sigma}\, dV + \int_{V(t)} \rho \mathbf{b}\, dV \tag{2.55}$$

Now we can collect all of the terms into a single integral,

$$\int_{V(t)} \left(\rho \frac{D\mathbf{v}}{Dt} - \nabla \cdot \boldsymbol{\sigma} - \rho \mathbf{b} \right) dV = \mathbf{0} \tag{2.56}$$

Finally, invoking the now-familiar argument that, because volume $V(t)$ is arbitrary, the integrand must vanish, we obtain a differential momentum balance equation:

$$\rho \frac{D\mathbf{v}}{Dt} = \nabla \cdot \boldsymbol{\sigma} + \rho \mathbf{b} \tag{2.57}$$

For fluid flow problems it is common to use Eq. (2.51) to replace the total stress by the pressure and the extra stress. Then the momentum balance equation becomes

$$\rho \frac{D\mathbf{v}}{Dt} = -\nabla p + \nabla \cdot \boldsymbol{\tau} + \rho \mathbf{b} \tag{2.58}$$

Both Eqs. (2.57) are (2.58) are general forms of the momentum balance that can be applied to any material.

2.3.3 INTEGRAL FORM

We can obtain a convenient integral form of the momentum balance by returning to Eq. (2.54). One can use the definition of the material derivative, Eq. (2.13), and the continuity equation, Eq. (2.29), to show that

$$\rho \frac{D\mathbf{v}}{Dt} = \frac{\partial}{\partial t}(\rho \mathbf{v}) + \nabla \cdot (\rho \mathbf{v}\mathbf{v}) \tag{2.59}$$

Integrate this over volume $V(t)$, and then use the divergence theorem to convert the second term into a surface integral.

$$\int_{V(t)} \rho \frac{D\mathbf{v}}{Dt} dV = \int_{V(t)} \frac{\partial}{\partial t}(\rho \mathbf{v}) dV + \int_{V(t)} \nabla \cdot (\rho \mathbf{v}\mathbf{v}) dV$$

$$= \int_{V(t)} \frac{\partial}{\partial t}(\rho \mathbf{v}) dV + \int_{S(t)} \mathbf{n} \cdot (\rho \mathbf{v}\mathbf{v}) dS \tag{2.60}$$

Note that $\rho \mathbf{v}$ is the momentum per unit volume of the fluid, so the new surface integral represents the momentum transported by fluid that is crossing the surface $S(t)$. Substitute this back into Eq. (2.54), and rearrange to obtain the desired integral form:

$$\underbrace{\int_V \frac{\partial}{\partial t}(\rho \mathbf{v}) dV}_{\text{rate of change inside}} = \underbrace{-\int_S \mathbf{n} \cdot (\rho \mathbf{v}\mathbf{v}) dS}_{\text{momentum entering surface}} + \underbrace{\int_S \mathbf{t} dS}_{\text{traction on surface}} + \underbrace{\int_V \rho \mathbf{b} dV}_{\text{body forces}} \tag{2.61}$$

Here we have dropped the time arguments from V and S, because this equation can be applied to any volume, including a control volume fixed in space.

2.3.4 THE NEWTONIAN FLUID

The mass and momentum balance equations, Eqs. (2.28) and (2.58), govern the dynamics of a continuum, but they are not sufficient to solve any problems because they contain more unknowns than there are equations. The unknowns include the density, three components of the velocity, and six components of stress. We have just four equations: the mass balance and three vector components from the linear momentum balance. To complete our description, we need a *constitutive equation* that describes the response of the material to applied stresses.

Constitutive relations may be developed from many sources: experimental observations and correlations, phenomenological theories, molecular theories, or inspired guesses. There are restrictions on the mathematical form that these relations may take, a point which we will take up in Chapter 6. A successful model, however, must use constitutive relations that accurately describe the material being modeled. In this section we introduce the constitutive relation for a Newtonian fluid, leaving the discussion of other constitutive equations for later.

The Newtonian constitutive relation is motivated by the experiment sketched in Fig. 2.3. A fluid is confined between two parallel plates, one of which is translated parallel to the other with a constant velocity. This generates the linear velocity profile shown in the figure, and it induces a shear stress τ_{yx}. For many fluids this stress is proportional to the derivative of the velocity:

$$\tau_{yx} = \mu \frac{\partial v_x}{\partial y} \tag{2.62}$$

The constant of proportionality μ is called the *shear viscosity*, and its units in the SI system are (newtons times seconds) per square meter, or pascals times seconds ($1 \, \text{Pa} = 1 \, \text{N/m}^2$).

This simple observation suggests that we might develop a constitutive relation in which the stress is proportional to the spatial derivatives of velocity. The complete set of velocity derivatives is given by the velocity gradient tensor $\nabla \mathbf{v}$, whose Cartesian components are $(\nabla \mathbf{v})_{ij} = \partial v_j / \partial x_i$. As a way to get a more natural indexing of the components, it is common to define a tensor $\mathbf{L}$, also called the velocity gradient tensor, whose components are

$$\mathbf{L} = \begin{bmatrix} \partial v_1/\partial x_1 & \partial v_1/\partial x_2 & \partial v_1/\partial x_3 \\ \partial v_2/\partial x_1 & \partial v_2/\partial x_2 & \partial v_2/\partial x_3 \\ \partial v_3/\partial x_1 & \partial v_3/\partial x_2 & \partial v_3/\partial x_3 \end{bmatrix} \tag{2.63}$$

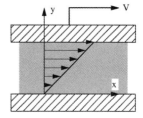

Figure 2.3: Simple shear flow experiment, used to define fluid viscosity.

or, in indicial notation,

$$L_{ij} = \frac{\partial v_i}{\partial x_j} \tag{2.64}$$

The relation between these two tensor is

$$\mathbf{L} = (\nabla \mathbf{v})^T \tag{2.65}$$

As with any tensor, we can separate $\mathbf{L}$ into a symmetric and an antisymmetric part:

$$\mathbf{L} = \underbrace{\frac{1}{2}(\mathbf{L} + \mathbf{L}^T)}_{\text{symmetric}} + \underbrace{\frac{1}{2}(\mathbf{L} - \mathbf{L}^T)}_{\text{antisymmetric}} \tag{2.66}$$

The symmetric part of the velocity gradient is called the *rate-of-deformation* tensor $\mathbf{D}$ and the antisymmetric part is the *vorticity* tensor $\mathbf{W}$,

$$\mathbf{D} = \frac{1}{2}(\mathbf{L} + \mathbf{L}^T) \tag{2.67}$$

$$\mathbf{W} = \frac{1}{2}(\mathbf{L} - \mathbf{L}^T) \tag{2.68}$$

so the velocity gradient can be written as

$$\mathbf{L} = \mathbf{D} + \mathbf{W} \tag{2.69}$$

In indicial notation these tensors are

$$D_{ij} = \frac{1}{2}\left(\frac{\partial v_i}{\partial x_j} + \frac{\partial v_j}{\partial x_i}\right) \tag{2.70}$$

$$W_{ij} = \frac{1}{2}\left(\frac{\partial v_i}{\partial x_j} - \frac{\partial v_j}{\partial x_i}\right) \tag{2.71}$$

Appendix B gives expressions for the components of $\mathbf{D}$ and $\mathbf{W}$ in other coordinate systems. Some texts define a rate of deformation tensor $\dot{\gamma}$ equal to $2\mathbf{D}$, and a vorticity tensor $\boldsymbol{\omega}$ equal to $-2\mathbf{W}$. Readers should be careful to check the definitions of these quantities when reading other texts.

The symmetric part $\mathbf{D}$ will be nonzero only if the material is deforming, and it seems sensible that this could cause some stress. The antisymmetric part $\mathbf{W}$ is nonzero when the material undergoes solid body rotation, and this should not induce any stress in the material. Thus, in a Newtonian fluid we expect the stress to be a function of $\mathbf{D}$, but independent of $\mathbf{W}$.

Basic theory shows that the most general way the extra stress τ can depend linearly on the rate of deformation (provided the material is isotropic) is

$$\tau = 2\mu \mathbf{D} + \lambda (\text{tr}\,\mathbf{D})\mathbf{I} \tag{2.72}$$

This general form introduces a second material constant λ, known as the dilatational viscosity. This property multiplies the quantity $\text{tr}\,\mathbf{D} = \nabla \cdot \mathbf{v}$, which will equal

2.3 MOMENTUM BALANCE

zero when the fluid has constant density; see Eq. (2.34). Substituting Eq. (2.72) into Eq. (2.58) yields the general form of the momentum balance for a Newtonian fluid:

$$\rho \frac{D\mathbf{v}}{Dt} = -\nabla p + \nabla \cdot (2\mu \mathbf{D} + \lambda (\text{tr } \mathbf{D})\mathbf{I}) + \rho \mathbf{b} \tag{2.73}$$

The property λ is difficult to measure experimentally. A common assumption, called the *Stokes hypothesis*, is $\lambda = -2\mu/3$ (e.g., Schlichting, 1968, pp. 57–58). This is valid for monatomic gases, but it may not be accurate for other fluids (Karim and Rosenhead, 1952). In most materials processing problems we can assume that the fluid density is constant, and λ is no longer required. Then the Newtonian constitutive equation becomes

$$\boldsymbol{\tau} = 2\mu \mathbf{D} \tag{2.74}$$

or, in indicial notation,

$$\tau_{ij} = 2\mu D_{ij} \tag{2.75}$$

For problems using Cartesian coordinates, we can use Eq. (2.70) to write τ directly in terms of the velocities:

$$\tau_{ij} = \mu \left(\frac{\partial v_i}{\partial x_j} + \frac{\partial v_j}{\partial x_i} \right) \tag{2.76}$$

Examples of Newtonian fluids include air, water, honey, and mercury.

An important special case of the momentum balance equation occurs when the material is a Newtonian fluid with constant density. One substitutes Eq. (2.74) into Eq. (2.58) to eliminate $\boldsymbol{\tau}$. This yields the *Navier–Stokes equation*:

$$\rho \frac{D\mathbf{v}}{Dt} = -\nabla p + \nabla \cdot (2\mu \mathbf{D}) + \rho \mathbf{b} \tag{2.77}$$

Because $\mathbf{D}$ is determined by the velocity field, Eqs. (2.77) and (2.33) provide a complete set of equations that, together with appropriate boundary conditions, may be used to solve many important fluid flow problems. There are now four equations (one from continuity and three components of the momentum balance) and four unknowns (pressure and three components of velocity).

If the viscosity is constant, Eq. (2.77) can be further simplified. We bring the viscosity outside the derivative, substitute Eq. (2.67) to eliminate $\mathbf{D}$, and use continuity equation, Eq. (2.33) to cancel some of the terms arising from $\nabla \cdot \mathbf{D}$. The result is

$$\rho \frac{D\mathbf{v}}{Dt} = -\nabla p + \mu \nabla^2 \mathbf{v} + \rho \mathbf{b} \tag{2.78}$$

Some texts refer to Eq. (2.78) as the Navier–Stokes equation, rather than the more general form in Eq. (2.77).

EXAMPLE 2.3.1 Drag Flow Between Parallel Plates

As an example of the use of the momentum balance equation, consider the flow between two large parallel plates. The plate surfaces are parallel to the x–y plane and we take $z = 0$ halfway between the plates, so that the plate surfaces are at

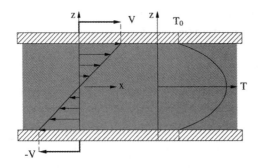

Figure 2.4: Velocity and temperature distributions for drag flow between parallel plates.

$z = \pm h$. The upper surface moves in the positive x direction with velocity V, and the lower surface moves in the negative x direction, so its x velocity is $-V$ (see Fig. 2.4). We further assume that:

- the velocity field is steady;
- the fluid is Newtonian, with constant density ρ and viscosity μ;
- the pressure p is uniform;
- the y- and z-direction velocities are zero everywhere;
- the x-direction velocity is a function of z only;
- the fluid does not slip where it contacts the plates; and
- body forces are negligible.

Some of these assumptions may seem quite arbitrary. For now we ask the reader to accept them, and we promise that in the next chapter we will provide a thorough treatment of how to evaluate and select assumptions of this type.

Any fluid flow solution must satisfy the continuity equation, so we start there. From Appendix B.2, for a fluid of constant density this equation is

$$\frac{\partial v_x}{\partial x} + \frac{\partial v_y}{\partial y} + \frac{\partial v_z}{\partial z} = 0 \tag{2.79}$$

The first term is zero because v_x depends only on z, and the second and third terms are zero because $v_y = v_z = 0$. In this problem the continuity equation is not very helpful. However, it is useful to know that our assumptions about the velocity field guarantee that the continuity equation will be satisfied.

To find $v_x(z)$ we will need a momentum balance equation. From Appendix B.6 we select the x component of the momentum balance equation in rectangular coordinates:

$$\rho \left(\frac{\partial v_x}{\partial t} + v_x \frac{\partial v_x}{\partial x} + v_y \frac{\partial v_x}{\partial y} + v_z \frac{\partial v_x}{\partial z} \right) = -\frac{\partial p}{\partial x} + \mu \left(\frac{\partial^2 v_x}{\partial x^2} + \frac{\partial^2 v_x}{\partial y^2} + \frac{\partial^2 v_x}{\partial z^2} \right) + \rho b_x \tag{2.80}$$

On the left-hand side, $\partial v_x/\partial t = 0$ because the flow is steady, $\partial v_x/\partial x = 0$ because v_x depends only on z, and v_y and v_z are both zero. On the right-hand side, $\partial p/\partial x = 0$ because p is constant, $\partial^2 v_x/\partial x^2$ and $\partial^2 v_x/\partial y^2$ are zero because v_x depends only on z, and we can drop ρb_x because body forces are negligible. This reduces the momentum balance equation to

$$0 = \mu \frac{d^2 v_x}{dz^2} \tag{2.81}$$

2.3 MOMENTUM BALANCE

Here we have changed the partial derivative to an ordinary derivative, because v_x is a function of z. Now divide by μ and integrate twice, finding

$$\frac{dv_x}{dz} = c_1 \tag{2.82}$$

$$v_x = c_1 z + c_2 \tag{2.83}$$

where c_1 and c_2 are constants of integration. We fix their values by using two boundary conditions, which come from the requirement that the fluid velocity must equal the plate velocity where the fluid touches the plates (the no slip condition). The boundary conditions are

$$v_x = V \quad \text{at } z = h \tag{2.84}$$

$$v_x = -V \quad \text{at } z = -h \tag{2.85}$$

Using these relations in Eq. (2.83), we find $c_1 = V/h$ and $c_2 = 0$, so the velocity distribution is

$$v_x = V \frac{z}{h} \tag{2.86}$$

This is a linear velocity profile that has zero velocity in the center (at $z = 0$) and matches the plate velocities $\pm V$ at $z = \pm h$. This profile is shown in Fig. 2.4.

Once the velocity distribution is known, we can evaluate the extra stresses. For a Newtonian fluid, each component of τ is proportional to the corresponding component of **D**. Examining Appendix B.3 for rectangular coordinates, we see that in this flow the only nonzero components of **D** are

$$D_{zx} = D_{xz} = \frac{1}{2}\left(\frac{\partial v_z}{\partial x} + \frac{\partial v_x}{\partial z}\right) \tag{2.87}$$

which, for this flow, are

$$D_{zx} = D_{xz} = \frac{V}{2h} \tag{2.88}$$

Following Eq. (2.75), we then find the stresses as

$$\tau_{zx} = \tau_{xz} = \frac{\mu V}{h} \tag{2.89}$$

and all the other components of τ are zero. Note that our result for τ is symmetric, as it must be.

As an aside, we note that this flow is the basis for the measurement of viscosity in a Couette viscometer. The total force F exerted on the upper plate is $\tau_{zx} A$, where A is the area of the plate. Thus, if we were to perform this experiment and carefully measure the force exerted on the top plate, the velocity, and the instrument geometry, we could rearrange Eq. (2.89) to find

$$\mu = \frac{Fh}{VA} \tag{2.90}$$

2.3.5 GRAVITATIONAL EFFECTS

The Modified Pressure

The motion of a fluid is affected by body forces, such as gravity, magnetic fields, and electrical fields. When gravity is the only body force acting and the fluid density is constant, there is a convenient trick to simplify the momentum balance equation. Let us use g to denote the acceleration that is due to gravity (9.82 m/s^2), and let us choose a Cartesian coordinate system in which z points up, that is, opposite to gravity. Now the body force per unit mass vector $\mathbf{b}$ has magnitude g and points in the negative z direction, so the body force term in the momentum balance will be

$$\rho \mathbf{b} = -\rho g \mathbf{e}_z \tag{2.91}$$

where $\mathbf{e}_z$ is the unit vector in the z direction. A convenient way to rephrase Eq. (2.91) is to introduce the variable h as the height, measured in the z direction, above some arbitrary reference plane. Note that $\partial h/\partial z = 1$, whereas $\partial h/\partial x = \partial h/\partial y = 0$. If we treat h as a scalar function of position (x, y, z), then the gradient of h is a unit vector pointing in the positive z direction, or

$$\nabla h = \mathbf{e}_z \tag{2.92}$$

Using this identity, we can rewrite Eq. (2.91) as

$$\rho \mathbf{b} = -\rho g \nabla h \tag{2.93}$$
$$= -\nabla(\rho g h) \tag{2.94}$$

where the last statement uses the fact that ρ and g are constants. This relationship can be used with any type of coordinate system, as long as we interpret $h(\mathbf{x})$ as the height of any point $\mathbf{x}$ above some reference level, measured in the $+z$ direction.

Substituting this result into the general form of the momentum balance equation, Eq. (2.58), we obtain

$$\rho \frac{D\mathbf{v}}{Dt} = -\nabla p + \nabla \cdot \boldsymbol{\tau} - \nabla(\rho g h) \tag{2.95}$$

Now it is convenient to combine the pressure gradient and body force terms, by defining a new variable $\hat{p}$ called the *modified pressure*,

$$\hat{p} \equiv p + \rho g h \tag{2.96}$$

so that the momentum balance equation becomes

$$\rho \frac{D\mathbf{v}}{Dt} = -\nabla \hat{p} + \nabla \cdot \boldsymbol{\tau} \tag{2.97}$$

That is, the momentum balance equation for a fluid with uniform density and a gravitational body force is identical to the momentum balance for a fluid with no body force, provided p is replaced by $\hat{p}$. This same substitution can be made in the Navier–Stokes equation, Eq. (2.78), and the Euler equation, Eq. (2.101).

Hydrostatics

To see how gravity affects the pressure distribution, consider the class of problems called *hydrostatics*, in which there is no fluid motion. With $\mathbf{v} = 0$ we have $D\mathbf{v}/Dt = 0$ and $\boldsymbol{\tau} = 0$ (at least for Newtonian fluids), and the momentum balance Eq. (2.97) reduces to

$$\nabla \hat{p} = 0 \tag{2.98}$$

2.3 MOMENTUM BALANCE

This has the simple, general solution of

$$\hat{p} = \text{constant} \tag{2.99}$$

where the constant is determined by fixing the value of $\hat{p}$ at one point. With the use of Eq. (2.96), this tells us that the actual pressure distribution is

$$p + \rho g h = \text{constant} \tag{2.100}$$

This is the pressure distribution in any constant-density fluid that is stationary and sees only gravity as a body force.

2.3.6 INVISCID AND IRROTATIONAL FLOWS

The Euler Equation

In some situations the viscosity of the fluid is small enough that it can be neglected, at least over most of the flow domain. The momentum balance for this case can be obtained either by dropping the $\nabla \cdot \tau$ term from the general momentum balance equation, Eq. (2.58), or by allowing μ to go to zero in the Navier–Stokes equation, Eq. (2.77). The result is called the *Euler equation*:

$$\rho \frac{D\mathbf{v}}{Dt} = -\nabla p + \rho \mathbf{b} \tag{2.101}$$

Irrotational and Potential Flows

An *irrotational flow* is one in which the vorticity is everywhere zero. This means that the vorticity tensor $\mathbf{W}$ equals zero at every point. Often this requirement is expressed in terms of the vorticity vector $\boldsymbol{\omega}$,

$$\boldsymbol{\omega} \equiv \nabla \times \mathbf{v} \tag{2.102}$$

and an irrotational flow has

$$\boldsymbol{\omega} = 0 \tag{2.103}$$

This is equivalent to requiring that the vorticity tensor $\mathbf{W}$ equal zero. You can think of an irrotational flow as one in which the fluid has no solid-body rotation at any point. If one placed a small ball in an irrotational flow, the ball would follow the flow field, but would not spin relative to the laboratory axes.

In any irrotational flow the velocity vector field $\mathbf{v}(\mathbf{x}, t)$ can be replaced by a scalar potential field $\phi(\mathbf{x}, t)$. To show this, we start with the general vector theorem that

$$\nabla \times \nabla \phi = 0 \tag{2.104}$$

for any scalar field ϕ. An irrotational field must have $\nabla \times \mathbf{v} = 0$, so we can identify the velocity vector with the gradient of the potential field:

$$\mathbf{v} = \nabla \phi \tag{2.105}$$

This can provide a useful simplification when solving flow problems, as one need only solve for the scalar field ϕ, and not for the vector field $\mathbf{v}$.

One strategy for solving potential flow problems is to combine Eq. (2.105) with the continuity equation for a fluid with constant density, Eq. (2.34). This gives

$$\nabla^2 \phi = 0 \tag{2.106}$$

If the proper boundary conditions are available, one can solve Eq. (2.106) for ϕ, use Eq. (2.105) to find the velocity, and then use the Euler Eq. (2.101) or one of the simpified forms derived below to find the pressure field.

A flow in which the velocity field satisfies Eq. (2.105) is called a *potential flow*. An important result, known as *Lagrange's theorem*, tells us that any flow of an incompressible, inviscid fluid that starts as a potential flow always remains as a potential flow (Richardson, 1989). Of course, a fluid at rest is experiencing a (not very interesting) potential flow, so many inviscid flows are potential flows.

Bernoulli's Equation, Potential Flows

To derive a special form of the momentum balance for inviscid potential flows, we start with a vector identity:

$$\nabla(\mathbf{u}_1 \cdot \mathbf{u}_2) = \mathbf{u}_2 \cdot \nabla \mathbf{u}_1 + \mathbf{u}_1 \cdot \nabla \mathbf{u}_2 + \mathbf{u}_2 \times (\nabla \mathbf{u}_1) + \mathbf{u}_1 \times (\nabla \mathbf{u}_2) \tag{2.107}$$

Now choose both $\mathbf{u}_1$ and $\mathbf{u}_2$ to be the velocity vector $\mathbf{v}$. Together with Eq. (2.102), this gives

$$\nabla(\mathbf{v} \cdot \mathbf{v}) = 2\mathbf{v} \cdot \nabla \mathbf{v} + 2\mathbf{v} \times \boldsymbol{\omega} \tag{2.108}$$

Substituting this into Euler Eq. (2.101) and recalling that $D\mathbf{v}/Dt = \partial \mathbf{v}/\partial t + \mathbf{v} \cdot \nabla \mathbf{v}$, we have

$$\rho \left[\frac{\partial \mathbf{v}}{\partial t} + \frac{1}{2} \nabla(\mathbf{v} \cdot \mathbf{v}) - \mathbf{v} \times \boldsymbol{\omega} \right] = -\nabla p + \rho \mathbf{b} \tag{2.109}$$

If the flow is irrotational then the term containing ω can be dropped. Let $v = \sqrt{\mathbf{v} \cdot \mathbf{v}}$ denote the magnitude of the velocity vector. Also assume that the body force comes only from gravity. Use Eq. (2.94) to replace the body force term in Eq. (2.109), and also use Eq. (2.105) to replace $\mathbf{v}$ in the transient term. This gives

$$\rho \frac{\partial(\nabla \phi)}{\partial t} + \frac{1}{2} \rho \nabla(v^2) = -\nabla p - \nabla(\rho g h) \tag{2.110}$$

We can interchange the order of differentiation in the first term. Because ρ and g are constant, every term is the gradient of something, and the momentum balance equation becomes

$$\nabla \left(\rho \frac{\partial \phi}{\partial t} + \frac{1}{2} \rho v^2 + p + \rho g h \right) = 0 \tag{2.111}$$

The quantity inside the parentheses must be independent of position, though it could depend on time. This gives us a general result for ideal, irrotational flows,

$$\rho \frac{\partial \phi}{\partial t} + \frac{1}{2} \rho v^2 + p + \rho g h = \zeta(t) \tag{2.112}$$

which is known as *Bernoulli's equation*. The quantity $\zeta(t)$ is called the Bernoulli constant, and for an irrotational flow it has a single value at each time for the entire flow field.

If the flow is steady, then Bernoulli's equation has a familiar form:

$$\frac{1}{2}\rho v^2 + p + \rho g h = \zeta \quad (2.113)$$

and ζ is independent of time and position. Again, this form of Bernoulli's equation applies at every point in a steady, irrotational, ideal flow.

Bernoulli's Equation: Streamline Form

We can also write versions of Bernoulli's equation for flows that are not irrotational, by integrating Euler Eq. (2.101) along a streamline. Let ds denote a unit of distance along the streamline and integrate from point 1 to point 2, where both points are on the same streamline. The transient equation for a constant-density fluid is

$$\frac{1}{2}\rho v_1^2 + p_1 + \rho g h_1 = \frac{1}{2}\rho v_2^2 + p_2 + \rho g h_2 + \rho \int_1^2 \frac{\partial v}{\partial t} ds \quad (2.114)$$

If the flow is steady, then we obtain a familiar form:

$$\frac{1}{2}\rho v^2 + p + \rho g h = \text{constant along a streamline} \quad (2.115)$$

Again, these equations apply only to two points on the same streamline.

2.4 ENERGY BALANCE

2.4.1 HEAT BUDGET

In most materials processing operations, the material is heated or cooled. Accordingly, we will need a governing equation for energy transport, or heat flow. To this end, we introduce the energy balance equation, sometimes referred to as the *first law of thermodynamics*. Simply stated, this equation represents a "budget" for the heat energy in our material volume $V(t)$:

$$\frac{d}{dt}(K + E) = Q + P \quad (2.116)$$

Here K is the kinetic energy, E is the internal energy, Q is the rate at which heat is added to the material by conduction or internal generation, and P is the power, that is, the rate of work done by mechanical forces. We now evaluate these terms one by one.

The kinetic energy is given by

$$K = \int_{V(t)} \frac{1}{2}\rho \mathbf{v} \cdot \mathbf{v} \, dV \quad (2.117)$$

Differentiate this expression with respect to time, and apply the Reynolds transport theorem together with the commutative property of the vector dot product, to obtain

$$\frac{dK}{dt} = \int_{V(t)} \rho \mathbf{v} \cdot \frac{D\mathbf{v}}{Dt} dV \quad (2.118)$$

Then substitute the momentum balance equation, Eq. (2.57), into Eq. (2.118) to obtain

$$\frac{dK}{dt} = \int_{V(t)} \mathbf{v} \cdot (\nabla \cdot \boldsymbol{\sigma} + \rho \mathbf{b}) \, dV \tag{2.119}$$

The internal energy E is written in terms of the *specific internal energy* ε, that is, the internal energy per unit mass:

$$E = \int_{V(t)} \rho \varepsilon \, dV \tag{2.120}$$

The time derivative is obtained in the usual way, using the Reynolds transport theorem:

$$\frac{dE}{dt} = \int_{V(t)} \rho \frac{D\varepsilon}{Dt} \, dV \tag{2.121}$$

The heating rate Q is analogous to the mechanical forces, having one component that enters through the surface $S(t)$ by *conduction*, and another component that is internally generated, perhaps by a chemical reaction or by induction heating. The heat transport by conduction across any surface is described by the heat flux vector $\mathbf{q}$, which has units of energy per unit area per unit time. The heat flows in the direction of the vector $\mathbf{q}$, so the dot product $\mathbf{n} \cdot \mathbf{q}$ is positive when heat is leaving the body. The rate of internal heat generation per unit mass is denoted by $\dot{R}$. Thus, the total heating rate in the body may be written as

$$Q = -\int_{S(t)} \mathbf{n} \cdot \mathbf{q} \, dS + \int_{V(t)} \rho \dot{R} \, dV \tag{2.122}$$

The minus sign appears because $\mathbf{n}$ points outward from the surface. Applying the divergence theorem to the first integral and combining terms gives the heating rate as

$$Q = \int_{V(t)} (-\nabla \cdot \mathbf{q} + \rho \dot{R}) \, dV \tag{2.123}$$

Finally, the rate of work done by the mechanical forces is obtained by integrating the dot product of the velocity with those forces:

$$P = \int_{S(t)} \mathbf{t} \cdot \mathbf{v} \, dS + \int_{V(t)} \rho \mathbf{v} \cdot \mathbf{b} \, dV \tag{2.124}$$

Introducing the stress tensor from Eq. (2.49) yields

$$P = \int_{S(t)} \mathbf{n} \cdot (\boldsymbol{\sigma} \cdot \mathbf{v}) \, dS + \int_{V(t)} \rho \mathbf{v} \cdot \mathbf{b} \, dV \tag{2.125}$$

Applying the divergence theorem to the first term yields

$$P = \int_{V(t)} [\boldsymbol{\sigma} : \nabla \mathbf{v} + \mathbf{v} \cdot (\nabla \cdot \boldsymbol{\sigma})] \, dV + \int_{V(t)} \rho \mathbf{v} \cdot \mathbf{b} \, dV \tag{2.126}$$

where the : symbol indicates a scalar product of two tensors, that is,

$$\boldsymbol{\sigma} : \nabla \mathbf{v} = \sigma_{ij} \frac{\partial v_i}{\partial x_j} \tag{2.127}$$

2.4 ENERGY BALANCE

To simplify this term, we separate $\nabla \mathbf{v}$ into its symmetric and antisymmetric parts by using Eq. (2.65) and (2.69). The scalar (double dot) product of a symmetric tensor with an antisymmetric tensor is always zero, so $\boldsymbol{\sigma} : \mathbf{W} = 0$ and we obtain the rate of mechanical work as

$$P = \int_{V(t)} [\boldsymbol{\sigma} : \mathbf{D} + \mathbf{v} \cdot (\nabla \cdot \boldsymbol{\sigma} + \rho \mathbf{b})] \, dV \tag{2.128}$$

Equations (2.119), (2.121), (2.123), and (2.128) may now be substituted into the energy budget, Eq. (2.116). Note that the kinetic energy terms in Eq. (2.119) cancel several terms from the power, Eq. (2.128). The result is

$$\int_{V(t)} \rho \frac{D\varepsilon}{Dt} dV = -\int_{V(t)} \nabla \cdot \mathbf{q} \, dV + \int_{V(t)} \boldsymbol{\sigma} : \mathbf{D} \, dV + \int_{V(t)} \rho \dot{R} \, dV \tag{2.129}$$

Because the control volume is arbitrary, the integrands must satisfy the differential equation

$$\rho \frac{D\varepsilon}{Dt} = -\nabla \cdot \mathbf{q} + \boldsymbol{\sigma} : \mathbf{D} + \rho \dot{R} \tag{2.130}$$

We next separate the stress tensor into the pressure and the extra stress tensor by using Eq. (2.51). The scalar product of the stress and rate-of-deformation tensors becomes

$$\begin{aligned}\boldsymbol{\sigma} : \mathbf{D} &= -p\mathbf{I} : \mathbf{D} + \boldsymbol{\tau} : \mathbf{D} \\ &= -p(\nabla \cdot \mathbf{v}) + \boldsymbol{\tau} : \mathbf{D}\end{aligned} \tag{2.131}$$

As discussed earlier, the term $\nabla \cdot \mathbf{v}$ represents the rate of change of volume per unit volume. Thus, the rate of work done by traction forces is the sum of work done by the pressure over a volume change (sometimes called the pV work) and work done to deform the material ($\boldsymbol{\tau} : \mathbf{D}$). Using Eq. (2.37), we find that

$$\boldsymbol{\sigma} : \mathbf{D} = -\rho p \frac{D\hat{V}}{Dt} + \boldsymbol{\tau} : \mathbf{D} \tag{2.132}$$

Substituting this into the energy balance gives

$$\rho \frac{D\varepsilon}{Dt} = -\nabla \cdot \mathbf{q} - \rho p \frac{D\hat{V}}{Dt} + \boldsymbol{\tau} : \mathbf{D} + \rho \dot{R} \tag{2.133}$$

The left-hand side of this equation is the rate of change of internal energy, and the terms on the right-hand side represent heat transferred by conduction, work done by compression, work done by shape change at constant volume, and internal heat generation, respectively.

2.4.2 CONSTITUTIVE EQUATION FOR HEAT CONDUCTION

Once again, we have more unknowns than equations. We now need constitutive relations to relate the heat flux $\mathbf{q}$ and internal energy ε to primitive variables, such as temperature. We start with the heat flux. A widely used constitutive relation that works well for most materials is *Fourier's law*, which states that the heat flux is proportional to the temperature gradient:

$$\mathbf{q} = -\mathbf{k} \cdot \nabla T \tag{2.134}$$

Here $\mathbf{k}$ is a material property, the *thermal conductivity tensor*, and T is temperature. The negative sign appears because heat flows in the opposite direction of the temperature gradient, from hot to cold. We can substitute Fourier's law into the energy balance, Eq. (2.133), to obtain

$$\rho \frac{D\varepsilon}{Dt} = \nabla \cdot (\mathbf{k} \cdot \nabla T) - \rho p \frac{D\hat{V}}{Dt} + \boldsymbol{\tau} : \mathbf{D} + \rho \dot{R} \tag{2.135}$$

If the material is isotropic (i.e., its properties are the same in all directions), then the conductivity tensor has the special form

$$\mathbf{k} = \begin{bmatrix} k & 0 & 0 \\ 0 & k & 0 \\ 0 & 0 & k \end{bmatrix} = k\mathbf{I} \tag{2.136}$$

where k is the scalar thermal conductivity, and Fourier's law becomes

$$\mathbf{q} = -k\nabla T \tag{2.137}$$

The conductivity term in Eq. (2.135) then simplifies to

$$\nabla \cdot (\mathbf{k} \cdot \nabla T) = \nabla \cdot (k\nabla T)$$
$$= \frac{\partial}{\partial x_i}\left(k\frac{\partial T}{\partial x_i}\right) \tag{2.138}$$

Careful measurements show that, for most materials, k varies slowly as a function of temperature. However, we sometimes ignore this effect and assume that k is constant. Then the conduction term of Eq. (2.138) simplifies further to $k\nabla^2 T$. Note that k can change radically when the material experiences a phase change such as solidification or melting, so constant conductivity may be a poor assumption in such cases.

2.4.3 CONSTITUTIVE EQUATIONS FOR INTERNAL ENERGY

Gases

The internal energy term $D\varepsilon/Dt$ is usually treated in one of two ways, depending on whether one is dealing with gases or with condensed phases (liquids and solids). The internal energy is a function of thermodynamic *state variables*, such as temperature, pressure, and specific volume. Gibbs observed that only two of these three quantities are independent. (See the collected works, Gibbs, 1961.)

Because gases fill a volume, it is convenient to select specific volume and temperature as the independent state variables. The derivative of the internal energy is then written as

$$\frac{D\varepsilon(T, \hat{V})}{Dt} = \left(\frac{\partial \varepsilon}{\partial T}\right)_{\hat{V}} \frac{DT}{Dt} + \left(\frac{\partial \varepsilon}{\partial \hat{V}}\right)_T \frac{D\hat{V}}{Dt} \tag{2.139}$$

2.4 ENERGY BALANCE

where the notation $(\ )_{\hat{V}}$ indicates differentiation with $\hat{V}$ held constant. The coefficients on the right-hand side are thermodynamic properties of the material, the specific heat at constant volume c_v and the latent heat of volume change L_v. These are defined (e.g., Prigogine and Defay, 1954, Chap. II) such that

$$c_v = \left(\frac{\partial \varepsilon}{\partial T}\right)_{\hat{V}}, \qquad L_v - p = \left(\frac{\partial \varepsilon}{\partial \hat{V}}\right)_T \qquad (2.140)$$

Substituting this into Eq. (2.135), we obtain

$$\rho c_v \frac{DT}{Dt} + \rho L_v \frac{D\hat{V}}{Dt} = \nabla \cdot (\mathbf{k} \cdot \nabla T) + \boldsymbol{\tau} : \mathbf{D} + \rho \dot{R} \qquad (2.141)$$

An *equation of state*, such as the ideal gas equation, must be introduced to relate $\hat{V}$, T, and p.

Condensed Phases

When dealing with condensed phases, we find it convenient to choose temperature and pressure as the state variables. We now introduce the *specific enthalpy H*, defined as

$$H(T, p) = \varepsilon + p\hat{V} \qquad (2.142)$$

Differentiating the enthalpy with respect to time yields

$$\frac{DH(T, p)}{Dt} = \frac{D\varepsilon}{Dt} + p\frac{D\hat{V}}{Dt} + \hat{V}\frac{Dp}{Dt} \qquad (2.143)$$

and using this expression to replace $D\varepsilon/Dt$ in Eq. (2.135) gives

$$\rho \frac{DH}{Dt} - \rho\hat{V}\frac{Dp}{Dt} = \nabla \cdot (\mathbf{k} \cdot \nabla T) + \boldsymbol{\tau} : \mathbf{D} + \rho \dot{R} \qquad (2.144)$$

Applying the chain rule for the derivative of enthalpy, we have

$$\frac{DH(T, p)}{Dt} = \left(\frac{\partial H}{\partial T}\right)_p \frac{DT}{Dt} + \left(\frac{\partial H}{\partial p}\right)_T \frac{Dp}{Dt} \qquad (2.145)$$

The coefficients in this expression give rise to two new material properties, the specific heat at constant pressure c_p, and the latent heat of pressure change L_p, defined such that

$$c_p = \left(\frac{\partial H}{\partial T}\right)_p, \qquad L_p + \hat{V} = \left(\frac{\partial H}{\partial p}\right)_T \qquad (2.146)$$

Using these new properties, we may write the energy balance equation for condensed phases as

$$\rho c_p \frac{DT}{Dt} + \rho L_p \frac{Dp}{Dt} = \nabla \cdot (\mathbf{k} \cdot \nabla T) + \boldsymbol{\tau} : \mathbf{D} + \rho \dot{R} \qquad (2.147)$$

As above, an equation of state is required to relate $\hat{V}$, T, and p.

Although we have advertised Eqs. (2.141) and (2.147) as energy balance equations for gases and condensed phases, respectively, both are completely general and can be applied to any material. The choice of one over the other is strictly a matter of convenience, and it depends on the form in which the equation of state is written. Note that the two specific heats are related by

$$c_p = c_v + L_v \left(\frac{\partial \hat{V}}{\partial T}\right)_p \tag{2.148}$$

2.4.4 SPECIAL CASES

An important special case is a material of constant density, in which case the work of volume change equals zero. An appropriate simplified energy equation is easily obtained from Eq. (2.141) by dropping the term $\rho L_v (D\hat{V}/Dt)$, because $\hat{V}$ is constant. Alternately, when the density is constant we have $c_p = c_v$, so dropping the term $\rho L_p(Dp/Dt)$ from Eq. (2.147) also gives an energy balance equation for constant-density materials.

When modeling solids or liquids that are *nearly* incompressible, one sometimes makes the approximation of constant density in the balance equations. For the energy equation, one must choose one of the two special forms mentioned in the previous paragraph. This amounts to choosing between c_p and c_v as the appropriate specific heat. In such cases the volume change term associated with $(D\hat{V}/Dt)$ is usually much larger than the pressure change term associated with (Dp/Dt) (Richardson, 1989, p. 47). Accordingly, the more accurate approximation is to use c_p as the specific heat.

If we specialize the energy balance equation for a material with a constant density and constant, isotropic thermal conductivity, we obtain

$$\rho c_p \frac{DT}{Dt} = k \nabla^2 T + \boldsymbol{\tau} : \mathbf{D} + \rho \dot{R} \tag{2.149}$$

The term involving the extra stress can be further simplified for the case of a Newtonian fluid, using Eq. (2.74). The result is

$$\boldsymbol{\tau} : \mathbf{D} = \mu \dot{\gamma}^2 \tag{2.150}$$

where $\dot{\gamma}$ is the scalar strain rate,

$$\dot{\gamma} \equiv \sqrt{2\mathbf{D} : \mathbf{D}} \tag{2.151}$$

EXAMPLE 2.4.1   Drag Flow Between Heated Parallel Plates

To demonstrate the use of the energy equation, we continue the example of Section 2.3.1 and find the temperature distribution. To the previous assumptions we add:

- the temperature distribution is steady,
- the upper and lower plates are both maintained at temperature T_0,
- T depends only on z,
- the fluid has constant heat capacity c_p and conductivity k, and
- there are no internal heat sources ($\dot{R} = 0$).

As with the other examples, we look to Appendix B.9 and write the general energy

2.4 ENERGY BALANCE

balance equation in rectangular coordinates:

$$\rho c_p \left(\frac{\partial T}{\partial t} + v_x \frac{\partial T}{\partial x} + v_y \frac{\partial T}{\partial y} + v_z \frac{\partial T}{\partial z} \right)$$
$$= k \left[\frac{\partial^2 T}{\partial x^2} + \frac{\partial^2 T}{\partial y^2} + \frac{\partial^2 T}{\partial z^2} \right] + 2\mu \left\{ \left(\frac{\partial v_x}{\partial x}\right)^2 + \left(\frac{\partial v_y}{\partial y}\right)^2 + \left(\frac{\partial v_z}{\partial z}\right)^2 \right\}$$
$$+ \mu \left\{ \left(\frac{\partial v_x}{\partial y} + \frac{\partial v_y}{\partial x}\right)^2 + \left(\frac{\partial v_y}{\partial z} + \frac{\partial v_z}{\partial y}\right)^2 + \left(\frac{\partial v_z}{\partial x} + \frac{\partial v_x}{\partial z}\right)^2 \right\} + \rho \dot{R}$$

(2.152)

With the present assumptions we can cancel any terms containing $\partial/\partial t$, $\partial/\partial x$, $\partial/\partial y$, v_y, and v_z. We may also change $\partial/\partial z$ to d/dz. This simplifies the energy balance equation to

$$-k \frac{d^2 T}{dz^2} = \mu \left(\frac{dv_x}{dz}\right)^2 \qquad (2.153)$$

We can differentiate our previous solution for the velocity distribution, Eq. (2.86), to find dv_x/dz, giving the energy balance equation as

$$\frac{d^2 T}{dz^2} = -\frac{\mu V^2}{k h^2} \qquad (2.154)$$

Integrate this equation twice,

$$\frac{dT}{dz} = -\frac{\mu V^2}{k h^2} z + c_3 \qquad (2.155)$$

$$T = -\frac{\mu V^2}{k h^2} \frac{z^2}{2} + c_3 z + c_4 \qquad (2.156)$$

where c_3 and c_4 are constants of integration. These constants are evaluated from boundary conditions at the upper and lower plates:

$$T = T_0 \quad \text{at} \quad z = \pm h \qquad (2.157)$$

Evaluating the constants and substituting them into Eq. (2.156), we find the temperature distribution to be

$$T - T_0 = \frac{\mu V^2}{2 k h^2}(h^2 - z^2) \qquad (2.158)$$

This is a parabolic temperature distribution with a maximum temperature in the center, as sketched in Fig. 2.4.

Once the temperature distribution is known, we can easily find the heat flux vector **q** by using Fourier's law, Eq. (2.137). From Appendix B.7 we see that the only nonzero component of **q** is

$$q_z = -k \frac{\partial T}{\partial z} \qquad (2.159)$$

Differentiating the temperature solution, Eq. (2.158), and substituting it in, we find the heat flux distribution to be

$$q_z = \frac{\mu V^2}{h^2} z \qquad (2.160)$$

> **Table 2.1** General Forms of the Balance Equations
>
> Mass balance:
> $$\frac{D\rho}{Dt} + \rho(\nabla \cdot \mathbf{v}) = 0$$
> Momentum balance:
> $$\rho \frac{D\mathbf{v}}{Dt} = -\nabla p + \nabla \cdot \boldsymbol{\tau} + \rho \mathbf{b}$$
> Energy balance:
> $$\rho c_p \frac{DT}{Dt} + \rho L_p \frac{Dp}{Dt} = \nabla \cdot (\mathbf{k} \cdot \nabla T) + \boldsymbol{\tau} : \mathbf{D} + \rho \dot{R}$$

That is, heat is flowing in the positive z direction above the center line, and in the negative z direction below the center line.

In this particular problem the temperature rise is due to viscous dissipation. Deforming the viscous fluid requires work, which is transformed into heat. The temperature distribution given in Eq. (2.158) provides the necessary temperature gradient for this heat to be conducted to the walls.

2.5 SUMMARY

The transport of mass, momentum, and energy in materials is governed by a set of fundamental balance equations. We have derived general forms of these equations that apply at every point in a continuum. The general equations are summarized in Table 2.1. Each of these equations is written out in rectangular, cylindrical, and spherical coordinates in Appendix B.

The balance equations alone seldom produce a well-posed problem. Usually one must introduce constitutive relations that describe the properties of the material. Two common constitutive relations were introduced here: the Newtonian fluid and Fourier's law for heat conduction. Well-posed problems also require boundary conditions, which reflect the interactions between the system being modeled and its surroundings.

The general balance equations may be specialized by making certain assumptions about the properties of the material, such as constant density. Some important special forms are summarized in Table 2.2. These are also written out in component form in Appendix B.

> **Table 2.2** Special Forms of the Balance Equations
>
> Mass balance, constant ρ:
> $$\nabla \cdot \mathbf{v} = 0$$
> Momentum balance, Newtonian fluid with constant ρ and μ:
> $$\rho \frac{D\mathbf{v}}{Dt} = -\nabla p + \mu \nabla^2 \mathbf{v} + \rho \mathbf{b}$$
> Energy balance, constant ρ and k, and Newtonian fluid:
> $$\rho c_p \frac{DT}{Dt} = k \nabla^2 T + \mu \dot{\gamma}^2 + \rho \dot{R}$$

The balance equations provide a firm basis from which to build models of materials processing. By starting with general forms and simplifying the equation for the problem at hand, one can be sure not to omit any terms or ignore any physical phenomena. The next chapter examines procedures for simplifying these governing equations.

BIBLIOGRAPHY

G. Astarita and G. Marrucci. *Principles of Non-Newtonian Fluid Mechanics*. McGraw-Hill, New York, 1974.

M. R. Barone and D. A. Caulk. Kinematics of flow in SMC. *Polymer Comp.* 6:105–109, 1985.

J. W. Gibbs. *The Scientific Papers of J. Willard Gibbs – Volume One: Thermodynamics*. Dover, New York, 1961.

S. M. Karim and L. Rosenhead. The second coefficient of viscosity of liquids and gases. *Rev. Modern Phys.* 24:108–116, 1952.

I. Prigogine and R. Defay. *Chemical Thermodynamics*. Longman, London, 1954. Translated by D. H. Everett.

S. M. Richardson. *Fluid Mechanics*. Hemisphere, New York, 1989.

H. Schlichting. *Boundary-Layer Theory*. McGraw-Hill, New York, 6th edition, 1968.

EXERCISES

1 ***Material and spatial descriptions of uniaxial elongation.*** A particular type of deformation, *uniaxial elongation*, is typical of tensile testing. A sample is stretched in the x_1 direction with an elongation rate $\dot{\varepsilon}$, and contracts in the x_2 and x_3 directions. The displacement functions for this flow are

$$x_1 = X_1 e^{\dot{\varepsilon} t}, \quad x_2 = X_2 e^{-\dot{\varepsilon} t/2}, \quad x_3 = X_3 e^{-\dot{\varepsilon} t/2}$$

(a) Find the velocity components v_1, v_2, and v_3 as functions of X_1, X_2, and X_3 (a material, or Lagrangian, description).

(b) Substitute the displacement functions to express the velocity components as functions of x_1, x_2, and x_3 (a spatial, or Eulerian, description).

(c) What is the rate-of-deformation tensor **D** for this flow?

2 ***Material and spatial descriptions of simple shear.*** The simple shear flow is widely used to measure the viscosity of liquids, and it occurs in practice when a fluid flows in a long channel or between closely spaced plates. The material moves in the x_1 direction, but the velocity varies with the x_2 coordinate. The velocity derivative $\partial v_1/\partial x_2$ is called the *shear rate*, $\dot{\gamma}$. The displacement functions for this flow are

$$x_1 = X_1 + X_2 \dot{\gamma} t, \quad x_2 = X_2, \quad x_3 = X_3$$

(a) Find the velocity components v_1, v_2, and v_3 as functions of X_1, X_2, and X_3 (a material, or Lagrangian, description).

(b) Substitute the displacement functions to express the velocity components as functions of x_1, x_2, and x_3 (a spatial, or Eulerian, description).

(c) What is the rate-of-deformation tensor **D** for this flow?

3. **The material derivative.** The water in a river has the following temperature distribution as a function of position and time:

$$T(x, y, t) = 25 + 3(x - 2t) + 7(y + t)^2$$

and the velocity of the river is

$$v_x = 2, \quad v_y = -1$$

Find the time derivative of temperature measured at $t = 0$ by an observer located at $x = 3$, $y = 3$ who:

(a) remains fixed at the given position,

(b) moves with the arbitrary velocity $V_x = 5$, $V_y = 0$, and

(c) translates with the fluid motion.

4. **Deriving the continuity equation.** Use a fixed differential control volume, as illustrated below, to derive the continuity equation for a fluid. Define new terms as needed.

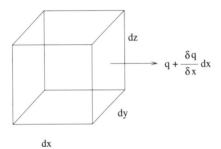

5. **One-dimensional continuity equation for fiber spinning.** For some models it is useful to have specialized balance equations. One example is the fiber spinning process. A slender, axisymmetric fiber is produced by stretching the fiber along its axis. The shape of the fiber can be characterized by its radius R as a function of position along the length z and time t. Because the fiber is slender, one usually assumes that the axial velocity v_z is not a function of r. However, v_z can depend on z and on t.

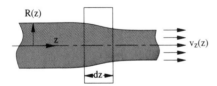

(a) Assuming that the material in the fiber has constant density, derive a one-dimensional continuity equation that relates the fiber radius $R(z, t)$ to the axial velocity $v_z(z, t)$. Do not assume a steady flow. Develop your equation by writing the integral form of the continuity equation for a control volume shown in the figure, which extends from z to $z + dz$ and is wider than the fiber width. Simplify the equation, and then let $dz \to 0$.

(b) Show that your resulting equation has the same form as the continuity equation for traffic flow from Chapter 1, Eq. (1.20).

6 *Velocity distribution in a lubricated squeeze flow.* There are not many fluid mechanics problems that can be solved by using only the continuity equation, but here is one. In compression molding, a mixture of fibers and polymer resin is squeezed between two mold halves, as shown in the figure below. The mold closing speed is s, and h is the instantaneous gap height. Under some conditions (Barone and Caulk, 1985) one can assume that the x-direction velocity does not change in the z direction, as the figure indicates. This is know as *plug flow*; the corresponding mathematical statement is $\partial v_x/\partial z = 0$.

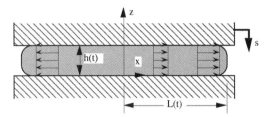

Your task is to solve for the velocity distribution $v_x(x, z)$ and $v_z(x, z)$ in a squeeze flow that obeys the plug flow assumption. You should also assume that the fluid is incompressible and that the flow is two dimensional in the x–z plane. That is, the mold and geometry of the problem prevent any motion in the y direction ($v_y = 0$), and there is no variation of v_x or v_z in the y direction. Follow these steps.

(a) Write the continuity equation, and specialize it for the assumptions given above.

(b) Take $\partial/\partial z$ of the continuity equation, and then interchange the order of differentiation. Show that $\partial^2 v_z/\partial z^2 = 0$. (This is simply a trick that works for this particular problem.)

(c) Integrate with respect to z and apply boundary conditions to find $v_z(x, z)$.

(d) Substitute this result back into the continuity equation to get an equation for $\partial v_x/\partial x$. Integrate with respect to x and apply a symmetry boundary condition at $x = 0$ to find $v_x(x, z)$.

7 *Stress and pressure in lubricated squeeze flow.* In the previous problem we used the continuity equation and some additional assumptions to solve for velocities in a two-dimensional squeezing flow. We found that the velocity field was

$$v_x = \frac{s}{h}x$$
$$v_y = 0$$
$$v_z = -\frac{s}{h}z$$

where s is the closing speed and h is the gap height. Now assume the fluid is Newtonian with viscosity μ. Assume that rate of deformation, stress, and pressure are not functions of position. Use the velocity field to answer the following questions.

(a) What is the rate-of-deformation tensor **D**?

(b) What is the extra stress tensor τ in the fluid?

(c) Suppose the edge of the material (a surface normal to the x axis) is exposed to atmospheric pressure p_{atm}. What is the surface traction vector **t** acting on this surface? What components of total stress σ can you specify based on this traction?

(d) Now find the pressure p in the fluid at the boundary. Warning: The right answer may not be obvious.

8 *Surface traction boundary conditions.* For each of the following surface traction boundary conditions, specify as many components of the total stress σ at the surface as you can, using only the boundary condition.

(a) The surface normal $\mathbf{n} = \{1, 0, 0\}^T$ and the surface traction is $\mathbf{t} = \{20, 0, 0\}^T$ MPa.

(b) $\mathbf{n} = \{1, 0, 0\}^T$ and $\mathbf{t} = \{0, 20, 0\}^T$ MPa.

(c) $\mathbf{n} = \{1, 0, 0\}^T$ and the surface is exposed to atmospheric pressure p_{atm}.

(d) $\mathbf{n} = \{-1, 0, 0\}^T$ and $\mathbf{t} = \{0, 20, 0\}^T$ MPa.

9 *Special forms of the balance equations.* For each special case given below, write out the requested balance equation, and then simplify the equation as much as possible within the stated assumptions.

(a) Write the continuity equation in cylindrical coordinates for an axisymmetric flow. (Axisymmetric means that the velocity and pressure fields are axially symmetric about the line $r = 0$).

(b) Write the r component of the equation of motion in cylindrical coordinates. Assume no body force, a Newtonian fluid of constant density, and motion in the θ direction only.

(c) Write the energy equation in rectangular coordinates. Assume constant density, Fourier's law but k not constant, steady state, and negligible dissipation. There is also a heat source that is due to a chemical reaction of $\dot{R} = H_R(dc/dt)$, where c is the degree of cure and H_R is the heat of reaction per unit mass.

10 *The Navier–Stokes equation.* Write out the x component of Eq. (2.77). Then, assuming constant viscosity and constant density, derive the x component of Eq. (2.78).

11 *Heat flux.* The temperature distribution in a body is given (with temperatures in degrees celsius and lengths in meters) by

$$T = -50\ln(x^2 + y^2)$$

and the thermal conductivity is $k = 0.2$ W/m K.

EXERCISES

(a) Find the components of the heat flux vector in Cartesian coordinates. Make a sketch of this vector field in the x–y plane in the vicinity of the origin.

(b) Transform the temperature distribution to cylindrical coordinates, and find the components of the heat flux vector in cylindrical coordinates.

12 *Velocity and temperature in a pressure-driven flow.* Repeat the solution from Sections 2.3.1 and 2.4.1 for velocity and temperature for a fluid between large parallel plates. However, this time let the plates be stationary ($V = 0$) and assume that the flow is driven by a uniform, known pressure gradient in the x direction, $-(\partial p/\partial x) = (\Delta p/L)$.

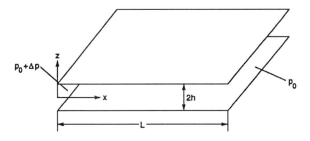

To recap, the other assumptions are:
- steady state,
- $v_y = v_z = 0$,
- v_x and T are functions of z only (often called *fully developed* velocity and temperature profiles),
- p is a function of x only,
- both plates are maintained at a uniform temperature T_o,
- ρ, c_p, k, and μ are constant (hence, the fluid is Newtonian),
- the fluid does not slip at the surface of the plates (at $z = \pm h$), and
- body forces are negligible and there are no heat sources.

Derive expressions for $v_x(z)$ and $T(z)$, and make sketches of your solutions.

13 *Viscous dissipation.* The viscous dissipation term in the energy equation will be important when we consider polymer flows. Consider the case of pressure-driven flow of a constant density Newtonian fluid in a very wide and long channel of height $2h$, where the velocity is given by

$$v_x = \frac{G}{2\mu}(h^2 - z^2), \quad v_y = 0, \quad v_z = 0$$

Here $-G = \partial p/\partial x$ is the pressure gradient and μ is the viscosity. In this flow p is a function of x only.

(a) Compute the rate-of-deformation tensor **D** for this flow.

(b) Compute the total stress tensor σ for this flow.

(c) Compute $\sigma : \mathbf{D}$. Using the results of parts (a) and (b) and the mass balance equation, show that for this flow,

$$\sigma : \mathbf{D} = \mu \left(\partial v_x / \partial z\right)^2$$

APPENDIX PROOFS REGARDING J

VOLUME CHANGE BETWEEN CONFIGURATIONS

Construct a short material vector $d\mathbf{X}$ by connecting two nearby material points in the reference configuration. In some other configuration this vector is $d\mathbf{x}$, and its components are

$$dx_i = \frac{\partial x_i}{\partial X_j} dX_j = F_{ij} dX_j \tag{2.161}$$

Now choose three material vectors $d\mathbf{X}^{(1)}, d\mathbf{X}^{(2)}$, and $d\mathbf{X}^{(3)}$, emanating from the same point but not coplanar. These vectors define the edges of a parallelepiped, whose volume in the reference state is the scalar triple product of the vectors:

$$dV_r = d\mathbf{X}^{(1)} \cdot \left[d\mathbf{X}^{(2)} \times d\mathbf{X}^{(3)}\right]$$
$$= dX_i^{(1)} dX_j^{(2)} dX_k^{(3)} \epsilon_{ijk} \tag{2.162}$$

Here ϵ_{ijk} represents the permutation symbol [Eq. (A.20)]. Similarly, the volume in the other configuration is

$$dV = d\mathbf{x}^{(1)} \cdot \left[d\mathbf{x}^{(2)} \times d\mathbf{x}^{(3)}\right]$$
$$= dx_i^{(1)} dx_j^{(2)} dx_k^{(3)} \epsilon_{ijk} \tag{2.163}$$

Into this equation substitute Eq. (2.161):

$$dV = F_{i\ell} dX_\ell^{(1)} F_{jm} dX_m^{(2)} F_{kn} dX_n^{(3)} \epsilon_{ijk}$$
$$= dX_\ell^{(1)} dX_m^{(2)} dX_n^{(3)} \epsilon_{ijk} F_{i\ell} F_{jm} F_{kn} \tag{2.164}$$

Now use the identity

$$\epsilon_{ijk} F_{i\ell} F_{jm} F_{kn} = \det \mathbf{F} \epsilon_{\ell mn} = J \epsilon_{\ell mn} \tag{2.165}$$

to obtain

$$dV = dX_\ell^{(1)} dX_m^{(2)} dX_n^{(3)} J \epsilon_{\ell mn} \tag{2.166}$$

Recognizing that the right-hand side contains Eq. (2.162), we obtain the desired result:

$$dV = J dV_r \tag{2.167}$$

APPENDIX: PROOFS REGARDING J

RATE OF CHANGE OF J

J is the determinant of the deformation gradient:

$$J = \det \mathbf{F} = \begin{vmatrix} \partial x_1/\partial X_1 & \partial x_1/\partial X_2 & \partial x_1/\partial X_3 \\ \partial x_2/\partial X_1 & \partial x_2/\partial X_2 & \partial x_2/\partial X_3 \\ \partial x_3/\partial X_1 & \partial x_3/\partial X_2 & \partial x_3/\partial X_3 \end{vmatrix} \tag{2.168}$$

The rule for differentiating a determinant is

$$\dot{J} = \begin{vmatrix} \partial \dot{x}_1/\partial X_1 & \partial \dot{x}_1/\partial X_2 & \partial \dot{x}_1/\partial X_3 \\ \partial x_2/\partial X_1 & \partial x_2/\partial X_2 & \partial x_2/\partial X_3 \\ \partial x_3/\partial X_1 & \partial x_3/\partial X_2 & \partial x_3/\partial X_3 \end{vmatrix} + \begin{vmatrix} \partial x_1/\partial X_1 & \partial x_1/\partial X_2 & \partial x_1/\partial X_3 \\ \partial \dot{x}_2/\partial X_1 & \partial \dot{x}_2/\partial X_2 & \partial \dot{x}_2/\partial X_3 \\ \partial x_3/\partial X_1 & \partial x_3/\partial X_2 & \partial x_3/\partial X_3 \end{vmatrix}$$

$$+ \begin{vmatrix} \partial x_1/\partial X_1 & \partial x_1/\partial X_2 & \partial x_1/\partial X_3 \\ \partial x_2/\partial X_1 & \partial x_2/\partial X_2 & \partial x_2/\partial X_3 \\ \partial \dot{x}_3/\partial X_1 & \partial \dot{x}_3/\partial X_2 & \partial \dot{x}_3/\partial X_3 \end{vmatrix} \tag{2.169}$$

From the chain rule,

$$\frac{\partial \dot{x}_1}{\partial X_i} = \frac{\partial \dot{x}_1}{\partial x_1} \frac{\partial x_1}{\partial X_i} \tag{2.170}$$

with similar results applying for $\partial \dot{x}_2/\partial X_i$ and $\partial \dot{x}_3/\partial X_i$. Also, if we multiply an entire row of a matrix by any factor, then the determinant of the matrix is multiplied by the same factor. Thus, Eq. (2.169) becomes

$$\dot{J} = \frac{\partial \dot{x}_1}{\partial x_1} \begin{vmatrix} \partial x_1/\partial X_1 & \partial x_1/\partial X_2 & \partial x_1/\partial X_3 \\ \partial x_2/\partial X_1 & \partial x_2/\partial X_2 & \partial x_2/\partial X_3 \\ \partial x_3/\partial X_1 & \partial x_3/\partial X_2 & \partial x_3/\partial X_3 \end{vmatrix} + \frac{\partial \dot{x}_2}{\partial x_2} \begin{vmatrix} \partial x_1/\partial X_1 & \partial x_1/\partial X_2 & \partial x_1/\partial X_3 \\ \partial x_2/\partial X_1 & \partial x_2/\partial X_2 & \partial x_2/\partial X_3 \\ \partial x_3/\partial X_1 & \partial x_3/\partial X_2 & \partial x_3/\partial X_3 \end{vmatrix}$$

$$+ \frac{\partial \dot{x}_3}{\partial x_3} \begin{vmatrix} \partial x_1/\partial X_1 & \partial x_1/\partial X_2 & \partial x_1/\partial X_3 \\ \partial x_2/\partial X_1 & \partial x_2/\partial X_2 & \partial x_2/\partial X_3 \\ \partial x_3/\partial X_1 & \partial x_3/\partial X_2 & \partial x_3/\partial X_3 \end{vmatrix} \tag{2.171}$$

Collecting terms and using indicial notation, we have

$$\dot{J} = J \frac{\partial \dot{x}_i}{\partial x_i} \tag{2.172}$$

The derivative denoted by the dot is actually the material derivative, so $\dot{x}_i$ is actually v_i, a component of the velocity vector. Recognizing that $(dv_i/dx_i) = \nabla \cdot \mathbf{v}$, we get the final result:

$$\frac{DJ}{Dt} = J \nabla \cdot \mathbf{v} \tag{2.173}$$

CHAPTER THREE

Scaling and Model Simplification

3.1 INTRODUCTION

The balance equations derived in the preceding chapter are the basis for all of the modeling work to come. When combined with constitutive equations and initial or boundary conditions, they provide the governing equations for most models. However, these governing equations are often difficult to solve in their general form, because they are multidimensional and nonlinear.

In this chapter we discuss a method for systematically simplifying the governing equations by determining which terms can be safely neglected *in a given problem*. We use the values of the physical parameters in the problem to estimate the relative importance of each term in the governing equations, using a particular procedure we call *scaling*. We then simplify the equations by neglecting terms that are "small" in comparison to terms that are "large." This produces governing equations that are often simpler than the original forms, but they still reflect all the important phenomena of the problem.

Scaling analysis produces other important results as well. Through scaling we learn the characteristic values of all of the problem variables. We also derive dimensionless parameters that have physical meaning for the particular problem. We can even determine whether the solution is likely to contain a boundary layer, and how large that layer will be. This makes scaling analysis a richer and more useful tool than traditional dimensional analysis. That is why we regard scaling as an essential step in model development.

3.1.1 SCALED VARIABLES

The value of any dimensional quantity, such as time, velocity, or stress, can only be said to be "small" or "large" in relation to some other quantity with the same units. Is a time of 5 minutes short or long? It may be short if we are growing a large single crystal for integrated circuit substrates, a process taking several days. It may be long if we are molding a thin-walled plastic housing for a cellular phone, where the cycle time is 15 seconds. Scaling provides a systematic means for determining the relative size and importance of the physical quantities in our problem.

3.1 INTRODUCTION

As a first step in scaling, we create a new, dimensionless version of each variable in the problem. A dimensionless variable can be obtained as simply as by dividing the original variable by a constant having the same units. We will represent most dimensionless variables by adding an asterisk to the original symbol. Hence, a dimensionless time t^* can be defined as

$$t^* = \frac{t}{t_c} \quad (3.1)$$

where t_c is some characteristic constant value of time.

In a process model, we choose the characteristic constant value (t_c in the example above) from the *problem parameters*. Problem parameters are physical quantities that are constant and known for any particular version of the process. For the pendulum model in Chapter 1, the parameters are m, L, g, and θ_0. In the traffic flow model the parameters are v_{max}, ρ_{max}, and L. Often the problem parameters provide an obvious choice for nondimensionalizing a variable. In the traffic flow problem we used L to nondimensionalize the distance x, and v_{max} to nondimensionalize the velocity v:

$$x^* = \frac{x}{L} \quad \text{and} \quad v^* = \frac{v}{v_{max}} \quad (3.2)$$

In other cases the choice is not so obvious, such as choosing a characteristic time from among combination of v_{max}, ρ_{max}, and L. In these cases we can use a symbol to represent the characteristic value, like t_c in Eq. (3.1), and let the scaling analysis guide us in determining how t_c is related to the known problem parameters.

The variables in Eq. (3.2) are not only dimensionless, they are scaled in a way appropriate to the problem. We will say that a variable is *properly scaled* if:

- it is dimensionless, and
- its magnitude ranges from zero to one over the entire time or space of the problem.

The magnitude requirement has to be satisfied only in a rough sense. In the traffic flow problem the variables x^* and v^* ranged exactly from zero to one, but they would have been adequately scaled if they ranged from zero to one-half, or from zero to 2π. However, a dimensionless variable that ranges from zero to 100 would *not* be properly scaled. We frequently say that a properly scaled variable is "of the order of one," writing this as $x^* = \mathcal{O}(1)$.

Most variables are scaled simply by dividing them by a constant, as in Eqs. (3.1) and (3.2). This is true if the variable takes on the value of zero somewhere in the problem. However, for some variables the dimensional value of zero never occurs in the problem, or zero has no special meaning for that variable. This is frequently the case with temperature. For variables of this type, the problem parameters usually provide both a maximum and a minimum value, so we can still produce a properly scaled variable. For instance, if we are heat treating a metal part by cooling from an initial temperature T_1 to a final temperature T_0, then a properly scaled temperature variable is

$$\theta = \frac{T - T_0}{T_1 - T_0} \quad (3.3)$$

With this definition, θ will have an initial value of one and a final value of zero.

3.2 BASIC SCALING ANALYSIS

3.2.1 SCALING RULES

The primary product of scaling analysis is a set of scaled governing equations. We say that an equation is *properly scaled* if it has the following attributes:

- all variables in the equation are properly scaled, and
- all dimensionless parameters in the equation are either $\mathcal{O}(1)$ or smaller.

We accomplish this scaling by using the following procedure.

1. Write the governing equations and boundary conditions, and simplify them as much as possible from considerations other than scaling. For example, the problem may be steady or axisymmetric, or Fourier's law for heat conduction may hold.
2. Scale the problem variables by dividing each variable by its characteristic value. If a variable, say ψ, takes on values in a characteristic range, then define an appropriate scaled variable as

$$\psi^* = \frac{\psi - \psi_{\min}}{\psi_{\max} - \psi_{\min}}$$

   For each equation there is usually one variable for which the relation between the problem parameters and the characteristic value is not obvious. For those variables choose a symbol to represent the unknown characteristic value.
3. Substitute the dimensionless variables into the governing equation and boundary conditions. In each term, separate the dimensional parameters from the dimensionless variables.
4. Choose one term that you think is important, and divide the entire equation by the group of parameters that multiply this term. Do not choose a term that contains an unknown characteristic value. Choosing the right term requires some judgment, but if you make the wrong choice here the analysis will reveal that later, and you can come back to this step and choose the right term. Each term in the equation is now dimensionless.
5. Express any unknown characteristic values in terms of the problem parameters by equating the dimensionless groups that contain them to one.
6. Compute the numerical values of any dimensionless parameters that remain in the equations. If the value of a parameter is small compared to one, then the term it multiplies can be neglected. (There are exceptions to this rule, which we come to later when we discuss boundary layers). If the value of any parameter is $\gg 1$, then that term is important but the equation is not properly scaled. Return to Step 4 and rescale accordingly.

In the next few sections, we illustrate the use of this procedure through several examples. We will explicitly identify each step in the procedure for these examples, but we will eventually drop this as the process becomes familiar.

3.2 BASIC SCALING ANALYSIS

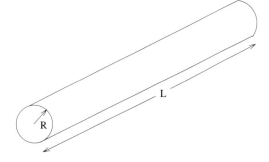

Figure 3.1: A thin, cylindrical rod, initially at temperature T_1 and quenched into a large bath of fluid at temperature T_0.

3.2.2 EXAMPLE: HEAT CONDUCTION IN A CYLINDRICAL ROD

It is easiest to see how scaling works in the context of an example. Consider the heat treatment of a thin solid cylindrical rod of radius R and length L. The rod is initially at temperature T_1, and it is quenched in a fluid at some lower temperature T_0. We want to estimate how long it will take the rod to reach the bath temperature T_0, using the procedure outlined in the preceding section. A sketch is shown in Fig. 3.1, where the sketch embodies the notion that the rod is long relative to its diameter; that is, $R \ll L$.

Step 1: Write and simplify the governing equations. The relevant governing equation is the energy balance equation. We will assume that the material follows Fourier's law of conduction, and that the pressure is approximately constant. Therefore, the most convenient form of the energy balance equation is

$$\rho c_p \left(\frac{\partial T}{\partial t} + \mathbf{v} \cdot \nabla T \right) = \nabla \cdot (\mathbf{k} \cdot \nabla T) + \boldsymbol{\tau} : \mathbf{D} + \rho \dot{R} \tag{3.4}$$

The following assumptions are made to simplify the governing equation.

1. The quenching medium is very large and maintains the surface of the rod at T_0, so we will solve only for the temperature in the rod.
2. The rod is solid and stationary, so $\mathbf{v} = 0$. This implies that $\dot{\gamma} = 0$, because if there are no velocities, there are no velocity gradients.
3. The thermal conductivity k is isotropic and constant.
4. There is no heat generation; $\dot{R} = 0$.
5. The temperature field is axisymmetric.

This leaves us with the following two-dimensional form of the equation:

$$\rho c_p \frac{\partial T}{\partial t} = k \left(\frac{1}{r} \frac{\partial}{\partial r} \left(r \frac{\partial T}{\partial r} \right) + \frac{\partial^2 T}{\partial z^2} \right) \tag{3.5}$$

The initial and boundary conditions are

$$\begin{aligned} T &= T_1 \quad \text{at } t = 0 \\ T &= T_0 \quad \text{at } r = R \\ T &= T_0 \quad \text{at } z = 0 \text{ and } z = L \end{aligned} \tag{3.6}$$

Step 2: Scale the problem variables. We now apply the scaling rules to determine which terms in Eq. (3.5) are most important in this problem. To this end, we define dimensionless variables as:

$$r^* = \frac{r}{R}$$
$$z^* = \frac{z}{L}$$
$$t^* = \frac{t}{t_c} \tag{3.7}$$
$$\theta = \frac{T - T_0}{T_1 - T_0}$$

Note that the characteristic time t_c is initially unknown.

Step 3: Substitute the dimensionless variables. We now change the governing equation from dimensional to dimensionless variables. Making use of the chain rule for differentiation, we find:

$$\frac{\partial}{\partial r} = \frac{dr^*}{dr}\frac{\partial}{\partial r^*} = \frac{1}{R}\frac{\partial}{\partial r^*}$$
$$\frac{\partial}{\partial z} = \frac{dz^*}{dz}\frac{\partial}{\partial z^*} = \frac{1}{L}\frac{\partial}{\partial z^*} \tag{3.8}$$
$$\frac{\partial}{\partial t} = \frac{dt^*}{dt}\frac{\partial}{\partial t^*} = \frac{1}{t_c}\frac{\partial}{\partial t^*}$$

Also,

$$\partial T = \partial[(T_1 - T_0)\theta] = (T_1 - T_0)\partial\theta \tag{3.9}$$

Using these relations to introduce the dimensionless variables into the governing equation yields

$$\left[\frac{\rho c_p (T_1 - T_0)}{t_c}\right]\frac{\partial \theta}{\partial t^*} = \left[\frac{k(T_1 - T_0)}{R^2}\right]\frac{1}{r^*}\frac{\partial}{\partial r^*}\left(r^*\frac{\partial \theta}{\partial r^*}\right) + \left[\frac{k(T_1 - T_0)}{L^2}\right]\frac{\partial^2 \theta}{\partial z^{*2}} \tag{3.10}$$

The boundary and initial conditions are scaled by substituting the dimensionless variables into Eqs. (3.6), giving

$$\theta = 1 \quad \text{at } t^* = 0$$
$$\theta = 0 \quad \text{at } r^* = 1 \tag{3.11}$$
$$\theta = 0 \quad \text{at } z^* = 0, 1$$

Step 4: Divide by one set of constants to make the equation dimensionless. It seems quite reasonable that conduction in the radial direction will be important, so we choose to divide the equation by $[k(T_1 - T_0)/R^2]$. This produces

$$\left\{\frac{\rho c_p R^2}{k t_c}\right\}\frac{\partial \theta}{\partial t^*} = \frac{1}{r^*}\frac{\partial}{\partial r^*}\left(r^*\frac{\partial \theta}{\partial r^*}\right) + \left\{\frac{R^2}{L^2}\right\}\frac{\partial^2 \theta}{\partial z^{*2}} \tag{3.12}$$

Note that all the quantities in braces are dimensionless.

3.2 BASIC SCALING ANALYSIS

Step 5: Determine the unknown characteristic values. By equating the dimensionless group containing t_c to one, we find that

$$t_c = \frac{\rho c_p R^2}{k} = \frac{R^2}{\alpha} \qquad (3.13)$$

where $\alpha \equiv k/(\rho c_p)$ is called the *thermal diffusivity*. This leaves

$$\frac{\partial \theta}{\partial t^*} = \frac{1}{r^*} \frac{\partial}{\partial r^*} \left(r^* \frac{\partial \theta}{\partial r^*} \right) + \left\{ \frac{R^2}{L^2} \right\} \frac{\partial^2 \theta}{\partial z^{*2}} \qquad (3.14)$$

Step 6: Compute numerical values for remaining dimensionless groups. The first two terms are $\mathcal{O}(1)$, since they contain only scaled variables and their derivatives. But what about the last term? We stated at the outset that the the rod is long compared with its radius, which means that

$$\frac{R^2}{L^2} \ll 1 \qquad (3.15)$$

The dimensionless variables in the last term are $\mathcal{O}(1)$, but they are multiplied by a dimensionless parameter that is much less than one, so the whole term must be small. Thus, the last term, which represents axial conduction, may be neglected in comparison to the term representing radial conduction. This leaves

$$\frac{\partial \theta}{\partial t^*} = \frac{1}{r^*} \frac{\partial}{\partial r^*} \left(r^* \frac{\partial \theta}{\partial r^*} \right) \qquad (3.16)$$

The solution to Eq. (3.16) with the initial and boundary conditions of Eq. (3.12) may be obtained by a separation of variables and is given by Carslaw and Jaeger (1959):

$$\theta = 1 - 2 \sum_{n=1}^{\infty} e^{-\beta_n^2 t^*} \frac{J_0(\beta_n r^*)}{\beta_n J_1(\beta_n)} \qquad (3.17)$$

Here J_0 and J_1 are Bessel functions of the first kind of order zero and one, respectively, and the β_n are the roots of

$$J_0(\beta) = 0 \qquad (3.18)$$

However, this solution is not really needed to achieve our goal of determining how long it takes the rod to reach T_0, because the cooling time is approximately the characteristic time, R^2/α. We found this from scaling the governing equation, without actually having to solving it.

In this problem, scaling also suggests that the temperature distribution should be the same at any axial location z along the rod, because our simplified governing equation, Eq. (3.16), does not contain any dependence on z. This is obviously untrue near the ends. Note that neglecting the terms for axial conduction left us unable to satisfy the boundary conditions on the end surfaces of the rod – the third of Eqs. (3.6). However, scaling does allow us to understand that "near the ends" means "within a few R's of the end." The difference between the actual temperature field and Eq. (3.17) near the ends of the rod is sometimes called an *end effect*. End effects are not some kind of mysterious physics; they are real details that were discarded to create a simplified model. Note that our estimate of cooling time is conservative near the ends, because the two-dimensional conduction that actually occurs there will cool the rod faster than the purely radial conduction in our simplified model.

Because we scaled the governing equation, the solution, Eq. (3.17), provides us with not just the temperature history of one particular cylindrical rod. It also provides the solution for *any* rod that has $R \ll L$, and it tells us how the dimensionless variables are related to the problem parameters. This allows us to extend and correlate observations. For example, we know from scaling that a rod of twice the diameter will take four times as long to cool under the same conditions.

When is a term small enough to neglect? One typically ignores terms that are two or more orders of magnitude smaller than the dominant terms. In this problem the axial conduction terms can be neglected when $(R^2/L^2) \leq 0.01$. Note that this only requires $R/L \leq 0.1$. In some cases one might even choose to neglect a term that was only one order of magnitude smaller than the dominant terms. Then we might be willing to neglect the axial conduction terms if we had $R/L \leq 0.3$.

3.2.3 IDENTIFYING DOMINANT TERMS: FLOW IN A TWO-DIMENSIONAL CONTRACTION

Step 4 in the scaling procedure requires that we identify one term as being important to the problem at hand. What happens if we guess incorrectly? Our procedure will eventually tells us. Here is an example that demonstrates this.

Suppose we want to model the flow of a fluid through the contraction shown in Fig. 3.2. We will assume that:

- the fluid is Newtonian with constant ρ and μ,
- the flow is steady and two dimensional in the x–y plane,
- body forces are negligible,
- the average velocity V in the upstream section is known, and
- the outlet pressure p_2 is known.

Note that we cannot specify both the inlet velocity V *and* the driving pressure difference $(p_1 - p_2)$ independently, so the inlet pressure p_1 is unknown. Finding an estimate for this quantity will be a goal of our model.

Step 1: The full formulation of this problem requires three equations: continuity and the x and y components of the Navier–Stokes equation, Eq. (2.77). However,

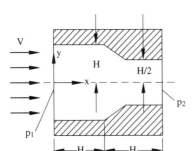

Figure 3.2: A contracting channel.

3.2 BASIC SCALING ANALYSIS

we will work with just the x component of Navier–Stokes and continuity, which in this problem contain all the interesting information. Simplifying these equations in view of the assumptions above yields

$$\frac{\partial v_x}{\partial x} + \frac{\partial v_y}{\partial y} = 0 \tag{3.19}$$

$$\rho\left(v_x \frac{\partial v_x}{\partial x} + v_y \frac{\partial v_x}{\partial y}\right) = -\frac{\partial p}{\partial x} + \mu\left(\frac{\partial^2 v_x}{\partial x^2} + \frac{\partial^2 v_x}{\partial y^2}\right) \tag{3.20}$$

Step 2: The length H is an obvious choice for nondimensionalizing the x and y coordinates, and we will use the same characteristic length for both variables.

$$x^* = \frac{x}{H}, \qquad y^* = \frac{y}{H} \tag{3.21}$$

We may use V as a characteristic value for v_x, but at this point the characteristic value for v_y is unknown. Let's call it V_y. We then have

$$v_x^* = \frac{v_x}{V}, \qquad v_y^* = \frac{v_y}{V_y} \tag{3.22}$$

The pressure p in this problem (as in all problems with incompressible fluids) has meaning only relative to some reference value, in this case p_2, so we nondimensionalize pressure much like we treated temperature in the previous example:

$$p^* = \frac{p - p_2}{\Delta p_c} \tag{3.23}$$

Here the characteristic pressure difference Δp_c is also initially unknown, and it will have to be discovered through the scaling procedure. We expect that Δp_c will be approximately equal to $(p_1 - p_2)$, so finding Δp_c through scaling allows us to estimate the inlet pressure p_1.

Step 3: Substituting the dimensionless variables first into Eq. (3.19) yields

$$\frac{V}{H}\frac{\partial v_x^*}{\partial x^*} + \frac{V_y}{H}\frac{\partial v_y^*}{\partial y^*} = 0 \tag{3.24}$$

and dividing through by V/H yields

$$\frac{\partial v_x^*}{\partial x^*} + \frac{V_y}{V}\frac{\partial v_y^*}{\partial y^*} = 0 \tag{3.25}$$

Because there are only two terms in this equation, both derivatives must be $\mathcal{O}(1)$, and we must have $V_y = V$. Note that this is a consequence of the fact that the length scales are the same in both the x and y directions.

Next, substituting the dimensionless variables into Eq. (3.20) yields

$$\left[\frac{\rho V^2}{H}\right]\left(v_x^* \frac{\partial v_x^*}{\partial x^*} + v_y^* \frac{\partial v_x^*}{\partial y^*}\right) = -\left[\frac{\Delta p_c}{H}\right]\frac{\partial p^*}{\partial x^*} + \left[\frac{\mu V}{H^2}\right]\left(\frac{\partial^2 v_x^*}{\partial x^{*2}} + \frac{\partial^2 v_x^*}{\partial y^{*2}}\right) \tag{3.26}$$

Step 4: Each of the three terms – inertial, pressure, and viscous forces – is multiplied by a group of problem parameters, here enclosed in brackets. Our next step is to choose one of these terms as "important" and divide through by its dimensional multiplier. We cannot choose the term with the unknown characteristic value (the pressure gradient in this case), so we must choose either inertial or viscous forces as being important. Let's start out by guessing that inertia is important. We multiply Eq. (3.26) by $H/(\rho V^2)$, and we get

$$\left(v_x^* \frac{\partial v_x^*}{\partial x^*} + v_y^* \frac{\partial v_x^*}{\partial y^*}\right) = -\left\{\frac{\Delta p_c}{\rho V^2}\right\} \frac{\partial p^*}{\partial x^*} + \left\{\frac{\mu}{\rho V H}\right\} \left(\frac{\partial^2 v_x^*}{\partial x^{*2}} + \frac{\partial^2 v_x^*}{\partial y^{*2}}\right) \quad (3.27)$$

Step 5: By setting the dimensionless group containing Δp_c equal to one, we determine that the characteristic pressure difference is

$$\Delta p_c = \rho V^2 \quad (3.28)$$

Also, the dimensionless collection of parameters that multiplies the viscous term is related to a classical dimensionless group, the Reynolds number, Re:

$$\text{Re} = \frac{\rho V H}{\mu} = \frac{V H}{\nu} \quad (3.29)$$

where $\nu = \mu/\rho$ is called the kinematic viscosity. Because the Reynolds number was formed by dividing the parameters multiplying the inertial terms by those multiplying the viscous terms, Re is often thought of as "the ratio of inertial forces to viscous forces." The final scaled equation is now

$$\left(v_x^* \frac{\partial v_x^*}{\partial x^*} + v_y^* \frac{\partial v_x^*}{\partial y^*}\right) = -\frac{\partial p^*}{\partial x^*} + \left\{\frac{1}{\text{Re}}\right\} \left(\frac{\partial^2 v_x^*}{\partial x^{*2}} + \frac{\partial^2 v_x^*}{\partial y^{*2}}\right) \quad (3.30)$$

Step 6: The inertial term and the pressure gradient term contain only dimensionless variables and their derivatives and, if the equation is properly scaled, these are "important" terms. We will also call such terms *major*, or *dominant*. The viscous term is multiplied by a dimensionless parameter, {1/Re}. The equation is *properly scaled* for a given problem if the value of this parameter is less than or equal to one.

For example, if we had values typical for casting of a liquid metal,

$$\mu = 3 \times 10^{-3} \text{ Pa s}, \quad \rho = 9 \times 10^3 \text{ kg/m}^3, \quad V = 0.20 \text{ m/s}, \quad H = 0.010 \text{ m} \quad (3.31)$$

then $\{1/\text{Re}\} = 1.7 \times 10^{-4}$, and the problem is properly scaled. For this the case our guess is correct: the major balance in the momentum equation is between fluid inertia and pressure, and viscosity plays a minor role. Also, the estimate for pressure provided by Eq. (3.28) is valid, and so we have successfully completed the scaling analysis.

What if the problem values are different? Suppose, for example, that we have values typical for forming of a molten polymer,

$$\mu = 8 \times 10^2 \text{ Pa s}, \quad \rho = 1 \times 10^3 \text{ kg/m}^3, \quad V = 0.20 \text{ m/s}, \quad H = 0.010 \text{ m} \quad (3.32)$$

so that $\{1/\text{Re}\} = 400$. Now the viscous terms in Eq. (3.30) are multiplied by a *large* number, and they seem to be larger than all the other terms. Can viscosity alone determine the flow pattern? No, not here. Instead, the large value of this parameter tells us that we guessed incorrectly in Step 4, and we need to redo the scaling. We go back to Eq. (3.27), but this time we know that the viscous terms are important, so we multiply the equation by $[H^2/(\mu V)]$ instead, and we get

$$\left\{\frac{\rho V H}{\mu}\right\}\left(v_x^* \frac{\partial v_x^*}{\partial x^*} + v_y^* \frac{\partial v_x^*}{\partial y^*}\right) = -\left\{\frac{\Delta p_c H}{\mu V}\right\}\frac{\partial p^*}{\partial x^*} + \left(\frac{\partial^2 v_x^*}{\partial x^{*2}} + \frac{\partial^2 v_x^*}{\partial y^{*2}}\right) \qquad (3.33)$$

Now we get a different estimate for the pressure difference,

$$\Delta p_c = \frac{\mu V}{H} \qquad (3.34)$$

and the scaled equation is

$$\{\text{Re}\}\left(v_x^* \frac{\partial v_x^*}{\partial x^*} + v_y^* \frac{\partial v_x^*}{\partial y^*}\right) = -\frac{\partial p^*}{\partial x^*} + \left(\frac{\partial^2 v_x^*}{\partial x^{*2}} + \frac{\partial^2 v_x^*}{\partial y^{*2}}\right) \qquad (3.35)$$

Note that the dimensionless parameter multiplying the inertial term is the inverse of the dimensionless parameter that multiplied the viscous term in our first try at scaling, so this parameter is guaranteed to be small. In fact we have $\text{Re} = 2.5 \times 10^{-3}$. Now we have a *properly* scaled equation, and Eq. (3.34) is the right estimate for the pressure difference. We have also learned that, for these parameter values, the important momentum balance is between viscous stresses and pressure, and that inertia plays a minor role.

In a comparison of Eq. (3.35) to Eq. (3.30), it might seem as if we have the same equation and have just moved the Reynolds number around. However, the two results are quite different, because they use very different scales for the pressure: Eqs. (3.34) and (3.28), respectively. Note that either scaling *could* be correct. It is only when we know the parameter values for a given problem that we can determine whether the scaling analysis is proper.

3.3 SMALL PARAMETERS AND BOUNDARY LAYERS

In the example of the cylindrical rod, neglecting the axial conduction terms made it impossible to satisfy some of the boundary conditions. This happens more frequently than one might at first guess. In the rod problem, neglecting the boundary conditions on the end caused us to lose some details from the model, but the main results were still correct. In other cases this is not true, and the simplified model leaves out important phenomena. In this section we show how to identify those cases, and how to model them properly.

Let us once again consider this phenomenon in the context of an example. Suppose we have a fluid flowing steadily through a wide, short duct, as illustrated in Fig. 3.3. It is easy to demonstrate by scaling that the velocity in the duct far from the perimeter is simply a constant V in the x direction. The incoming fluid is assumed to be at constant temperature T_0, and there is a heated screen at the exit that raises the exit temperature to T_1. (This problem is somewhat artificial, but it simplifies the analysis and is convenient to illustrate the point of this section.) We assume the fluid has

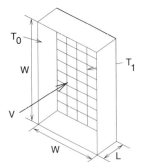

Figure 3.3: Conceptual sketch of one-dimensional flow in a duct.

constant material properties and no internal heat generation, and that the process is at steady state. With these assumptions the energy equation reduces to

$$\rho c_p V \frac{\partial T}{\partial x} = k \left(\frac{\partial^2 T}{\partial x^2} + \frac{\partial^2 T}{\partial y^2} + \frac{\partial^2 T}{\partial z^2} \right) \tag{3.36}$$

Note that the viscous dissipation is zero because there are no velocity gradients.

We now introduce dimensionless variables:

$$\{x^*, y^*, z^*\} = \left\{ \frac{x}{L}, \frac{y}{W}, \frac{z}{W} \right\} \tag{3.37}$$

$$\theta = \frac{T - T_0}{T_1 - T_0} \tag{3.38}$$

Substituting into Eq. (3.36) and dividing through by k/L^2 yields

$$\left\{ \frac{\rho c_p V L}{k} \right\} \frac{\partial \theta}{\partial x^*} = \frac{\partial^2 \theta}{\partial x^{*2}} + \left\{ \frac{L^2}{W^2} \right\} \left(\frac{\partial^2 \theta}{\partial y^{*2}} + \frac{\partial^2 \theta}{\partial z^{*2}} \right) \tag{3.39}$$

When we said that the duct was "short," we meant $L \ll W$. Thus, as we did when we considered the quenching of the slender rod, we can neglect the latter two terms, which represent conduction perpendicular to the flow, and we are left with

$$\left\{ \frac{\rho c_p V L}{k} \right\} \frac{\partial \theta}{\partial x^*} = \frac{\partial^2 \theta}{\partial x^{*2}} \tag{3.40}$$

The dimensionless group in braces on the left-hand side is called the Péclet number, Pe; it represents the ratio of heat advection to heat conduction.

$$\text{Pe} = \frac{\rho c_p V L}{k} = \frac{V L}{\alpha} \tag{3.41}$$

Here α is again the thermal diffusivity. The scaled boundary conditions are

$$\theta(x^* = 0) = 0 \tag{3.42}$$

$$\theta(x^* = 1) = 1 \tag{3.43}$$

It is often useful to examine the behavior of a scaled equation at extreme values of the parameters. Consider first the case in which the Péclet number is small, Pe $\ll$ 1. The left-hand side of Eq. (3.40) is neglected, leaving

$$\frac{\partial^2 \theta}{\partial x^{*2}} = 0 \tag{3.44}$$

3.3 SMALL PARAMETERS AND BOUNDARY LAYERS

for which the solution (satisfying both boundary conditions) is simply

$$\theta = x^* \tag{3.45}$$

This is the solution for steady-state conduction between the two ends, as if the fluid velocity were zero.

The behavior at large Péclet number is more interesting. When Pe $\gg$ 1, we follow the same procedure as in the previous fluid flow example, rewriting the equation as

$$\frac{\partial \theta}{\partial x^*} = \frac{1}{\text{Pe}} \frac{\partial^2 \theta}{\partial x^{*2}} \tag{3.46}$$

because we want the largest term in the equation to be $\mathcal{O}(1)$. If we now neglect the term multiplied by the small parameter, we obtain the simplified form

$$\frac{\partial \theta}{\partial x^*} = 0 \tag{3.47}$$

whose solution is

$$\theta = \text{constant} \tag{3.48}$$

Now we can satisfy *either* the boundary condition at $x^* = 0$ ($\theta = 0$), *or* the boundary condition at $x^* = 1$ ($\theta = 1$), but not *both* at once. What has happened here? The difficulty lies in the fact that the small parameter multiplied the term containing the highest-order derivative in the governing equation. When we eliminated that term, we could no longer satisfy all of the boundary conditions. This problem did not arise for Pe $\ll$ 1, because we dropped the convection term with its first derivative, but retained the conduction term with its second derivative.

There is a fundamental physical issue with neglecting these high-order derivatives, which becomes clear when we consider the solution of the full equation, Eq. (3.40). This solution, which satisfies both boundary conditions, is

$$\theta = \frac{\exp(\text{Pe}\, x^*) - 1}{\exp(\text{Pe}) - 1} \tag{3.49}$$

Graphs of this solution for various values of Pe appear in Fig. 3.4. It is clear from the figure that when the Péclet number is large, a *boundary layer* exists near the exit. That is, there is a small region near the boundary where the solution differs dramatically from that in the rest of the domain. The fact that the small parameter multiplies the highest derivative term can be thought of as a warning that a boundary layer will appear.

Boundary layers arise when a particular physical phenomenon affects the solution locally, within the boundary layer, but is unimportant over the rest of the domain. In Fig. 3.4, at Pe $= 100$ the solution from $x^* = 0$ to approximately $x^* = 0.95$ is essentially the constant value predicted by Eq. (3.48), whereas the solution is very different in the boundary layer, which extends approximately from $x^* = 0.95$ to $x^* = 1$. We next show how scaling analysis can be used to estimate this boundary layer thickness.

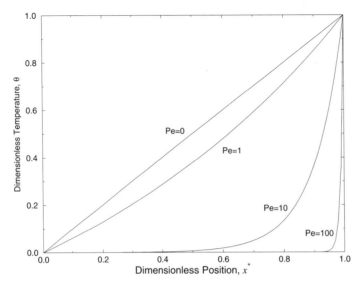

Figure 3.4: Plot of Eq. (3.49) for various values of Pe. Note the boundary layer that appears at the exit for large values of Pe.

3.3.1 BOUNDARY LAYERS AND SCALING

The scaled equation can actually be used to estimate the size of the boundary layer, without solving the equation. One way to interpret the form of the governing equation at a large Pe is to say that advection dominates conduction over most of the domain. However, within the boundary layer these effects are comparable. Reversing the latter statement is also helpful: the boundary layer is that portion of the domain where conduction and advection effects have the same magnitude.

We can use this idea to estimate the thickness δ of the boundary layer. This is done by *rescaling* the governing equation. That is, we accept Eq. (3.47) as the governing equation *outside* the boundary layer, and we derive a new, rescaled equation that applies *inside* the boundary layer. In this new equation, we use a new dimensionless variable ξ, defined as

$$\xi = \frac{x}{\delta} \tag{3.50}$$

That is, our length scale is based on the boundary layer thickness δ. This length is δ is initially unknown.

The scaled equation comes out much as before, with δ appearing wherever we saw L before:

$$\left\{\frac{\rho c_p V \delta}{k}\right\} \frac{\partial \theta}{\partial \xi} = \frac{\partial^2 \theta}{\partial \xi^2} \tag{3.51}$$

However, advection and conduction must balance each other within the boundary layer, so we set the dimensionless parameter in the braces to unity,

$$\left\{\frac{\rho c_p V \delta}{k}\right\} = 1 \tag{3.52}$$

3.3 SMALL PARAMETERS AND BOUNDARY LAYERS

This gives the boundary layer thickness as

$$\delta = \frac{k}{\rho c_p V} \tag{3.53}$$

Equation (3.52) can also be rewritten in terms of the original Péclet number as

$$\left\{\frac{\rho c_p V L}{k}\right\}\left\{\frac{\delta}{L}\right\} = \text{Pe}\left\{\frac{\delta}{L}\right\} = 1 \tag{3.54}$$

Thus, the characteristic dimensionless boundary layer thickness is $(\delta/L) = (1/\text{Pe})$. Often the solution in the boundary layer falls off exponentially from the boundary value to the "free stream" value. In that case, we expect the solution to be essentially equal to the free stream value at three to five times δ. Look back now at Fig. 3.4 and confirm this observation.

How do we develop and solve models when boundary layers are present? Very few problems admit an analytical solution to the full equations, like the example shown here. An approach that is sometimes possible is to develop two separate sets of governing equations, one inside the boundary layer (using δ as the length scale and keeping the high-order derivative terms) and one outside the boundary layer (using L as the length scale and discarding the high-order derivative terms). These two solutions must then be matched at the surface of the boundary layer by appropriate boundary conditions. This approach can be executed in a very formal mathematical way, in which case it is the method of *matched asymptotic expansions*. The inner and outer solutions can be found analytically or numerically.

Another approach is to keep all the terms in the equation and solve the problem numerically, perhaps by using a commercial code. However, simply keeping all the terms does not guarantee success. If the boundary layer is not resolved by the numerical grid or mesh, then the numerical solution will behave poorly. The computed solution may oscillate in the vicinity of the boundary layer, the solution may be unstable, or the code may never converge to a solution. Scaling analysis is very helpful in these cases, because it not only reveals the likely presence of a boundary layer, but the estimate of the boundary layer thickness tells us in advance how fine the grid has to be to get a useful solution.

3.3.2 BOUNDARY LAYERS IN NUMERICAL SOLUTIONS (ADVANCED)

The example in the previous section provides us with an opportunity to explore the effect of boundary layers in a numerical analysis. Let us start by deriving a numerical scheme for solving Eq. (3.40), using central finite differences. We approximate the derivatives by using a uniform mesh with spacing Δx^*, and we write

$$\frac{\partial \theta}{\partial x^*} \simeq \frac{\theta_{n+1} - \theta_{n-1}}{2\Delta x^*} \tag{3.55}$$

$$\frac{\partial^2 \theta}{\partial x^{*2}} \simeq \frac{\theta_{n+1} - 2\theta_n + \theta_{n-1}}{\Delta x^{*2}} \tag{3.56}$$

Substituting these expressions into Eq. (3.40) and rearranging yields

$$-\left(\frac{\text{Pe}\,\Delta x^*}{2}+1\right)\theta_{n-1}+2\theta_n+\left(\frac{\text{Pe}\,\Delta x^*}{2}-1\right)\theta_{n+1}=0 \quad (3.57)$$

Recognizing that Δx^* is simply $\Delta x/L$, we define the combination $\text{Pe}\,\Delta x^*$ as

$$\text{Pe}\,\Delta x^* = \frac{VL}{\alpha}\frac{\Delta x}{L} = \frac{V\Delta x}{\alpha} = \text{Pg} \quad (3.58)$$

where Pg is called the *grid Péclet number*. Rewriting Eq. (3.57) in terms of Pg, we have

$$-\left(\frac{\text{Pg}}{2}+1\right)\theta_{n-1}+2\theta_n+\left(\frac{\text{Pg}}{2}-1\right)\theta_{n+1}=0 \quad (3.59)$$

To complete the problem specification, we add the boundary conditions that

$$\theta_1 = 0 \quad (3.60)$$
$$\theta_N = 1 \quad (3.61)$$

This is a very rare instance in which the difference equation can be solved explicitly. The solution is

$$\theta_i = \frac{1-\left(\frac{1+\text{Pg}/2}{1-\text{Pg}/2}\right)^i}{1-\left(\frac{1+\text{Pg}/2}{1-\text{Pg}/2}\right)^N} \quad (3.62)$$

Numerical solutions to Eq. (3.59) are shown in Fig. 3.5 for several values of Pg. If Pg < 2, the numerical solution closely matches the analytical solution, but a nonphysical oscillation appears at the right-hand boundary for Pg > 2. This oscillation appears when the term multiplying θ_{n+1} in Eq. (3.59) changes sign. A physical interpretation is more revealing. Recall that the dimensionless thickness of the

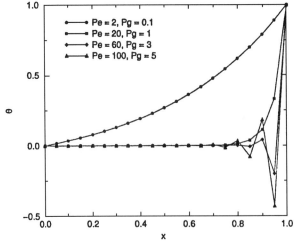

Figure 3.5: Numerical solution of Eq. (3.59) for several values of Pg, using a central difference scheme. The solution oscillates near the downstream boundary for Pg > 2.

3.3 SMALL PARAMETERS AND BOUNDARY LAYERS

boundary layer was computed as (1/Pe). When Pg exceeds two we have

$$\frac{VL}{\alpha}\frac{\Delta x}{L} > 2 \tag{3.63}$$

$$\frac{\Delta x}{L} > \frac{2}{\text{Pe}} \tag{3.64}$$

Thus, oscillations occur whenever the grid spacing exceeds twice the size of the boundary layer. The solution oscillates because the grid is not fine enough to resolve the boundary layer!

Oscillations are a common occurrence when problems are solved on coarse grids. However, there are times when the cost of resolving the boundary layer is prohibitive. Therefore, numerical schemes have been developed that will produce a smooth solution without resolving the boundary layer. One of the simplest methods is called *upwinding* or *upwind differencing*, in which the first derivative term is approximated as a backward difference:

$$\frac{\partial \theta}{\partial x^*} \sim \frac{\theta_n - \theta_{n-1}}{\Delta x^*} \tag{3.65}$$

The difference equation then becomes

$$-(\text{Pg}+1)\theta_{n-1} + (2+\text{Pg})\theta_n + (-1)\theta_{n+1} = 0 \tag{3.66}$$

Notice that the coefficient of the θ_{n+1} term is always negative in this algorithm. The solution of Eq. (3.66) is

$$\theta_i = \frac{1 - \left(1 + \frac{\text{Pg}}{2}\right)^i}{1 - \left(1 + \frac{\text{Pg}}{2}\right)^N} \tag{3.67}$$

This solution will not display the oscillatory behavior of the central difference scheme for any value of Pg, because none of the terms involving Pg changes sign.

As usual, however, anything that appears to be too good to be true probably is. The solutions with upwinding are physically plausible, but they are incorrect. Figure 3.6 compares the two numerical solutions to the exact solution for Pg = 3. Upwinding eliminates the oscillations, but it gives a much thicker boundary layer than the exact solution. Even though the unstable central difference solution produces physically impossible values near the boundary, it actually appears to be more accurate than the solution using the upwind difference algorithm.

In order to understand the behavior of the upwind difference scheme, we may ask the question, "For what differential equation is Eq. (3.66) a proper central difference representation?" A little manipulation of Eq. (3.66) shows that the answer to this question is

$$\text{Pe}\frac{\partial \theta}{\partial x^*} = \left(1 + \frac{\text{Pg}}{2}\right)\frac{\partial^2 \theta}{\partial x^{*2}} \tag{3.68}$$

Thus we see that we have actually stabilized the numerical scheme by adding artificial or "numerical" diffusion. The same phenomenon appeared in Section 1.3.2 when we used Lax's method to solve the one-dimensional wave equation.

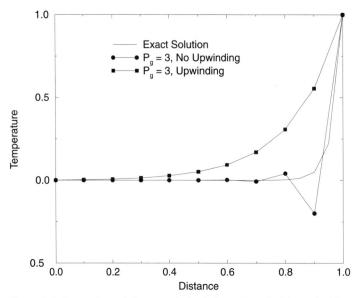

Figure 3.6: Comparison of the numerical solution of Eqs. (3.59) and (3.66) for several values of Pg.

There are better upwind difference formulations that suffer less from the artificial diffusion than what we encountered here (cf. Tannehill et al., 1997). Nevertheless, all of these schemes introduce some artificial diffusion in order to stabilize the solution when boundary layers are not resolved. It is important for the analyst to understand the behavior of these schemes, especially when using commercial programs that make it easy to select such a formulation. If possible, is always best to use a grid that resolves the boundary layer, so that upwinding schemes are not needed.

3.4 CLASSICAL DIMENSIONLESS GROUPS

One should always use scaling analysis to determine the proper dimensionless groups for any given problem. However, dimensionless groups have a long history, and many important groups have acquired standard names and standard definitions. Scaling analysis often reproduces these groups, such as the Péclet number in Section 3.3. Scaling may also produce inversions or combinations of the classical groups, such as the inverse of the Reynolds number in the first (inertia-dominant) scaling in Section 3.2.3.

These classical dimensionless groups appear in many problems, and it is useful to know what they are so that one can "speak the language." Table 3.1 lists many of the dimensionless groups that appear in materials processing problems, and it gives a brief interpretation of their meaning. In this table V and L are characteristic scales for velocity and length, and ΔT is an imposed temperature difference. All other quantities are either material properties or specified by the boundary conditions. Note that each dimensionless group represents a ratio of two phenomena, arising from terms in the governing equations.

Table 3.1 Classical Dimensionless Groups

Name of No.	Expression	Physical Meaning
Biot	$\mathrm{Bi} = hL/k$	Ratio of heat advection at the surface of a solid to heat conduction inside
Boussinesq	$\mathrm{Bo} = g\beta \Delta T_c L^3 / \alpha_0^2$	Ratio of heat advected by buoyancy to conducted heat
Brinkman	$\mathrm{Br} = \mu V^2 / k \Delta T$	Ratio of viscous dissipation to heat conduction from an imposed temp. difference
Capillary	$\mathrm{Ca} = \mu V / \Gamma$	Ratio of viscous stresses to surface tension Γ
Fourier	$\mathrm{Fo} = \alpha t / L^2$	Comparison of the penetration depth of a temp. change over time t to a characteristic dimension of a body
Froude	$\mathrm{Fr} = V^2 / gL$	Ratio of inertia to gravity
Graetz	$\mathrm{Gz} = V H^2 / \alpha L$	Ratio of heat advection in a direction with characteristic length L, to heat conduction in a direction having characteristic length H
Grashof	$\mathrm{Gr} = g\beta \Delta T_c L^3 / v_0^2$	Ratio of advective flow caused by buoyancy to viscosity
Lewis	$\mathrm{Le} = \alpha / D$	Ratio of thermal diffusion to mass diffusion
Nusselt	$\mathrm{Nu} = hL/k$	Ratio of heat conduction to heat advection, when both take place in a fluid
Nahme	$\mathrm{Na} = V^2 (\partial \mu / \partial T) / k$	Ratio of temp. change from viscous heating to the temp. needed to significantly change the viscosity (also called the Griffith number)
Pearson	$\mathrm{Pn} = (\partial \mu / \partial T) \Delta T / \mu$	Ratio of an imposed operating temp. difference to the temp. difference required to change the viscosity of the fluid
Péclet	$\mathrm{Pe} = VL / \alpha$	Ratio of heat advection to heat conduction, when both phenomena are characterized by the same length scale L (cf. the Graetz number)
Prandtl	$\mathrm{Pr} = c_p \mu / k = v / \alpha$	Ratio of momentum and thermal diffusivities in a fluid (note that the Prandtl no. is a material property)
Rayleigh	$\mathrm{Ra} = g\beta \Delta T_c L^3 / v_0 \alpha_0$	Ratio comparing advection from a buoyancy-driven flow to the product of viscosity and heat conduction
Reynolds	$\mathrm{Re} = \rho V L / \mu$	Ratio of inertial to viscous forces in a fluid
Schmidt	$\mathrm{Sc} = \mu / \rho D = v / D$	Ratio of momentum diffusivity to mass diffusivity
Sherwood	$\mathrm{Sh} = h_D L / D$	Ratio of mass transfer at the surface of a body (characterized by a mass transfer coefficient h_D) to mass transfer by diffusion within the body
Stefan	$\mathrm{Ste} = c_p \Delta T / L_f$	Ratio of sensible heat to latent heat in problems involving melting or solidification (L_f is the latent heat of fusion.)

Adapted from Lee and Castro (1989).

3.5 NONDIMENSIONALIZATION FOR NUMERICAL SOLUTIONS (ADVANCED)

In practice, it is quite common to use a computer code to solve the equations of a process model numerically. Numerical solutions are attractive because they can retain all of the terms in the governing equations, treat complicated geometries, and incorporate material properties that vary as a function of temperature, strain rate, and so on. When these codes are used, it is usually best to enter the data in dimensionless form.

Well-written software for fluid flow, heat transfer, structural analysis, and the like does not make any assumptions about units for the data. Rather, it relies on the user to enter the data in a consistent set of units, and to interpret the results accordingly. For example, suppose we have a code that solves fluid flow problems. If the coordinates of the grid points are entered in units of meters, the density is entered in kilograms per cubic meter, and the viscosity in (newtons times seconds) per square meter, then the calculated velocities will be in meters per second and the calculated stresses will be in newtons per square meter.

Although some simple problems can be solved numerically by entering the data values in some familiar unit system, in other cases this will create problems with roundoff errors and convergence. As a user, you can help a code do its best by keeping all of the values it works with close to unity. This sounds like scaling the variables and governing equations is exactly what is needed, but in fact there are a few important differences. For this reason we speak of *nondimensionalization* for numerical solutions, to distinguish this activity from scaling analysis.

3.5.1 ISOTROPIC LENGTH SCALES

In scaling analysis we are free to choose different length scales along different coordinate axes, and in fact this may allow us to eliminate some terms from the governing equations. However, this is generally not a good idea when nondimensionalizing for numerical solutions.

To see why, consider again the problem of heat conduction in the long, slender cylindrical rod from Section 3.2.2. The dimensional governing equation,

$$\rho c_p \frac{\partial T}{\partial t} = k \frac{1}{r} \frac{\partial}{\partial r}\left(r \frac{\partial T}{\partial r}\right) + k \frac{\partial^2 T}{\partial z^2} \tag{3.69}$$

was defined on the domain $r \in [0, R]$, $z \in [0, L]$. We chose dimensionless variables

$$r^* = r/R, \quad z^* = z/L, \quad t^* = tR^2/\alpha, \quad \theta = \frac{T - T_0}{T_1 - T_0} \tag{3.70}$$

which transformed the equation to

$$\frac{\partial \theta}{\partial t^*} = \frac{1}{r^*}\left(r^* \frac{\partial \theta}{\partial r^*}\right) + \left\{\frac{R^2}{L^2}\right\} \frac{\partial^2 \theta}{\partial z^{*2}} \tag{3.71}$$

This scaled problem occupies the domain $r^* \in [0, 1]$, $z^* \in [0, 1]$.

Assume now that we have a code that solves Eq. (3.69) with the appropriate boundary conditions. We could give the code the scaled problem to solve, define a mesh or grid on the $r^* \in [0, 1]$, $z^* \in [0, 1]$, enter the initial values of θ, and enter values of one for ρc_p and k. The output temperatures would be values of θ, and the

3.5 NONDIMENSIONALIZATION FOR NUMERICAL SOLUTIONS (ADVANCED)

output times would be values of t^*. However, this approach has a fatal flaw. The scaled equation, Eq. (3.71), has a multiplier of $\{R^2/L^2\}$ on the second term on the right-hand side. There is no way to enter this factor into our code, because the corresponding term in Eq. (3.69) is multiplied by k, the same factor that multiplies the first term on the right-hand side. In effect, there is no way to tell the code the length of the domain L. (This is not strictly true. Some codes permit the definition of an anisotropic thermal conductivity, which could capture the difference between the axial and radial conduction. However, this makes the solutions even more complicated to relate to the original problem.)

The cure for this problem is simple: use the same physical length to nondimensionalize both r and z. That is, choose the dimensionless variables to be

$$r^* = r/R, \quad \hat{z} = z/R, \quad t^* = tR^2/\alpha, \quad \theta = (T - T_0)/(T_1 - T_0) \tag{3.72}$$

The only difference between Eqs. (3.72) and (3.70) is the way that z is made dimensionless. Now the dimensionless energy balance equation becomes

$$\frac{\partial \theta}{\partial t^*} = \frac{1}{r^*}\left(r^* \frac{\partial \theta}{\partial r^*}\right) + \frac{\partial^2 \theta}{\partial \hat{z}^2} \tag{3.73}$$

The new dimensionless variable $\hat{z}$ is no longer properly scaled, because it ranges from zero to L/R, rather than from zero to one. However, the new dimensionless governing equation, Eq. (3.73) has exactly the same form as the original dimensional equation, Eq. (3.69). This allows us to use the code written for Eq. (3.69) to solve dimensionless form of Eq. (3.73). The correspondence between the internal variables in the code and the dimensionless values entered or interpreted by the user are shown in Table 3.2. For a complete problem setup, one must also transform the boundary conditions by using the same dimensionless variables.

This type of nondimensionalization uses an *isotropic* length scale, in the sense that the ratio between physical and dimensionless lengths (in this case the length scale R) is the same along all coordinate axes. When other vector quantities appear in the problem, such as velocity, they must also be nondimensionalized isotropically in order to preserve the form of the governing equation.

Table 3.2 Correspondence Between Software Program Variables and the User's Input Data or Interpretation of the Output, for Analysis of Heat Conduction in a Slender Rod

Program Variable	User's Value
T	θ
t	t^*
r	r^*
z	$\hat{z}$
ρc_p	1
k	1

3.5.2 PARAMETER VARIATION

We have seen in this chapter that the dimensionless solution to a model usually depends on one or more dimensionless parameter groups. A second use for nondimensionalization in numerical solutions is to provide a convenient way to change these parameters.

To put this in the context of an example, consider the flow through the two-dimensional contraction from Section 3.2.3. Suppose that we have a computational fluid dynamics code that solves the governing equations, which we now write out in full:

$$\frac{\partial v_x}{\partial x} + \frac{\partial v_y}{\partial y} = 0 \tag{3.74}$$

$$\rho\left(v_x \frac{\partial v_x}{\partial x} + v_y \frac{\partial v_x}{\partial y}\right) = -\frac{\partial p}{\partial x} + \mu\left(\frac{\partial^2 v_x}{\partial x^2} + \frac{\partial^2 v_x}{\partial y^2}\right) \tag{3.75}$$

$$\rho\left(v_x \frac{\partial v_y}{\partial x} + v_y \frac{\partial v_y}{\partial y}\right) = -\frac{\partial p}{\partial y} + \mu\left(\frac{\partial^2 v_y}{\partial x^2} + \frac{\partial^2 v_y}{\partial y^2}\right) \tag{3.76}$$

This problem had only one length scale H and one velocity scale V, so the original scaling is isotropic. Happily, we do not need to redo the scaling. For the case in which the Reynolds number is $\mathcal{O}(1)$ or greater, the dimensionless variables are

$$x^* = \frac{x}{H}, \quad y^* = \frac{y}{H}, \quad v_x^* = \frac{v_x}{V}, \quad v_y^* = \frac{v_y}{V}, \quad p^* = \frac{p - p_2}{\rho V^2} \tag{3.77}$$

and the dimensionless governing equations are

$$\frac{\partial v_x^*}{\partial x^*} + \frac{\partial v_y^*}{\partial y^*} = 0 \tag{3.78}$$

$$\left(v_x^* \frac{\partial v_x^*}{\partial x^*} + v_y^* \frac{\partial v_x^*}{\partial y^*}\right) = -\frac{\partial p^*}{\partial x^*} + \left\{\frac{1}{\text{Re}}\right\}\left(\frac{\partial^2 v_x^*}{\partial x^{*2}} + \frac{\partial^2 v_x^*}{\partial y^{*2}}\right) \tag{3.79}$$

$$\left(v_x^* \frac{\partial v_y^*}{\partial x^*} + v_y^* \frac{\partial v_y^*}{\partial y^*}\right) = -\frac{\partial p^*}{\partial y^*} + \left\{\frac{1}{\text{Re}}\right\}\left(\frac{\partial^2 v_y^*}{\partial x^{*2}} + \frac{\partial^2 v_y^*}{\partial y^{*2}}\right) \tag{3.80}$$

Note that Eqs. (3.78)–(3.80) have the same form as Eqs. (3.74)–(3.76).

Now suppose that we wish to calculate a set of solutions for different values of the Reynolds number, $\text{Re} = \rho V H/\mu$. In the physical situation, increasing the Re might correspond to increasing the value of the inlet velocity V. If one solved the problem by using dimensional variables, the average velocities would be different for each solution and comparison would be difficult. A convenient alternative is to pose the problem in dimensionless terms and change only the parameter Re for each run of the code. All of the velocities in the results are then values of v_x^* and v_y^*, so the solutions for different cases can readily be compared with one another. The factor of μ in dimensional Eqs. (3.75) and (3.76) occupies the same position as the factor of $\{1/\text{Re}\}$ in dimensionless Eqs. (3.79) and (3.80), so the easy way to change the Reynolds number is to provide the code with a viscosity value that is numerically equal to $\{1/\text{Re}\}$. The complete correspondence between internal program variables and user-supplied variables for this problem is given in Table 3.3.

Table 3.3 Correspondence Between Software Program Variables and the User's Input Data or Interpretation of the Output, for Analysis of Fluid Flow in a Two-Dimensional Contraction

Program Variable	User's Value
x	x^*
y	y^*
v_x	v_x^*
v_y	v_y^*
p	p^*
ρ	1
μ	$1/\mathrm{Re}$

3.6 SUMMARY

Scaling is a powerful tool in the development of process models. It shows which terms in the general governing equations can be neglected, and provides a rational basis for doing so. It gives order-of-magnitude estimates for the problem variables in terms of known problem parameters. It reveals the dimensionless groups that govern the problem, in a physically meaningful form. It can detect the presence of boundary layers and estimate their thickness. Best of all, scaling provides all of these results by using a straightforward procedure, without solving any partial differential equations. For this reason, a scaling analysis should always be done when a new process model is developed.

BIBLIOGRAPHY

E. Buckingham. Model experiments and the forms of empirical equations. *ASME Trans.* 37:263–296, 1916.

H. S. Carslaw and J. C. Jaeger. *Conduction of Heat in Solids.* Clarendon Press, Oxford, 2nd edition, 1959.

S. J. Kline. *Similitude and Approximation Theory.* McGraw-Hill, New York, 1965.

C.-C. Lee and J. M. Castro. Model simplification. In C. L. Tucker III, editor, *Fundamentals of Computer Modeling for Polymer Processing,* pages 69–112. Carl Hanser Verlag, Munich, 1989.

C. C. Lin and L. A. Segel. *Mathematics Applied to Deterministic Problems in the Natural Sciences.* SIAM, Philadelphia, PA, 1988.

J. C. Tannehill, D. A. Anderson, and R. H. Pletcher. *Computational Fluid Mechanics and Heat Transfer.* Taylor and Francis, Washington, DC, 2nd edition, 1997.

EXERCISES

1 *Heat conduction in a slab.* Consider heat conduction in a thin square slab, of thickness δ and having edge length L. Assume $\delta \ll L$. The initial temperature is T_1, and at time zero the slab is quenched into a medium at temperature T_0

with a heat transfer coefficient h. We want to scale this problem and estimate the cooling time.

(a) Reduce the governing equations as far as possible by making appropriate assumptions, but without using scaling data. Be sure to state your assumptions as you delete terms.

(b) Introduce appropriate length, time, and temperature scales, and rewrite the equations in dimensionless form. Identify any new characteristic values and eliminate terms that should be small.

(c) Scale the boundary condition.

(d) At this point, you should have found a new dimensionless group, the Biot number:

$$\text{Bi} = \frac{h\delta}{k}$$

Examine the two extremes of $\text{Bi} \gg 1$ and $\text{Bi} \ll 1$ and sketch the temperature profiles through the section for the two cases. Remember that when the Bi is large, you need to divide through to have $(1/\text{Bi})$ in the equation instead. Look carefully at the boundary condition to see what it is telling you.

(e) Estimate the cooling times for these two cases.

2 *Typical material properties.* When scaling analyses are done, it can be very useful to have some typical values of material properties at hand. These can be used to estimate the values of the dimensionless parameters, even when the exact property values are unknown.

Use any reliable source to fill in the table of properties below. Express all values in SI units. Give the names of the metal and the polymer. For each data item give the source of your information (book, authors, publisher, year, and page number).

Material	ρ	c_p	k	μ
Air				
Water				
A liquid metal				
A molten polymer				

3 *Cooling length of sheet products.* Frequently in materials processing, a sheet or slab of material is produced continuously and must be cooled after it leaves some processing device. This situation arises in polymer extrusion and in continuous casting of metals, for instance. The situation is sketched below. Assume

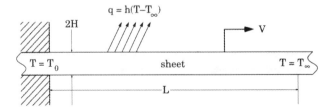

that the sheet has a uniform temperature T_0 as it leaves the upstream device, and that it moves with a constant velocity V. The thickness of the sheet is $2H$. Cooling is accomplished by exposing the sheet to a coolant (usually air or water), and it may involve either free or forced convection. The cooling is characterized by a heat transfer coefficient h and a coolant temperature T_∞. The goal of our analysis is to determine how the cooling length L depends on the other problem parameters.

(a) Write the energy equation, and simplify it by assuming a two-dimensional temperature distribution, steady-state behavior, constant properties, and no heat sources. Also write the boundary conditions.

(b) What are the physical parameters for this problem? Which parameter is initially unknown?

(c) Define appropriate dimensionless variables for this problem.

(d) Scale the energy equation, and simplify it by assuming that $H \ll L$.

(e) Scale the boundary conditions. Identify a dimensionless group that describes whether the heat transfer coefficient h is small or large.

(f) Estimate the cooling length L in terms of known problem parameters, assuming that $hH/k \gg 1$.

4 *Viscous flow in a narrow gap.* The flow of a fluid in a narrow gap occurs in polymer extrusion and molding and in metal die casting. To keep the analysis simple, consider only steady two-dimensional flow in the x–z plane. Assume that the fluid is incompressible and Newtonian with a constant viscosity, and neglect body forces. The flow geometry is shown below. Assume that the gap is narrow ($H \ll L$); V is the average velocity of the fluid entering the gap.

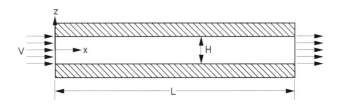

(a) Write the balance equations that govern this problem, simplifying them for the assumptions given above.

(b) List the problem parameters. Scale the problem variables. For this problem use different characteristic values for v_x and v_z, and for x and z. State which characteristic values are unknown at this point.

(c) Scale the continuity equation, and find an estimate for v_z.

(d) Scale the x-direction equation of motion, and simplify it on the basis that $H \ll L$. Assume that viscous stresses and the pressure gradient dominate the x-direction momentum balance. Derive an estimate for the pressure gradient $\partial p/\partial x$. What parameter describes the importance of the inertial terms? Is this a classical dimensionless group?

(e) Rescale the x-direction momentum equation and get a new estimate for the pressure gradient $\partial p/\partial x$, this time assuming that the inertial terms are much larger than the viscous terms. What is the parameter describing the importance of the viscous terms? If you neglected viscosity, would any boundary conditions have to be ignored?

(f) Using the properties from Problem 2, calculate the relative importance of viscosity and inertia for a liquid metal, and also for a molten polymer, in this geometry. Take $H = 3$ mm, $L = 150$ mm, and $V = 75$ mm/s.

5 **Startup of drag flow between parallel plates.** A fluid is confined between two large parallel plates, which are separated by a gap of height H. The pressure in the fluid is uniform. Initially the fluid and the plates are at rest. At time $t = 0$ the top plate begins to move in the x direction with velocity V. Initially the inertia of the fluid is important, and the velocity has time-dependent profiles that are similar to the profile on the left in the sketch below. After a long time the velocity reaches a steady-state profile like the one on the right. We are interested in determining how much time is required to achieve the steady-state velocity profile.

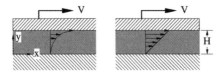

Assume that the flow is two dimensional ($v_z = 0$ and $\partial/\partial z = 0$), v_x depends on y and t but does not vary in the x direction, $v_y = 0$, $\partial p/\partial x = 0$, and body forces are negligible.

(a) Write the x-direction Navier–Stokes equation, and simplify it by using the assumptions given above.

(b) Use scaling to find the characteristic time for this problem in terms of the known parameters. This is approximately the time required to reach steady state.

(c) Calculate the numerical value of this characteristic time for a spin-coating process in which the fluid has approximately the properties of water ($\rho = 1 \times 10^3$ kg/m^3, $\mu = 1 \times 10^{-3}$ Pa s), the thickness H is 0.1 mm, and the velocity is 10 m/s.

6 **Velocities in spin coating.** Spin coating is a process used to place a thin, uniform layer of polymeric materials onto a flat substrate. It is widely used in the manufacture of solid-state electronics, and it is also used to put the protective coating on the tops of compact disks. The solid substrate is rotated at an angular velocity Ω, and a solution of polymer and solvent is dropped onto the substrate. The fluid quickly acquires the angular velocity of the substrate, and it is slung outward by the action of centrifugal force. After a very short time the situation is like the sketch: the fluid layer is nearly uniform in

thickness, and the thickness decreases slowly in time as more fluid flows off the edge of the substrate. The goal of the process is to create a fluid layer of uniform and controlled thickness. In practice the fluid motion is halted by drying (i.e., the solvent evaporates, leaving a thin layer of polymer). Here we will ignore evaporation and obtain an estimate for the radial velocity of the fluid.

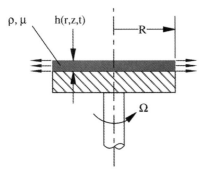

Assume that the flow is axisymmetric, and that the fluid moves in the θ direction with the velocity of the rotating solid disk, so $v_\theta = \Omega r$. Assume that the fluid film thickness $h(r, t)$ is initially known, and use H as the characteristic value of this thickness. The disk has radius R, and we are interested in thin coatings so $H \ll R$. Assume that the fluid has constant properties ρ and μ. Use V_r and V_z as characteristic velocities in the r and z directions respectively; both of these quantities are initially unknown. We can also assume that $\partial p/\partial r = 0$, because the upper surface of the fluid is at constant pressure.

(a) Write the continuity equation, simplify it for the given assumptions, and use scaling to obtain a relationship between V_r and V_z.

(b) Write the r-direction momentum balance equation, and simplify it for the given assumptions.

(c) Scale the r-direction momentum balance equation, using $t_c = R/V_r$ as a characteristic time for the unsteady inertial term. Eliminate terms that are small when $V_r \ll \Omega R$, and find an estimate for the radial velocity V_r in terms of known problem parameters.

7 *Heating a thin slab with radio-frequency energy.* A thin slab of material is heated with radio-frequency energy, which dissipates heat at a known rate $\rho \dot{R}$ per unit volume. The heating goes on for a prescribed time t_c. The slab is initially at T_0. It is always in contact with the surrounding air, which is also at T_0, through a heat transfer coefficient h. That is, the thermal boundary condition at the surface of the slab is

$$-k\frac{\partial T}{\partial n} = h(T - T_0)$$

where $\partial T/\partial n$ stands for the derivative normal to the surface of the part. We wish to know the maximum temperature and to have some idea of the temperature uniformity in the slab.

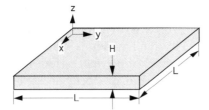

(a) Simplify the energy equation for this problem by using the assumptions stated above. Also assume that the thermal properties are constant.

(b) List the problem parameters. Scale the problem variables. State which characteristic values are unknown at this point.

(c) Scale the energy equation and simplify it under the assumption that $H \ll L$. Also assume that the heat conduction terms are small. Identify a characteristic temperature difference. How will the temperature solution vary with z if heat conduction is neglected altogether?

(d) Produce a new scaled energy equation, this time assuming that the heat conduction term balances the heat source term, and that the transient term is small. Identify a new characteristic temperature difference. How will the temperature solution vary with t if the transient term is neglected?

(e) Write the initial and boundary conditions for this problem in dimensionless form. Do any new dimensionless parameters appear? If so, what is their physical significance?

CHAPTER FOUR

Heat Conduction and Materials Processing

The preceding chapters presented fundamental tools needed to model materials processing operations. We now begin to apply these concepts, considering problems that involve heat conduction. Heat conduction is the predominant phenomenon in many important applications, including heat treatment of steel, metal casting, welding, and polymer injection molding. Heat conduction is also a good place to start, because it involves just one equation, the energy balance, and one variable, the temperature. In this chapter we introduce several simple problems, building in complexity, and then provide an extensive treatment of heat conduction with phase change. The emphasis will be on how to approach problems, though we will also pick up some handy solution techniques along the way.

Energy Balance Equation

The general energy balance equation was derived in Chapter 2, where we saw that the equation can take several possible forms. In particular, to represent the thermodynamic behavior of a material, we must choose two of the three fundamental variables – temperature, pressure, and specific volume – to regard as independent. For condensed phases (liquids and solids) it is usually most convenient to choose temperature and pressure as the independent variables. Accordingly, we start with the following general form of the energy balance equation:

$$\rho \left(c_p \frac{DT}{Dt} + L_p \frac{Dp}{Dt} \right) = \nabla \cdot (\mathbf{k} \cdot \nabla T) + \boldsymbol{\tau} : \mathbf{D} + \rho \dot{R} \tag{4.1}$$

where c_p is the specific heat at constant pressure, L_p is the latent heat of pressure change, $\mathbf{k}$ is the thermal conductivity tensor, $\dot{R}$ is the rate of internal heat generation, and $\mathbf{D}$ and $\boldsymbol{\tau}$ are the rate-of-deformation and extra stress tensors, respectively.

For the examples in this chapter, we will confine ourselves to cases in which all material properties are constant. We will also make the following assumptions.

1. Pressure variations are small enough that $L_p(Dp/Dt)$ is negligible.
2. The thermal conductivity is isotropic, so that $\mathbf{k} = k\mathbf{I}$, and k is constant.

Table 4.1 Phenomenological Names for Terms in the Energy Equation

Term	Phenomenological Name
$\rho c_p \, \partial T / \partial t$	Transient or unsteady transport
$\rho c_p \mathbf{v} \cdot \nabla T$	Advection or convective transport
$k \nabla^2 T$	Conduction
$\boldsymbol{\tau} : \mathbf{D}$	Viscous dissipation
$\rho \dot{R}$	Internal generation or production

With these restrictions the energy balance equation becomes

$$\rho c_p \left(\frac{\partial T}{\partial t} + \mathbf{v} \cdot \nabla T \right) = k \nabla^2 T + \boldsymbol{\tau} : \mathbf{D} + \rho \dot{R} \qquad (4.2)$$

Here the material derivative has been expanded to remind the reader of the terms it includes. The various terms in this equation are often referred to by the phenomenological names listed in Table 4.1.

In this chapter we further restrict our attention to problems in which the body in question is not moving. Then both $\mathbf{v}$ and $\mathbf{D}$ are zero, and the energy balance equation reduces to

$$\rho c_p \frac{\partial T}{\partial t} = k \nabla^2 T + \rho \dot{R} \qquad (4.3)$$

This equation describes heat conduction in a solid, or in a fluid undergoing no motion, possibly having internal heat generation. For most of the problems in this section we also assume that there is no internal heat generation, and we are left with what is usually called the *heat conduction equation*:

$$\rho c_p \frac{\partial T}{\partial t} = k \nabla^2 T \qquad (4.4)$$

This will be the energy equation for most problems in this chapter. In Chapter 7 we will return to heat transfer problems in moving bodies and flowing fluids.

Boundary Conditions

The heat conduction equation requires a boundary condition at every point on the boundary of the solution domain. One must specify either the temperature, the heat flux normal to the surface, or some combination of the two. Note that the complete heat flux vector $\mathbf{q}$ is not specified on the boundary, only the component of $\mathbf{q}$ normal to the boundary. This is equal to $\mathbf{q} \cdot \mathbf{n}$, where $\mathbf{n}$ is a unit vector normal to the boundary and (by convention) pointing outward. Using Fourier's law of heat conduction, on the boundary we have

$$\mathbf{q} \cdot \mathbf{n} = -k \nabla T \cdot \mathbf{n} \qquad (4.5)$$

The right-hand side of Eq. (4.5) involves the projection of the temperature gradient vector ∇T in the direction of $\mathbf{n}$. That is, $\nabla T \cdot \mathbf{n}$ is the rate that temperature changes

Table 4.2 Common Names and Forms for Boundary Conditions on the Heat Conduction Equation

Name	Boundary Condition
Dirichlet	$T = T_s$
Neumann	$-k\, \partial T/\partial n = q_s$
Mixed	$-k\, \partial T/\partial n = h(T - T_\infty)$
Radiation	$-k\, \partial T/\partial n = \sigma\epsilon(T^4 - T_\infty^4)$

Here $-k(\partial T/\partial n)$ is shorthand for the heat flux leaving the surface, $\mathbf{q} \cdot \mathbf{n}$, where $\mathbf{n}$ is the outward-directed unit normal vector at the surface.

with position as one moves in the direction specified by **n**. We customarily write the right-hand side of Eq. (4.5) in shorthand as $-k\, \partial T/\partial n$. The common types of boundary conditions are tabulated in Table 4.2.

The *mixed* boundary condition is also sometimes called a *Robin*, *third kind*, or *heat transfer coefficient* boundary condition, and the quantity h is called a convective heat transfer coefficient. This boundary condition is used to model heat transfer behavior where a solid body is in contact with a moving fluid. On the boundary, the fluid temperature matches the solid temperature, while far away from the boundary the fluid has a temperature T_∞. The heat flux entering or leaving the fluid depends on T_∞, but also on the thermal properties of the fluid and the nature of the fluid flow. Fluids with high heat capacity and thermal conductivity transfer heat more readily than fluids with low c_p and k, and turbulent fluid flow also enhances the heat transfer relative to laminar flow. A good deal of effort has been expended to learn how to predict h given the details of the fluid properties and flow, and many theoretical calculations and correlation formulas are available (cf. Incropera and DeWitt, 1996). For some purposes a rough estimate of h is quite useful, and Table 4.3 gives the order of magnitude of h for a variety of situations.

Table 4.3 Typical Values of the Convective Heat Transfer Coefficient h

Condition	h (W/m²K)
Gases, natural convection	5–30
Flowing gases	10–300
Flowing liquids (nonmetallic)	170–5,500
Flowing liquid metals	5,500–30,000
Boiling liquids	1,000–30,000
Condensing vapors	3,000–30,000

Adapted from Rohsenow and Choi (1961, p. 102).

4.1 STEADY HEAT CONDUCTION IN SOLIDS

There are a number of materials processing applications in which the temperature field can be considered to be steady, and we consider those first. Examples include temperature distributions in furnace walls and pipes, in which the usual goal is to compute heat loss rates.

For steady conduction with constant k and without internal heat generation, the heat conduction equation reduces to Laplace's equation for the temperature:

$$\nabla^2 T = 0 \tag{4.6}$$

4.1.1 A HOMOGENEOUS WALL

Consider steady heat conduction through the wall section shown in Fig. 4.1. The temperature on one side of the wall is maintained at T_1, while the temperature on the other side is maintained at T_2. We assume that the wall is thin relative to its lateral dimensions, that is, $H/L \ll 1$. We would like to know the temperature distribution and heat flux through the wall. Writing out Eq. (4.6) in Cartesian coordinates, we have

$$\frac{\partial^2 T}{\partial x^2} + \frac{\partial^2 T}{\partial y^2} + \frac{\partial^2 T}{\partial z^2} = 0 \tag{4.7}$$

Now introduce the following dimensionless variables:

$$x^* = \frac{x}{H}, \quad y^* = \frac{y}{L} \tag{4.8}$$

$$z^* = \frac{z}{L}, \quad \theta = \frac{T - T_1}{T_2 - T_1} \tag{4.9}$$

Inserting these variables into Eq. (4.7) yields

$$\frac{\partial^2 \theta}{\partial x^{*2}} + \left\{ \frac{H^2}{L^2} \right\} \left(\frac{\partial^2 \theta}{\partial y^{*2}} + \frac{\partial^2 \theta}{\partial x^{*2}} \right) = 0 \tag{4.10}$$

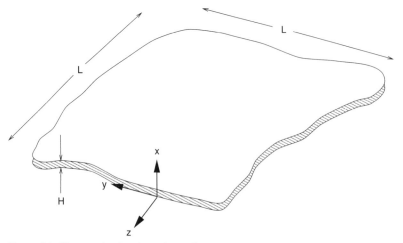

Figure 4.1: Heat conduction through a wall segment.

4.1 STEADY HEAT CONDUCTION IN SOLIDS

We may neglect the latter two terms for the case in which $H/L \ll 1$, and we solve the resulting equation to find

$$\theta = c_1 x^* + c_2 \tag{4.11}$$

The scaled boundary conditions on the upper and lower boundaries become

$$\theta(x^* = 0) = 0 \tag{4.12}$$
$$\theta(x^* = 1) = 1 \tag{4.13}$$

Using these conditions to evaluate the constants c_1 and c_2 gives the dimensionless solution:

$$\theta = x^* \tag{4.14}$$

In dimensional form this is

$$T = T_1 + (T_2 - T_1)\frac{x}{H} \tag{4.15}$$

Once the temperature solution is known, we can differentiate it to compute the heat flux:

$$q_x = -k\frac{dT}{dx} = -\frac{k}{H}(T_2 - T_1) \tag{4.16}$$

An analogy is often made between heat flow by conduction and current flow in electrical circuits. The rate of heat flow (heat flux times area) is analogous to the current, and the temperature difference is analogous to the potential (voltage) difference. We therefore introduce a "resistance" to heat flow $\mathcal{R}$, defined by analogy to Ohm's law, as

$$q_x A = -\frac{(T_2 - T_1)}{\mathcal{R}} \tag{4.17}$$

where A is the surface area. The negative sign is incorporated because of our definition of the direction of heat flux. The resistance should always be positive. For the one-dimensional wall section above, the resistance of the wall is

$$\mathcal{R} = \frac{H}{Ak} \tag{4.18}$$

This equation satisfies our intuition that we may increase the resistance to heat flow either by increasing the thickness of material or by decreasing its thermal conductivity.

Let us now modify this problem slightly by changing the boundary condition at $x = 0$ to a mixed condition:

$$k\frac{\partial T}{\partial x} = h(T - T_1) \tag{4.19}$$

(There is no minus sign on the left-hand side because the outward normal at the lower surface points in the negative x direction). The solution is readily found (and is left as an exercise):

$$T = \frac{h(T_2 - T_1)x}{hH + k} + \frac{hHT_1 + kT_2}{hH + k} \quad (4.20)$$

The heat flux is computed by differentiating this solution to find

$$q_x = -\frac{T_2 - T_1}{\frac{H}{k} + \frac{1}{h}} \quad (4.21)$$

Referring to the definition of resistance to heat flow, Eq. (4.17), we find the important result that, for this case, the total resistance $\mathcal{R}_T$ is

$$\mathcal{R}_T = \frac{H}{Ak} + \frac{1}{Ah} \quad (4.22)$$

This shows that the heat transfer coefficient h at the surface contributes a resistance to heat flow given by

$$\mathcal{R} = \frac{1}{Ah}$$

and that the total resistance is the sum of the wall and surface resistances acting in series. This latter result is a consequence of the fact that, at steady state, the heat fluxes through the wall and the surface are equal.

For a cylindrical wall with inner radius R_i, outer radius R_o, and length L, the thermal resistance is

$$\mathcal{R} = \frac{\ln(R_o/R_i)}{2\pi L k}$$

This result is left as an exercise.

4.1.2 A COMPOSITE WALL

Another application of the same principle concerns a composite wall, such as that shown in Fig. 4.2. The heat conduction equation must be satisfied in both layers, so we have

$$\frac{d^2 T_A}{dx^2} = 0, \quad 0 \leq x \leq L_A \quad (4.23)$$

$$\frac{d^2 T_B}{dx^2} = 0, \quad L_A \leq x \leq L_A + L_B \quad (4.24)$$

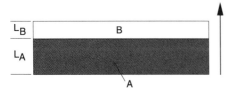

Figure 4.2: A composite wall made of two materials, A and B.

to which the solutions are

$$T_A = c_1^A x + c_2^A, \quad 0 \leq x \leq L_A \tag{4.25}$$

$$T_B = c_1^B x + c_2^B, \quad L_A \leq x \leq L_B \tag{4.26}$$

The boundary conditions at the upper and lower surfaces permit the resolution of two of the constants:

$$T_A(x = 0) = T_1 = c_2^A \tag{4.27}$$

$$T_B(x = L_A + L_B) = T_2 = c_1^B(L_A + L_B) + c_2^B \tag{4.28}$$

At this point we have

$$T_A = T_1 + c_1^A x \tag{4.29}$$

$$T_B = T_2 + c_1^B(x - L_A - L_B) \tag{4.30}$$

The other two constants are determined by two matching conditions at the interface, $x = L_A$. The temperature must be continuous there, so the first matching condition is

$$T_A = T_B \quad \text{at } x = L_A \tag{4.31}$$

In addition, the heat flux leaving A must be equal to the flux entering B. The fluxes at $x = L_A$ are given by

$$k_A \frac{dT_A}{dx} = k_A c_1^A \tag{4.32}$$

$$k_B \frac{dT_B}{dx} = k_B c_1^B \tag{4.33}$$

Equating these, we conclude that $c_1^B = (k_A/k_B) c_1^A$. Combining this with Eqs. (4.29)–(4.31), we find that

$$c_1^A = \frac{T_2 - T_1}{k_A \left(\frac{L_A}{k_A} + \frac{L_B}{k_B} \right)} \tag{4.34}$$

From this result and Eq. (4.29) we find that the total resistance to heat flow for the composite wall is simply the sum of the individual resistances in series:

$$\mathcal{R}_T = \frac{L_A}{A k_A} + \frac{L_B}{A k_B} \tag{4.35}$$

Once again, this occurs because the fluxes must match in both materials. This procedure of summing the resistances can be applied to any number of components and interfaces.

4.2 TRANSIENT HEAT CONDUCTION

Many materials processing operations involve transient heat flow. These include quenching of steel during heat treatment and cooling of polymers in molds, among many other applications. The purpose of these operations is usually to solidify the material, melt it, or induce some other phase transformation. In this section we consider several basic transient problems. The solutions to these problems provide important insights into transient conduction phenomena, and some of them will be useful building blocks for solutions to more complex problems. The solution methods themselves are also useful tools that can be applied to other problems.

4.2.1 SEMI-INFINITE BODY

Consider the cooling of a semi-infinite slab, as depicted in Fig. 4.3. Obviously no real material is infinite in extent, but we will develop expressions to help us understand when this might be a reasonable approximation. The material occupies the domain $x > 0$, and it is initially at temperature T_1. At time $t = 0$ the surface at $x = 0$ is instantaneously reduced to temperature T_0. We continue to consider the case in which the material properties are constant and there is no heat generation. The temperature must depend only on x and t, so Eq. (4.4) reduces to

$$\rho c_p \frac{\partial T}{\partial t} = k \frac{\partial^2 T}{\partial x^2} \tag{4.36}$$

The boundary and initial conditions are

$$T(x > 0, t = 0) = T_1 \tag{4.37}$$
$$T(x = 0, t) = T_0 \tag{4.38}$$
$$T(x \to \infty, t) = T_1 \tag{4.39}$$

Because the medium is semi-infinite, this problem has no natural length scale, nor does it have a natural time scale. In such cases, it is often possible to find a *similarity variable*, which is a combination of the two independent variables x and t, that reduces the partial differential equation to an ordinary differential equation. That is, we want to transform the problem for $T(x, t)$ into a problem for $T(\zeta)$, where $\zeta(x, t)$ is the similarity variable. In this problem scaling provides no more simplification, so we will proceed by using the dimensional variables.

We start by making an (educated) guess for the similarity variable,

$$\zeta = Cxt^m \tag{4.40}$$

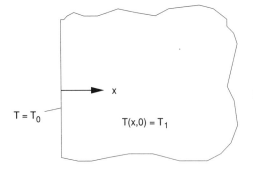

Figure 4.3: A model of a semi-infinite medium.

4.2 TRANSIENT HEAT CONDUCTION

where C and m are constants that we need to choose. Transforming the various derivatives in Eq. (4.36) by using the chain rule, we have

$$\frac{\partial T}{\partial t} = \frac{dT}{d\zeta}\frac{\partial \zeta}{\partial t} = \frac{dT}{d\zeta}mCxt^{m-1} = \frac{dT}{d\zeta}m\zeta t^{-1} \tag{4.41}$$

$$\frac{\partial T}{\partial x} = \frac{dT}{d\zeta}\frac{\partial \zeta}{\partial x} = \frac{dT}{d\zeta}Ct^m \tag{4.42}$$

$$\frac{\partial^2 T}{\partial x^2} = \frac{d}{d\zeta}\left(\frac{\partial T}{\partial x}\right)\frac{\partial \zeta}{\partial x} = \frac{d^2 T}{d\zeta^2}C^2 t^{2m} \tag{4.43}$$

Substituting these results into Eq. (4.36) and combining the material properties by defining the thermal diffusivity $\alpha = k/\rho c_p$ yields

$$m\zeta t^{-1}\frac{dT}{d\zeta} = \alpha C^2 t^{2m}\frac{d^2 T}{d\zeta^2} \tag{4.44}$$

We can satisfy our requirement that this becomes an ordinary differential equation in ζ if we choose $2m = -1$, thus eliminating from the equation any explicit reference to x or t. We may also choose C arbitrarily without affecting this result, so we select $C^2 = 1/4\alpha$. Thus our similarity variable becomes

$$\zeta = x/\sqrt{4\alpha t} \tag{4.45}$$

and the energy balance equation, Eq. (4.36) is transformed to

$$\frac{d^2 T}{d\zeta^2} + 2\zeta\frac{dT}{d\zeta} = 0 \tag{4.46}$$

The transformed boundary conditions are

$$T(\zeta = 0) = T_0 \tag{4.47}$$

$$T(\zeta \to \infty) = T_1 \tag{4.48}$$

The first of these equations comes from the condition at $x = 0$, whereas the latter corresponds to *both* the initial condition *and* to the boundary condition at $x = \infty$. If the initial and far-field boundary conditions were not the same, then no similarity solution would exist.

To solve Eq. (4.46), we first define a new variable $y = \partial T/\partial \zeta$. Equation (4.46) then becomes

$$\frac{dy}{d\zeta} = -2\zeta y \tag{4.49}$$

which we integrate to find y as

$$y = C_1 \exp(-\zeta^2) \tag{4.50}$$

Replacing y by $\partial T/\partial \zeta$ and integrating again yields

$$T = C_2 + C_1 \int \exp(-\zeta^2)\, d\zeta \tag{4.51}$$

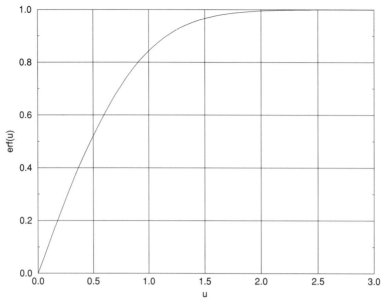

Figure 4.4: Plot of the Gaussian error function, erf (u).

There is no analytical solution to this integral, but it appears in a variety of different problems, and its functional behavior is well known. This integral is the kernel of the *Gaussian error function*, frequently called just the error function, which is defined as

$$\text{erf}(\zeta) \equiv \frac{2}{\sqrt{\pi}} \int_0^\zeta \exp(-u^2)\, du \qquad (4.52)$$

The error function is plotted in Fig. 4.4. The factor of $2/\sqrt{\pi}$ gives the handy property that $\text{erf}(\infty) = 1$, and from the definition we can see that $\text{erf}(0) = 0$. Intermediate values can be calculated numerically, and erf is also a tabulated function. A table of values appears in Table A.1, and Appendix A.7 gives some useful approximations for erf (u). Note that erf $(-u) = -\text{erf}(u)$. The expression $(1 - \text{erf}(u))$ occurs frequently enough that it is given its own name, the *complementary error function*, and its own symbol, so that

$$\text{erfc}(u) \equiv 1 - \text{erf}(u) \qquad (4.53)$$

Returning to the semi-infinite slab problem, we now rewrite Eq. (4.51) in terms of the error function,

$$T = C_2 + C_3 \,\text{erf}(\zeta) \qquad (4.54)$$

and then use the boundary conditions to determine the constants. This gives

$$\frac{T - T_1}{T_0 - T_1} = 1 - \text{erf}\left(\frac{x}{\sqrt{4\alpha t}}\right) = \text{erfc}\left(\frac{x}{\sqrt{4\alpha t}}\right) \qquad (4.55)$$

A plot of this solution is shown in Fig. 4.5 for various times. Note that the curves are parameterized by αt, because time appears only in this product combination in the solution. The horizontal axis in Fig. 4.5 uses arbitrary dimensional units for x. If we

4.2 TRANSIENT HEAT CONDUCTION

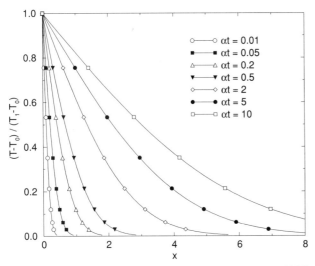

Figure 4.5: Temperature solution in a semi-infinite body, Eq. (4.55), for various values of (αt).

were to make this axis dimensionless by plotting θ versus $x/\sqrt{4\alpha t}$, the solutions for all times t would collapse onto a single curve. This is the character we exploited in finding a similarity solution.

We can understand a little more about this solution by considering the asymptotic limits. The value of erf (2) is 0.995, so for practical purposes any value of ζ greater than 2 is "infinite." In a body of finite thickness the semi-infinite body solution will be valid for very short times, and it will remain valid as long as the center of the body remains at the initial temperature. For a slab of thickness $2H$ this means that $x = H$ must correspond to $\zeta \geq 2$, or that $H/\sqrt{4\alpha t} \geq 2$. Thus, the semi-infinite body solution will be valid for

$$t \leq \frac{H^2}{16\alpha} \tag{4.56}$$

For example, suppose that we have a steel plate that is 50 mm thick. The thermal diffusivity of steel is approximately 6 mm²/s, so the plate can be considered to be semi-infinite for

$$t \leq \frac{(25 \text{ mm})^2}{16 (6 \text{ mm}^2/\text{s})} = 6.5 \text{ s} \tag{4.57}$$

If the plate were 100 mm thick, it could be approximated as a semi-infinite body for *four* times as long, or up to 26 s, because the penetration depth for heat conduction grows in proportion to the square root of time.

We assumed at the outset that the surface temperature could be changed from T_1 to T_0 instantaneously at $t = 0$. An analysis of Eq. (4.55) shows that this requires the heat flux $-k \, (\partial T/\partial x)$ to be infinite at $x = 0$ and $t = 0$. Mathematically, the solution has a singularity at this point. (As a sidelight, this singularity does not invalidate the solution, because it is *integrable*. That is, the amount of heat transferred across the outer surface over any *finite* time interval is finite, because of the way that heat flux decreases with time.) However, as a practical matter, we can expect that the actual behavior at *very* short times will differ from Eq. (4.55). A more realistic model is to use a convective heat transfer coefficient boundary condition at $x = 0$. The solution to this problem is (Carslaw and Jaeger, 1959, Art. 2.7)

$$\frac{T - T_1}{T_0 - T_1} = \mathrm{erfc}\left(\frac{x}{\sqrt{4\alpha t}}\right) - \left[\exp\left(\frac{xh}{k} + \frac{h^2 \alpha t}{k^2}\right) \mathrm{erfc}\left(\frac{x}{\sqrt{4\alpha t}} + \frac{h\sqrt{\alpha t}}{k}\right)\right] \tag{4.58}$$

The term in brackets represents the difference in the solution introduced by a finite heat transfer coefficient. This term will be small if the argument to the exponential function is small, so this solution approaches the solution with fixed surface temperature, Eq. (4.55), when $h\sqrt{\alpha t}/k$ is large, say greater than 10. Hence, Eq. (4.58) should be used for very short times or for small heat transfer coefficients, whereas for large enough heat transfer coefficients or for long times, Eq. (4.55) is quite adequate.

4.2.2 SLABS OF FINITE THICKNESS

Governing Equations and Scaling

It is of somewhat more practical interest to consider slabs of finite thickness. To this end, we consider again the slab shown in Fig. 4.1, but this time we examine transient heat transfer. It is convenient in this case to define the slab thickness as $2H$, instead of H. We consider the case in which the slab is quenched from an initially constant temperature T_0 into a medium at temperature T_∞ with the same heat transfer coefficient h at both surfaces. The energy equation for this problem is

$$\frac{\partial T}{\partial t} = \alpha \left(\frac{\partial^2 T}{\partial x^2} + \frac{\partial^2 T}{\partial y^2} + \frac{\partial^2 T}{\partial z^2} \right) \tag{4.59}$$

and the initial and boundary conditions are

$$T(0 \leq x \leq H, 0) = T_0 \tag{4.60}$$

$$\frac{\partial T(0, t)}{\partial x} = 0 \tag{4.61}$$

$$k\frac{\partial T(H, t)}{\partial x} = -h[T(H, t) - T_\infty] \tag{4.62}$$

Here we are taking the total thickness to be $2H$, with $x = 0$ at the midplane. In writing these equations we have continued to make the assumptions that material properties are constant, and that there is no internal heat generation.

We now scale the variables, introducing appropriate temperature and length scales. The problem parameters provide no obvious time scale, so we nondimensionalize time by using an unknown characteristic value t_c.

$$x^* = x/H \tag{4.63}$$

$$y^* = y/L \tag{4.64}$$

$$z^* = z/L \tag{4.65}$$

$$\theta = (T - T_\infty)/(T_0 - T_\infty) \tag{4.66}$$

$$t^* = t/t_c \tag{4.67}$$

4.2 TRANSIENT HEAT CONDUCTION

Substituting these variables into the energy equation gives

$$\frac{\partial \theta}{\partial t^*} = \left\{\frac{\alpha t_c}{H^2}\right\} \left(\frac{\partial^2 \theta}{\partial x^{*2}} + \left\{\frac{H^2}{L^2}\right\} \frac{\partial^2 \theta}{\partial y^{*2}} + \left\{\frac{H^2}{L^2}\right\} \frac{\partial^2 \theta}{\partial z^{*2}}\right) \quad (4.68)$$

Since $H \ll L$, we can neglect the last two terms. Only two terms remain, so they must have the same magnitude. Thus, we set the remaining parameter group equal to unity and so identify the characteristic time as

$$t_c = \frac{H^2}{\alpha} \quad (4.69)$$

The scaled energy equation is now

$$\frac{\partial \theta}{\partial t^*} = \frac{\partial^2 \theta}{\partial x^{*2}} \quad (4.70)$$

whereas the dimensionless versions of the boundary and initial conditions, Eqs. (4.60)–(4.62), are

$$\theta(x^*, 0) = 1 \quad (4.71)$$

$$\frac{\partial \theta(0, t^*)}{\partial x^*} = 0 \quad (4.72)$$

$$\frac{\partial \theta(1, t^*)}{\partial x^*} = -\left\{\frac{hH}{k}\right\} \theta \quad (4.73)$$

The boundary condition, Eq. (4.73) contains a second dimensionless group, the *Biot number* Bi, which is defined as

$$\text{Bi} = \frac{hH}{k} \quad (4.74)$$

The Biot number represents the relative importance of convective heat transfer at the surface relative to heat conduction within the body. Not surprisingly, the nature of the solution depends on whether the Biot number is large, small, or in between. We now consider each of these cases in turn.

Small Biot Number

When the Biot number is very small compared to unity, the boundary condition, Eq. (4.73), becomes

$$\frac{\partial \theta(1, t)}{\partial x^*} = -\text{Bi}\,\theta \approx 0 \quad (4.75)$$

When combined with the no-flux condition at the center line, Eq. (4.72), this tells us that the temperature gradient within the slab is zero. This seems to imply that there is no heat loss from the slab. Zero heat loss is obviously not a very interesting solution, but it is also not what actually happens. Instead, Eq. (4.75) is telling us that transporting the heat away from the surface is much more difficult than conducting heat from the interior to the surface. This seems to create a difficulty in solving the problem: the temperature gradients within the body are what we usually calculate when solving the energy equation, but here they are precisely the quantity that is negligible.

To reconcile this matter we must use the energy balance equation in its integral form, where the integral is taken over the entire body. For the present assumptions we need only the internal energy and the heat conduction terms, so the energy balance equation, Eq. (2.129), can be written as

$$\int_V \rho c_p \frac{\partial T}{\partial t} dV = -\int_V \nabla \cdot \mathbf{q}\, dV \qquad (4.76)$$

The right-hand side is the conduction term, which our scaled boundary condition tells us is unimportant within the body. However, this term originated from the consideration of applied surface fluxes; see the discussion around Eq. (2.123). We can recover the earlier form by applying the divergence theorem, Eq. (A.59), to the right-hand side of Eq. (4.76) and obtain

$$\int_V \rho c_p \frac{\partial T}{\partial t} dV = -\int_S \mathbf{q} \cdot \mathbf{n}\, dS \qquad (4.77)$$

Here S is the external surface of the body. On this surface the value of $\mathbf{q} \cdot \mathbf{n}$ is given by the boundary condition as $h(T - T_\infty)$, so we have

$$\int_V \rho c_p \frac{\partial T}{\partial t} dV = -\int_S h(T - T_\infty)\, dS \qquad (4.78)$$

The boundary condition, Eq. (4.75), has told us that the temperature gradients within the body are small, so we may conclude that at any time the temperature inside the slab is nearly uniform, say at temperature $\bar{T}(t)$. With this approximation the temperatures can be brought outside the integral signs, the integrals can be evaluated, and the energy balance equation becomes

$$\rho c_p V \frac{\partial \bar{T}}{\partial t} = -hA(\bar{T} - T_\infty) \qquad (4.79)$$

where A and V are the surface area and volume of the body, respectively. Equation (4.79) is readily solved to find

$$\bar{T} - T_\infty = c_1 \exp\left(-\frac{hA}{\rho c_p V} t\right) \qquad (4.80)$$

The constant of integration is evaluated by applying the initial condition, Eq. (4.60), so our final solution for small Biot number is

$$\frac{\bar{T} - T_\infty}{T_0 - T_\infty} = \exp\left(-\frac{hA}{\rho c_p V} t\right) \qquad (4.81)$$

Note that the solution depends on the surface area to volume ratio A/V but is independent of other aspects of the body's geometry, including its overall shape. We have found the transient temperature solution for a body of *any* shape, provided that the Biot number is small. The ratios of surface area to volume for several simple shapes are tabulated in Table 4.4.

Large Biot Number

Now consider the other limiting case, where $\text{Bi} \gg 1$. For this case we multiply the surface boundary condition in Eq. (4.73) by $1/\text{Bi}$ so that the dimensionless parameter will be $\mathcal{O}(1)$ or smaller, and we have

$$\theta(x^* = 1, t^*) = \frac{1}{\text{Bi}} \frac{\partial \theta}{\partial x^*} \approx 0 \qquad (4.82)$$

Table 4.4 Surface Area to Volume Ratio for Some Simple Geometries

Geometry	Characteristic Dimension	A/V
Slab	thickness $= 2H$	$1/H$
Cylinder	radius $= R$	$2/R$
Sphere	radius $= R$	$3/R$
Square rod	edge $= H$	$4/H$

This boundary condition corresponds to a fixed temperature at the surface. That is, when the heat transfer coefficient h is very large, any heat conducted from the interior of the body to the surface is easily removed, without raising the surface temperature much above T_∞. Rather than solve this system, we will proceed to solve the more general case of finite Bi and then extract the fixed temperature condition as a special case.

Arbitrary Biot Number

The problem defined by Eqs. (4.70)–(4.73) does not have a similarity solution, because the initial condition written in terms of ζ does not match one of the boundary conditions. Instead we must find some other approach. Another common tool for solving partial differential equations is to seek a solution by *separation of variables*. We guess that the temperature, which is a function of position and time, can be written as the product of two other functions,

$$\theta(x^*, t^*) = \Psi(t^*)\Phi(x^*) \tag{4.83}$$

where Ψ is a function of t^* only and Φ is a function of x^* only. Substituting this form into the governing equation, Eq. (4.70), yields

$$\Psi'\Phi = \Psi\Phi'' \tag{4.84}$$

where the prime indicates differentiation with respect to the appropriate argument, either t^* or x^*. Dividing through by $\Psi\Phi$ yields

$$\frac{\Psi'}{\Psi} = \frac{\Phi''}{\Phi} = -\lambda^2 \tag{4.85}$$

We know that λ must be a constant, because the first expression depends only on t^* whereas the second expression depends only on x^*; $-\lambda^2$ is sometimes referred to as the *separation constant*.

This leaves us with two ordinary differential equations to solve:

$$\Psi' + \lambda^2\Psi = 0 \tag{4.86}$$

$$\Phi'' + \lambda^2\Phi = 0 \tag{4.87}$$

These may be integrated to find

$$\Psi = c_0 \exp(-\lambda^2 t^*) \tag{4.88}$$

$$\Phi = c_1 \sin \lambda x^* + c_2 \cos \lambda x^* \tag{4.89}$$

Returning these solutions to Eq. (4.83) yields

$$\theta = c_0 \exp(-\lambda^2 t^*)(c_1 \sin \lambda x^* + c_2 \cos \lambda x^*) \tag{4.90}$$

Now we can see that the constant c_0 from Eq. (4.88) is redundant, so we can choose $c_0 = 1$ without losing any generality in the solution.

We now proceed to evaluate the constants c_1, c_2, and λ by using the boundary and initial conditions. The boundary condition of zero slope at the center line, Eq. (4.71), implies that $c_1 = 0$. The boundary condition, Eq. (4.73), at the surface gives us

$$\frac{\partial \theta}{\partial x^*} = -\text{Bi}\,\theta \quad \text{at } x^* = 1 \tag{4.91}$$

Substituting Eq. (4.90) into this expression, we find

$$-c_2 \lambda \sin \lambda \, \exp(-\lambda^2 t^*) = -\text{Bi}\exp(-\lambda^2 t^*) c_2 \cos \lambda$$

$$\lambda \tan \lambda = \text{Bi} \tag{4.92}$$

This last expression is a transcendental equation for λ. The function on the left-hand side is plotted in Fig. 4.6. One could find the roots by drawing a horizontal line at any particular value of Bi and noting the points of intersection with the plotted function. We see from the figure that there are an infinite number of roots, with the nth root λ_n falling in the interval $n\pi \leq \lambda_n \leq (2n+1)\pi/2$. The exact values of the λ_n's for any given Biot number could be determined, for example, by a simple root-finding program.

Now we have an infinite number of λ_n's, each with its own value of c_2, which we will call b_n. So there is an infinity of versions of Eq. (4.90). Which one is the solution? Actually, they all are, and any linear combination of them is also a solution to the energy equation. To construct a general solution, we sum this infinite set of functions, so the temperature is now given by

$$\theta = \sum_{n=0}^{\infty} b_n \exp(-\lambda_n^2 t^*) \cos \lambda_n x^* \tag{4.93}$$

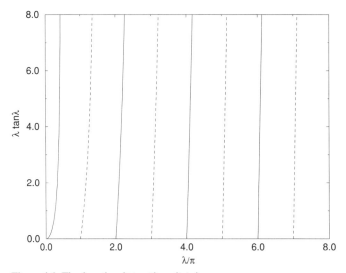

Figure 4.6: The function $(\lambda \tan \lambda)$ vs. (λ/π).

To evaluate the constants b_n we enforce the initial condition:

$$\theta(x^*, 0) = 1 = \sum_{n=0}^{\infty} b_n \cos \lambda_n x^* \tag{4.94}$$

This is a Fourier series representation of the initial condition. The different cosine functions are orthogonal on the interval $[0, 1]$. To obtain the coefficients b_n, we multiply this equation by $\cos \lambda_m x^*$ and integrate from zero to one. This gives

$$\int_0^1 \cos \lambda_m x^* \, dx = \int_0^1 \sum_{n=0}^{\infty} b_n \cos \lambda_m x^* \cos \lambda_n x^* \, dx \tag{4.95}$$

Thanks to the orthogonality property, the terms in the infinite sum are all zero except for the term $n = m$. Evaluating the integral, we find the constants that match the initial condition for our problem to be

$$b_m = \frac{2 \sin \lambda_m}{\lambda_m + \sin \lambda_m \cos \lambda_m} \tag{4.96}$$

Using this result in Eq. (4.93), we finally obtain the temperature solution for a slab at the arbitrary Biot number as

$$\theta = \sum_{n=0}^{\infty} \frac{2 \sin \lambda_n}{\lambda_n + \sin \lambda_n \cos \lambda_n} \exp(-\lambda_n^2 t^*) \cos \lambda_n x^* \tag{4.97}$$

Although this solution may appear formidable at first, each λ_n is close to $(2n - 1)\pi/2$, so as long as t^* is not too close to zero, the exponential terms go to zero very quickly for all but the first few terms in the series. Thus, we can get a very accurate approximation to the solution by summing just a few terms from the series. Solutions are shown in Fig. 4.7 for several values of the Biot number at three different depths in the slab. The computer code used to generate the data for these figures is given in the Appendix to this chapter.

Note that for $\text{Bi} = 0.01$, the temperature is essentially constant in the slab at all times. For $\text{Bi} = 100$, we can observe that the temperature in the slab reaches T_∞ ($\theta = 0$) everywhere after the dimensionless time reaches one.

Constant Surface Temperature

We now return to the case where $\text{Bi} \to \infty$. For this case we replace the heat transfer coefficient boundary condition, Eq. (4.91), with a fixed temperature boundary condition, $\theta(1, t) = 0$. Then Eq. (4.92) for λ is replaced by

$$\cos \lambda = 0 \tag{4.98}$$

This has the solutions $\lambda = \pi/2, 3\pi/2, 5\pi/2, \ldots$, or $\lambda_n = (2n + 1)\pi/2$. The temperature solution for $\text{Bi} \to \infty$ is then

$$\theta = \sum_{n=0}^{\infty} \frac{4(-1)^n}{(2n+1)\pi} \exp\left\{-\left[\frac{(2n+1)\pi}{2}\right]^2 t^*\right\} \cos \frac{(2n+1)\pi}{2} x^* \tag{4.99}$$

Again this series converges rapidly. For long times the first term dominates the series, and we can approximate the temperature as a single cosine function with an

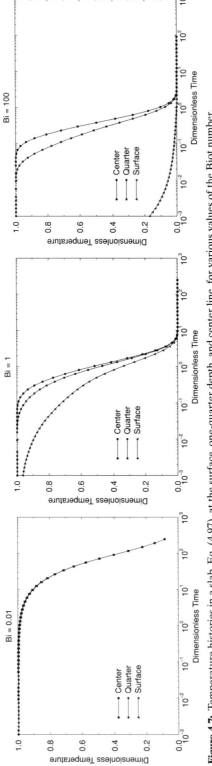

Figure 4.7: Temperature histories in a slab, Eq. (4.97), at the surface, one-quarter depth, and center line, for various values of the Biot number.

4.2 TRANSIENT HEAT CONDUCTION

exponentially decreasing amplitude:

$$\theta \simeq \frac{4}{\pi} \exp[-(\pi/2)^2 t^*] \cos\left(\frac{\pi x^*}{2}\right) \tag{4.100}$$

In fact this equation is a good approximation as long as $t^* > 0.2$. As a simple approximation for very short times, say $t^* < 0.05$, we can use the semi-infinite body solution in Eq. (4.5), provided we account for the different location of $x = 0$ in that equation.

More on Separation of Variables

Before leaving this problem we make a few additional comments on the method of separation of variables. Equation (4.70) was solved by using a product form such as Eq. (4.83). This led to an infinite series solution, Eq. (4.93), each term being the product of a space function and a time function. The key point here is that the *infinite sum* must satisfy the boundary conditions of the problem. This is most easily done when the boundary conditions are *homogeneous*, that is, when the dependent variable or its derivative equals zero on every boundary.

This was exactly the case in the slab problem for a large Biot number, in which the boundary conditions were $\theta = 0$ on the surface and $\partial\theta/\partial x^* = 0$ on the midplane. If each term in the infinite series solution satisfied these conditions, then any sum of terms also satisfied them. We could then adjust the coefficients to match the initial temperature distribution.

If a problem does not have homogeneous boundary conditions, it may still be possible to solve it by using a separation of variables. For instance, suppose we wanted to solve a one-dimensional transient conduction problem in a slab where the two surfaces are held at different temperatures. The governing equation and initial conditions would be the same as before, and the boundary conditions would be

$$\theta(x^* = -1, t^*) = 0 \tag{4.101}$$

$$\theta(x^* = +1, t^*) = \theta_1 \tag{4.102}$$

where θ_1 is a constant. This problem is not immediately amenable to solution by separation of variables. However, it can be solved analytically by first partitioning the solution into a transient part $\hat{\theta}$ and a steady part $\mathcal{S}$.

$$\theta(x^*, t^*) = \hat{\theta}(x^*, t^*) + \mathcal{S}(x^*) \tag{4.103}$$

We then require $\mathcal{S}$ to satisfy the steady one-dimensional heat conduction equation and the nonhomogeneous boundary conditions,

$$0 = \frac{\partial^2 \mathcal{S}}{\partial x^{*2}} \tag{4.104}$$

$$\mathcal{S}(x^* = -1) = 0 \tag{4.105}$$

$$\mathcal{S}(x^* = +1) = \theta_1 \tag{4.106}$$

while $\hat{\theta}$ satisfies the transient conduction equation and the homogeneous boundary conditions,

$$\frac{\partial \hat{\theta}}{\partial t^*} = \frac{\partial^2 \hat{\theta}}{\partial x^{*2}} \tag{4.107}$$

$$\hat{\theta}(-1) = 0 \tag{4.108}$$

$$\hat{\theta}(+1) = 0 \tag{4.109}$$

The solution for S is the usual linear steady-state solution:

$$S = \theta_1 \frac{1 + x^*}{2} \tag{4.110}$$

Now $\hat{\theta}$ can be found by the separation of variables method as before. The results for $\hat{\theta}$ depend on the initial conditions. Some specific examples are given by Carslaw and Jaeger (1959, p. 99ff).

4.2.3 OTHER ANALYTICAL SOLUTIONS FOR HEAT CONDUCTION

Heat conduction is a mature subject, and most problems that can be solved analytically have been solved at some time or other. The classic text on the subject is that by Carslaw and Jaeger (1959). If you are working on a problem and you suspect that an analytical solution might exist, this is the place to look for it.

Important additional one-dimensional solutions include problems of conduction in an infinite cylinder, and in a sphere. By proper transformation to cylindrical or spherical coordinates, these problems can be cast as one dimensional. The Laplacian takes on different forms in these coordinate systems, and the solution looks somewhat different, but the overall behavior and scaling dependence remains the same. These solutions are treated well by Carslaw and Jaeger (1959).

We note that we have discussed only one-dimensional problems thus far. The reason for this is that our emphasis is on understanding the phenomena, and the relation between scaling and solution behavior. There are also relatively few analytical solutions for problems in more than one dimension, unless the geometry is very regular. Most other cases must be treated numerically.

4.3 CONDUCTION WITH PHASE CHANGE

Some of the most important heat transfer problems in materials processing involve a change of phase. Most metal processing begins with solidification of a liquid alloy cast into a cold mold. There are similar applications in the crystalization of polymers from the melt, and the melting of polymers is also an important problem. In this section we show how to analyze the heat transfer in these problems, and we develop some basic analytical solutions.

The key new material property that we must introduce is the latent heat of fusion, L_f. This is the energy per unit mass (i.e., joules per kilogram) that must be removed from the material to transform liquid at the melting temperature T_m into solid at T_m. For the time being we will focus on pure substances, which solidify at a single

4.3 CONDUCTION WITH PHASE CHANGE

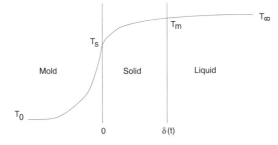

Figure 4.8: Schematic drawing of one-dimensional solidification of a pure substance in a thick mold, with no interfacial resistance to heat flow.

temperature T_m. Eutectic alloys also have this property and can be analyzed the same way. However, many metal alloys solidify over a range of temperatures and are not represented well by the models developed here. The solidification of alloys requires a more complicated treatment, which we will discuss in Section 8.3.

Consider the solidification of a pure material near the wall of a mold, as depicted in Fig. 4.8. As discussed in the previous section, for short times the mold and the solidifying material can be considered to be semi-infinite. Thus, as shown in Fig. 4.8, we assume that the mold surface is at $x = 0$, the mold extends to $x = -\infty$, and the solidifying material extends to $x = +\infty$, and we model the heat transfer as being one dimensional. Other basic assumptions are that there are no heat sources, the material properties are constant in each phase (mold, solid, and liquid), and that the material is not moving.

We will need boundary conditions at each interface between the different phases. The mold–solid interface can provide a noticeable resistance to heat transfer, either because a small air gap opens there, or because contaminants coat the surface of the mold. This can be modeled by using a boundary condition of the third kind (i.e., a heat transfer coefficient) at the mold–solid interface. However, this boundary condition precludes the possibility of finding analytical solutions when the other thermal resistances are finite. Thus for now we shall assume that the thermal resistance of the interface is negligible, in which case the temperature is continuous across that interface.

The Stefan Condition

The boundary conditions at the liquid–solid interface require careful treatment, because the position of this boundary is unknown. In fact, the position of the boundary must be found as part of the solution. This type of problem is known as a *moving boundary problem*, because one of the boundaries – the solid–liquid interface – moves with time in a manner that is determined by the heat flow solution itself. If the problem is steady but the position of one or more boundaries must be found as part of the solution, then we have what is called a *free boundary problem*.

Free and moving boundary problems require an additional boundary condition to fix the position or motion of the boundary. In a heat transfer problem, one typically knows either the temperature or the heat flux at each point on the boundary. As indicated in Fig. 4.8, the temperature at the interface is the melting point of the material T_m. Because the position of this boundary is unknown, however, we require another boundary condition to resolve its position and motion in time.

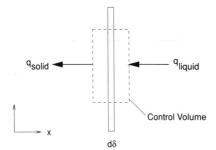

Figure 4.9: A control volume containing a moving liquid–solid interface.

To derive this boundary condition, consider a thin control volume that contains the liquid–solid interface, as shown in Fig. 4.9. The x coordinate of the interface is δ, and the instantaneous velocity of the interface is $d\delta/dt$. As the interface moves, liquid is transformed to solid, releasing latent heat L_f per unit mass at the interface. For a unit area of interface, the rate of heat evolution from the phase change is $\rho_s L_f \, d\delta/dt$. This heat must be conducted away from the interface into the solid, with a flux given by $k_s \left[\partial T/\partial x\right]_s$. In addition, any other heat conducted into the interface from the liquid, $k_\ell \left[\partial T/\partial x\right]_\ell$, must also conduct into the solid. (Subscripts s and ℓ denote properties of the solid and liquid, respectively). Combining these quantities into an energy balance for the control volume yields the *Stefan condition*:

$$k_s \left.\frac{\partial T}{\partial x}\right|_s - k_\ell \left.\frac{\partial T}{\partial x}\right|_\ell = \rho_s L_f \frac{d\delta}{dt} \tag{4.111}$$

This is the new boundary condition that is needed to account for the heat of fusion, and it completes the mathematical model.

The Stefan condition can be generalized to multiple dimensions. In this case, we must account for the possibility that the interface is curved, in which case the area of the liquid–solid interface changes as the interface moves. The net heat q_{cond} conducted away from a surface element of area A on the interface is given by

$$q_{\text{cond}} = A(k_s \nabla T_s \cdot \mathbf{n} - k_\ell \nabla T_\ell \cdot \mathbf{n}) \tag{4.112}$$

where $\mathbf{n}$ is a unit vector normal to the interface and pointing from the solid into the liquid. This heat must balance the heat generated by the phase change q_{freeze} in the corresponding volume V swept out by the moving interface,

$$q_{\text{freeze}} = \rho_s L_f \frac{dV}{dt} - \Gamma \frac{dA}{dt} \tag{4.113}$$

where Γ is the surface energy per unit area (i.e., the surface tension). Note that the signs of the two terms are opposite because the solidification reaction liberates heat, whereas the creation of surface absorbs heat. Before combining the two expressions, we note that we may use the chain rule to write

$$\frac{dA}{dt} = \frac{dA}{dV}\frac{dV}{dt} = 2\kappa_m \frac{dV}{dt} \tag{4.114}$$

where κ_m is the mean curvature, a geometric property of the surface (see Appendix A.6). The volume swept out by the moving surface element is proportional

4.3 CONDUCTION WITH PHASE CHANGE

to the surface area A and the normal velocity V_n:

$$\frac{dV}{dt} = A\mathbf{V} \cdot \mathbf{n} = AV_n \tag{4.115}$$

Substituting these two expressions into Eq. (4.113), and then equating q_freeze to q_cond, we have

$$(\rho_s L_f - 2\Gamma \kappa_m)V_n = (k_s \nabla T_s \cdot \mathbf{n} - k_\ell \nabla T_\ell \cdot \mathbf{n}) \tag{4.116}$$

Unless the radii of curvature are very small (approaching atomic dimensions), the contribution of the curvature term to this expression is usually negligible in comparison to the latent heat. Equation (4.116) can be adapted to systems in which the material is moving with velocity $\mathbf{v}$, by replacing V_n with $(\mathbf{V} - \mathbf{v}) \cdot \mathbf{n}$.

4.3.1 SCALING AND LIMITING CASES

Before considering the complete problem with conduction in the mold, solid, and liquid, as depicted in Fig. 4.8, we find it useful to consider some limiting cases. For the moment we restrict our attention to those cases in which the liquid is poured into the mold at its melting point, that is, where there is no superheat. Because $T = T_m$ everywhere in the liquid, there is no temperature gradient and no heat flux in the liquid, and the liquid remains at T_m. This simplifies the analysis because we need consider heat flow only in the solid and the mold. We will treat problems with superheat in Section 4.3.2.

We wish to solve for the temperatures in the mold and the solid, which satisfy the following energy balance equations:

$$\frac{\partial T}{\partial t} = \alpha_m \frac{\partial^2 T}{\partial x^2}, \quad -\infty < x \leq 0 \tag{4.117}$$

$$\frac{\partial T}{\partial t} = \alpha_s \frac{\partial^2 T}{\partial x^2}, \quad 0 < x \leq \delta(t) \tag{4.118}$$

where the subscript m on the thermal properties refers to the mold. We also have the boundary conditions

$$T(x \to -\infty) = T_0 \tag{4.119}$$

$$T[x = \delta(t)] = T_m \tag{4.120}$$

We also specify the mold–solid interface temperature:

$$T(x = 0) = T_s$$

This is not a boundary condition, because the temperature T_s is unknown, but using this symbol will simplify the math. We also assume that T_s does not vary in time, a notion that will be justified a posteriori.

Finally, there are additional boundary conditions on the heat fluxes at the interfaces:

$$k_m \left.\frac{\partial T}{\partial x}\right|_{x=0-} = k_s \left.\frac{\partial T}{\partial x}\right|_{x=0+} \tag{4.121}$$

$$k_s \left.\frac{\partial T}{\partial x}\right|_{x=\delta(t)} = \rho_s L_f \frac{d\delta}{dt} \tag{4.122}$$

The first of these expresses the continuity of the heat flux at the mold–solid interface, and the second is the Stefan condition, specialized for the case with no superheat and a planar interface ($\kappa_m = 0$).

Before proceeding to the solution, we will scale the governing equations and identify some limiting cases. There is neither a natural length scale nor a natural time scale, so we will choose undefined values L and t_c respectively, and proceed. The dimensionless variables are

$$x^* = \frac{x}{L}, \quad t^* = \frac{t}{t_c}, \quad \theta = \frac{T - T_0}{T_m - T_0}, \quad \delta^* = \frac{\delta}{L} \qquad (4.123)$$

The scaled equations are then

$$\frac{\partial \theta}{\partial t^*} = \frac{\alpha_m t_c}{L^2} \frac{\partial^2 \theta}{\partial x^{*2}}, \quad -\infty < x^* \leq 0 \qquad (4.124)$$

$$\frac{\partial \theta}{\partial t^*} = \frac{\alpha_s t_c}{L^2} \frac{\partial^2 \theta}{\partial x^{*2}}, \quad 0 < x^* \leq \delta^* \qquad (4.125)$$

These equations suggest two different time scales, one based on the mold thermal diffusivity, and the other based on the solid thermal diffusivity. The limiting cases that we will identify shortly correspond to cases in which these time scales are very different. At this point we choose $t_c = L^2/\alpha_s$, so the scaled equations become

$$\frac{\partial \theta}{\partial t^*} = \frac{\alpha_m}{\alpha_s} \frac{\partial^2 \theta}{\partial x^{*2}}, \quad -\infty < x^* \leq 0 \qquad (4.126)$$

$$\frac{\partial \theta}{\partial t^*} = \frac{\partial^2 \theta}{\partial x^{*2}}, \quad 0 < x^* \leq \delta^* \qquad (4.127)$$

The scaled boundary conditions are

$$\theta(x^* \to -\infty) = 0 \qquad (4.128)$$

$$\theta(x^* = 0) = \theta_s \qquad (4.129)$$

$$\theta(x^* = \delta^*) = 1 \qquad (4.130)$$

where θ_s is the (constant) dimensionless surface temperature. The scaled heat flux boundary condition at the mold–solid interface is

$$\frac{k_m}{k_s} \frac{\partial \theta}{\partial x^*}\bigg|_{x^*=0-} = \frac{\partial \theta}{\partial x^*}\bigg|_{x^*=0+} \qquad (4.131)$$

whereas the Stefan condition scales as follows:

$$k_s \frac{\partial T}{\partial x}\bigg|_{x=\delta(t)} = \rho_s L_f \frac{d\delta}{dt} \qquad (4.132)$$

$$\frac{k_s(T_m - T_0)}{L} \frac{\partial \theta}{\partial x^*}\bigg|_{x^*=\delta^*} = \rho_s L_f \frac{\alpha_s}{L} \frac{d\delta^*}{dt*} \qquad (4.133)$$

$$\frac{\partial \theta}{\partial x^*}\bigg|_{x^*=\delta^*} = \frac{L_f}{c_{ps}(T_m - T_0)} \frac{d\delta^*}{dt*} \qquad (4.134)$$

$$\frac{\partial \theta}{\partial x^*}\bigg|_{x^*=\delta^*} = \frac{1}{\text{Ste}} \frac{d\delta^*}{dt*} \qquad (4.135)$$

4.3 CONDUCTION WITH PHASE CHANGE

Here we have found a new dimensionless parameter, the *Stefan number:*

$$\text{Ste} = \frac{c_{ps}(T_m - T_0)}{L_f} \tag{4.136}$$

The Stefan number is a measure of the relative importance of sensible heat compared with latent heat. A large Stefan number, $\text{Ste} \gg 1$, means that the sensible heat required to cool the solid from T_m to T_0 is large compared with the latent heat of fusion, and the latent heat can be ignored. A small Stefan number, $\text{Ste} \ll 1$, means that the latent heat is a dominant factor in the energy balance. Metals usually have a Stefan number much less than one, whereas in crystallizing polymers the Stefan number can be of the order of one or even larger. For the remainder of this chapter we will assume that $\text{Ste} \leq 1$.

Mold Control: $(\alpha_m/\alpha_s) \ll 1$

In the sand casting of metals, the thermal diffusivity of the mold material is typically much less than that of the metal. For example, for aluminum cast in a typical sand mold, $(\alpha_m/\alpha_s) \approx 0.007$. When (α_m/α_s) is small, then we find in Eq. (4.126) a small parameter multiplying the highest spatial derivative in the equation, and we anticipate that there is a boundary layer near the interface where the temperature changes rapidly. Another way to express the same idea is to consider the ratio of the characteristic time scales in the two media:

$$\frac{t_c(\text{solid})}{t_c(\text{mold})} = \frac{L^2}{\alpha_s} \frac{\alpha_m}{L^2} = \frac{\alpha_m}{\alpha_s} \tag{4.137}$$

When this ratio is much smaller than one, conduction is very rapid in the solid compared with that in the mold. Consequently, the solid is nearly isothermal, and the heat flow is controlled by conduction in the mold. The temperature distribution in this case is sketched in Fig. 4.10. There must be some temperature gradient in the solid, but in this special case it is small compared with the temperature gradient in the mold.

If the solid is nearly isothermal then its temperature must be approximately T_m, because of the boundary condition at $\delta(t)$. Therefore we set $T_s = T_m$, or $\theta_s = 1$. The heat transfer problem in the mold is then exactly the case of the semi-infinite body with constant surface temperature, which we solved in Section 4.2.1. We can therefore

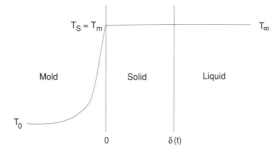

Figure 4.10: Schematic view of the temperature distribution in the solid and in the mold for the case in which $(\alpha_m/\alpha_s) \ll 1$.

immediately write the mold temperature distribution as

$$\theta = \operatorname{erfc}\left(\frac{-x^*}{2\sqrt{\frac{\alpha_m}{\alpha_s}t^*}}\right) \tag{4.138}$$

We can recover information about the motion of the interface by noting that the heat flux entering the solid from the liquid must equal the heat flux leaving the solid and entering the mold. (This comes from an energy balance on the isothermal solid.) Equating the fluxes, we find

$$k_m \left.\frac{\partial T}{\partial x}\right|_{x=0} = \rho_s L_f \frac{d\delta}{dt} \tag{4.139}$$

or, in dimensionless form,

$$\left.\frac{\partial \theta}{\partial x^*}\right|_{x=0} = \frac{k_s}{k_m} \frac{1}{\text{Ste}} \frac{d\delta^*}{dt^*} \tag{4.140}$$

Substituting the solution from Eq. (4.138) for θ and using Eq. (A.79) to take the derivative gives

$$\frac{2}{\sqrt{\pi}} \frac{1}{2\sqrt{\frac{\alpha_m}{\alpha_s}t^*}} = \frac{k_s}{k_m} \frac{1}{\text{Ste}} \frac{d\delta^*}{dt^*} \tag{4.141}$$

This equation can be rearranged to give an expression for the growth rate of the solid:

$$\frac{d\delta^*}{dt^*} = \frac{\text{Ste}}{\sqrt{\pi t^*}} \frac{\sqrt{k_m \rho_m c_{pm}}}{\sqrt{k_s \rho_s c_{ps}}} \tag{4.142}$$

(Here we have used the fact that $k_m/\sqrt{\alpha_m} = \sqrt{k_m \rho_m c_{pm}}$.) To find the solid layer thickness, we integrate this expression in time and use the initial condition that $\delta^* = 0$ at $t^* = 0$:

$$\delta^* = \frac{2\,\text{Ste}}{\sqrt{\pi}} \frac{\sqrt{k_m \rho_m c_{pm}}}{\sqrt{k_s \rho_s c_{ps}}} \sqrt{t^*} \tag{4.143}$$

In dimensional form this is

$$\delta = \frac{2(T_m - T_0)}{\rho_s L_f \sqrt{\pi}} \sqrt{k_m \rho_m c_{pm}} \sqrt{t} \tag{4.144}$$

Equation (4.144) allows us to see more clearly the individual roles of the metal and mold properties. The conductivity of the solid does not appear because we have assumed the solid to be isothermal, but the heat of fusion and the solid density do appear. The mold properties are represented by the combined property $(k_m \rho_m c_{pm})$. This quantity is called the *heat diffusivity*, which is not to be confused with the thermal diffusivity $\alpha_m = (k_m/\rho_m c_{pm})$. Heat diffusivity is the property that governs the rate at which a material can absorb heat by conduction. For mold-controlled solidification, the freezing rate is governed by the rate at which heat is absorbed by the mold, so it is the heat diffusivity of the mold that appears here.

4.3 CONDUCTION WITH PHASE CHANGE

Equation (4.144) can be extended to estimate the solidification time of sand castings of various shapes. Recall that when we considered the cooling of a slab at low Biot number, the energy balance gave us a result in terms of the surface area to volume ratio that was valid for an object of arbitrary shape. We can make the same argument here, because the solid is approximately isothermal. In this case, we generalize Eq. (4.144) to determine the final solidification time t_f of a casting in which $(\alpha_m/\alpha_s) \ll 1$, by setting $\delta = V/A$ in Eq. (4.144) and solving for t_f. The result is

$$t_f = \frac{\pi}{4k_m \rho_m c_{pm}} \left(\frac{\rho_s L_f}{T_m - T_0}\right)^2 \left(\frac{V}{A}\right)^2 \qquad (4.145)$$

Thus the solidification time increases with the square of (V/A). This scaling relation is sometimes referred to as *Chvorinov's rule*, and (V/A) is called the "modulus" of the geometrical section. Several commercially available computer programs predict the solidification pattern of complex-shaped castings by using local values of the modulus.

Solid Control: $(\alpha_s/\alpha_m) \ll 1$

There are also important examples at the other extreme, where $(\alpha_s/\alpha_m) \ll 1$. One example is the solidification of polymers, which are relatively poor conductors, in metal molds. In this case, the *mold* is nearly isothermal, and we expect a boundary layer conduction solution in the solid. The approximate temperature distribution for this case is shown schematically in Fig. 4.11.

The mold is nearly isothermal and its far-field temperature is T_0, so we set the surface temperature to $T_s = T_0$. We must then solve the following scaled equations for the temperature in the solid:

$$\frac{\partial \theta}{\partial t^*} = \frac{\partial^2 \theta}{\partial x^{*2}}, \quad 0 < x^* \leq \delta^* \qquad (4.146)$$

$$\theta(x^* = 0) = 0 \qquad (4.147)$$

$$\theta(x^* = \delta^*) = 1 \qquad (4.148)$$

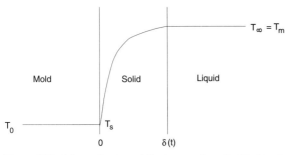

Figure 4.11: Schematic view of the temperature distribution in the solid and mold for the solid-control case, in which $(\alpha_m/\alpha_s) \gg 1$.

We must also satisfy the Stefan condition at $\delta(t)$,

$$k_s \frac{\partial T}{\partial x}\bigg|_{x=\delta} = \rho_s L_f \frac{d\delta}{dt} \tag{4.149}$$

which becomes, in dimensionless form,

$$\frac{\partial \theta}{\partial x^*}\bigg|_{x^*=\delta^*} = \frac{1}{\text{Ste}} \frac{d\delta^*}{dt^*} \tag{4.150}$$

This is a one-dimensional problem with no natural length scale, so we try an error function solution by using the similarity variable we discovered earlier:

$$\theta = c_1 + c_2 \, \text{erf}\left(\frac{x^*}{2\sqrt{t^*}}\right) \tag{4.151}$$

We know that this temperature distribution satisfies the heat conduction equation, Eq. (4.146). If it can also satisfy all three boundary conditions, then it is indeed the solution to the problem.

The boundary condition at $x^* = 0$ implies that $c_1 = 0$, and the temperature boundary condition at $x^* = \delta^*$ becomes

$$1 = c_2 \, \text{erf}\left(\frac{\delta^*}{2\sqrt{t^*}}\right) \tag{4.152}$$

The left-hand side of this equation is constant, whereas the right-hand side contains the dimensionless time variable t^*. The only way that the boundary condition can be satisfied is for δ^* to be proportional to $\sqrt{t^*}$. We write this as

$$\delta^* = 2\phi\sqrt{t^*} \tag{4.153}$$

where ϕ is a constant that must be determined. This gives $c_2 = 1/\text{erf}(\phi)$, and the temperature solution is now

$$\theta = \frac{1}{\text{erf}(\phi)} \, \text{erf}\left(\frac{x^*}{2\sqrt{t^*}}\right) \tag{4.154}$$

Finally, to evaluate ϕ we use the remaining boundary condition, the Stefan condition. Differentiating the temperature solution, Eq. (4.154) – see Eq. (A.79) – and substituting it into the Stefan condition from Eq. (4.150) yields

$$\frac{2}{\sqrt{\pi}} \frac{\exp(-\phi^2)}{\text{erf}(\phi)} \frac{1}{2\sqrt{t^*}} = \frac{1}{\text{Ste}} \frac{\phi}{\sqrt{t^*}} \tag{4.155}$$

We rearrange this to find a transcendental equation for ϕ:

$$\phi \exp(\phi^2) \text{erf}(\phi) = \frac{\text{Ste}}{\sqrt{\pi}} \tag{4.156}$$

Although this equation looks daunting, it can be solved graphically or numerically to find ϕ. The left-hand side is well behaved, as shown in Fig. 4.12, and for most cases of interest ϕ lies between 0.2 and 1. To find the solution to Eq. (4.156) graphically, first find the right-hand side of the equation, $\text{Ste}/\sqrt{\pi}$, on the ordinate, then draw a horizontal line until the curve is intersected, and read down to find the corresponding

4.3 CONDUCTION WITH PHASE CHANGE

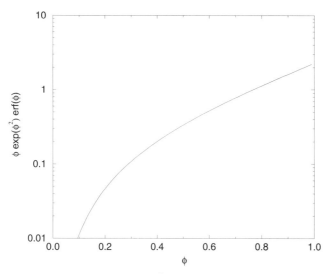

Figure 4.12: The function $\phi \exp(\phi^2) \text{erf}(\phi)$. Numerical values are given in Table A.1

value of ϕ that solves the equation. The function $\phi \exp(\phi^2) \text{erf}(\phi)$ is tabulated in Table A.1, so one can also find the value of ϕ by interpolating within the table.

After solving for ϕ, we can determine the solidification time t_f for a slab of total thickness $2H$ (cooled from both sides) by redimensionalizing and inverting the expression for δ:

$$t_f = H^2/4\phi^2 \alpha_s \tag{4.157}$$

Note that one *cannot* apply this solution to bodies of arbitrary geometry because the solution relies on one-dimensional heat flow in the solid, which is not generally applicable.

4.3.2 ANALYSIS OF THE GENERAL CASE

Let us now analyze the case when both the solid and the mold present a significant resistance to heat flow. The configuration is illustrated in Fig. 4.8. Our approach is to solve the conduction equation in each phase and then determine the unknown constants in the solution by matching the boundary conditions between the two solutions. Because of the different material properties in each phase, the scaled equations are no more convenient than their dimensional counterparts, so we solve this problem in dimensional form.

In the mold, we must satisfy the energy equation and the boundary conditions

$$\begin{aligned}
\frac{\partial T}{\partial t} &= \alpha_m \frac{\partial^2 T}{\partial x^2}, & -\infty < x \leq 0 \\
T &= T_0, & x \to -\infty \\
T &= T_s, & x = 0
\end{aligned} \tag{4.158}$$

Note that the surface temperature T_s is not yet known, but we again assume that it is constant. (There is no guarantee that T_s is constant. However, if it is not constant, we will be unable to find a solution that satisfies all of the boundary and initial conditions.) The solution to this problem was derived earlier as Eq. (4.138), which we now write in dimensional form with a surface temperature of T_s:

$$T = T_s + (T_0 - T_s)\,\text{erf}\left(\frac{-x}{2\sqrt{\alpha_m t}}\right), \quad -\infty < x \leq 0 \tag{4.159}$$

We will need the heat flux at the surface of the mold, so we compute it here for future use:

$$k_m \left.\frac{\partial T}{\partial x}\right|_{x=0} = \frac{(T_s - T_0)\sqrt{k_m \rho_m c_{pm}}}{\sqrt{\pi t}} \tag{4.160}$$

Next we consider the solid, which lies between $x = 0$ and $x = \delta(t)$. We must satisfy the following differential equation and boundary conditions there:

$$\begin{aligned}\frac{\partial T}{\partial t} &= \alpha_s \frac{\partial^2 T}{\partial x^2}, & 0 \leq x \leq \delta(t) \\ T &= T_s, & x = 0 \\ T &= T_m, & x = \delta(t)\end{aligned} \tag{4.161}$$

The Stefan condition, written earlier, will also be used in a moment. Let us again assume that the temperature follows an error function solution in the solid, and see where that leads:

$$T = c_1 + c_2\,\text{erf}\left(\frac{x}{2\sqrt{\alpha_s t}}\right) \tag{4.162}$$

The boundary condition at $x = 0$ gives $c_1 = T_s$. Substituting the boundary condition at $x = \delta(t)$, we have

$$T_m = T_s + c_2\,\text{erf}\left(\frac{\delta(t)}{2\sqrt{\alpha_s t}}\right) \tag{4.163}$$

Now the left-hand side of Eq. (4.163) is constant, as is T_s, so either $\delta(t)$ is proportional to $\sqrt{t}$ or there is no solution. Accordingly, we choose

$$\delta(t) = 2\phi\sqrt{\alpha_s t} \tag{4.164}$$

where ϕ is a constant to be determined in the analysis. The corresponding interface velocity is then

$$\frac{d\delta}{dt} = \frac{\phi\sqrt{\alpha_s}}{\sqrt{t}} = \frac{2\phi^2 \alpha_s}{\delta} \tag{4.165}$$

Equation (4.163) then becomes

$$T_m = T_s + c_2\,\text{erf}(\phi) \tag{4.166}$$

and we may solve for c_2. The temperature in the solid is then

$$T = T_s + \frac{T_m - T_s}{\text{erf}(\phi)}\,\text{erf}\left(\frac{x}{2\sqrt{\alpha_s t}}\right), \quad 0 \leq x \leq \delta(t) \tag{4.167}$$

4.3 CONDUCTION WITH PHASE CHANGE

As we did for the mold, we now compute the fluxes at both solid interfaces:

$$k_s \frac{\partial T}{\partial x}\bigg|_{x=0} = \frac{(T_m - T_s)\sqrt{k_s \rho_s c_{ps}}}{\text{erf}(\phi)\sqrt{\pi t}} \tag{4.168}$$

$$k_s \frac{\partial T}{\partial x}\bigg|_{x=\delta} = \frac{(T_m - T_s)\sqrt{k_s \rho_s c_{ps}}}{\text{erf}(\phi)\sqrt{\pi t}} \exp(-\phi^2) \tag{4.169}$$

Finally, we solve for the temperature in the liquid, where we must satisfy the following equations:

$$\begin{aligned}\frac{\partial T}{\partial t} &= \alpha_\ell \frac{\partial^2 T}{\partial x^2}, \quad \delta(t) \leq x < \infty \\ T &= T_m, \quad x = \delta(t) \\ T &= T_\infty, \quad x \to \infty\end{aligned} \tag{4.170}$$

This set of equations also has an error-function-type solution for $\delta(t) \leq x < \infty$, given by

$$T - T_m = \frac{1}{\text{erf}\left(\phi\sqrt{\frac{\alpha_s}{\alpha_\ell}}\right) - 1}\left[T_\infty \text{erf}\left(\phi\sqrt{\frac{\alpha_s}{\alpha_\ell}}\right) - T_m + (T_m - T_\infty)\text{erf}\left(\frac{x}{2\sqrt{\alpha_\ell t}}\right)\right] \tag{4.171}$$

The heat flux at the liquid–solid interface, found by differentiating this solution, is then given by

$$k_\ell \frac{\partial T}{\partial x}\bigg|_{x=\delta} = \frac{(T_\infty - T_m)\sqrt{k_\ell \rho_\ell c_{p\ell}}}{\left[\text{erf}\left(\phi\sqrt{\frac{\alpha_s}{\alpha_\ell}}\right) - 1\right]\sqrt{\pi t}} \exp\left(-\phi^2 \frac{\alpha_s}{\alpha_\ell}\right) \tag{4.172}$$

It now remains to determine the constants T_s and ϕ. We do this by matching the heat fluxes in the mold and solid at $x = 0$, and by enforcing the Stefan condition at $x = \delta(t)$. At the mold–solid interface, equate the heat flux results from Eqs. (4.160) and (4.168), and find

$$\frac{(T_s - T_0)\sqrt{k_m \rho_m c_{pm}}}{\sqrt{\pi t}} = \frac{(T_m - T_s)\sqrt{k_s \rho_s c_{ps}}}{\text{erf}(\phi)\sqrt{\pi t}} \tag{4.173}$$

Note that the time dependence $1/\sqrt{t}$ appears on both sides and can be cancelled. This tells us that our assumption of a constant interface temperature was correct. We may now solve this equation to find the interface temperature:

$$T_s = \frac{T_0\sqrt{k_m \rho_m c_{pm}}\,\text{erf}(\phi) + T_m\sqrt{k_s \rho_s c_{ps}}}{\sqrt{k_m \rho_m c_{pm}}\,\text{erf}(\phi) + \sqrt{k_s \rho_s c_{ps}}} \tag{4.174}$$

Finally, we use the fluxes at the liquid–solid interface, Eqs. (4.169) and (4.172), in the Stefan condition (4.111) to find

$$\frac{(T_m - T_s)\sqrt{k_s \rho_s c_{ps}}}{\text{erf}(\phi)\sqrt{\pi t}}\exp(-\phi^2) - \frac{(T_m - T_\infty)\sqrt{k_\ell \rho_\ell c_{p\ell}}}{\left[\text{erf}\left(\phi\sqrt{\frac{\alpha_s}{\alpha_\ell}}\right) - 1\right]\sqrt{\pi t}}\exp\left(-\phi^2\frac{\alpha_s}{\alpha_\ell}\right)$$

$$= \rho_s L_f \phi \sqrt{\frac{\alpha_s}{t}} \tag{4.175}$$

Substituting the result of Eq. (4.174) for T_s and performing some rearrangement of terms yields a transcendental equation for ϕ:

$$\left(\phi \exp(\phi^2) + \frac{c_{ps}(T_m - T_\infty)}{L_f \sqrt{\pi}} \frac{\exp\left[(1 - \frac{\alpha_s}{\alpha_\ell})\phi^2\right]}{\left[\text{erfc}\left(\phi\sqrt{\frac{\alpha_s}{\alpha_\ell}} - 1\right)\right]} \sqrt{\frac{k_\ell \rho_\ell c_{p\ell}}{k_s \rho_s c_{ps}}}\right)$$

$$\cdot \left(\text{erf}(\phi) + \sqrt{\frac{k_s \rho_s c_{ps}}{k_m \rho_m c_{pm}}}\right) = \frac{c_{ps}(T_m - T_0)}{L_f \sqrt{\pi}} \tag{4.176}$$

This equation is admittedly messy, but it yields readily to graphical or numerical solution.

A special case that is often considered is casting with no superheat, $T_\infty = T_m$. For this case Eq. (4.176) reduces to

$$\phi \exp(\phi^2) \left(\text{erf}(\phi) + \sqrt{\frac{k_s \rho_s c_{ps}}{k_m \rho_m c_{pm}}}\right) = \frac{c_{ps}(T_m - T_0)}{L_f \sqrt{\pi}} \tag{4.177}$$

Equation (4.177) is especially important because it can be used to estimate the solidification time of a section with finite thickness. If there is no superheat, then the solution for the temperature in the liquid is a constant temperature at the melting point. There is no heat flux into the liquid, and the solidification time for a slab of total thickness $2H$ can be found from the expression for the interface position:

$$t_f = \frac{H^2}{4\phi^2 \alpha_s} \tag{4.178}$$

Note that the solution *with* superheat, Eq. (4.176), *cannot* be used to find the solidification time of a finite-thickness section. The temperature solution underlying this equation extends into the liquid and involves a temperature reduction from T_∞ to T_m over a distance of the order of $\sqrt{\alpha_\ell t}$ ahead of the solid–liquid interface. For a finite-thickness slab the temperature solutions growing from the two sides "collide" in the center well before the center of the slab is solid, and the solutions are not valid from that point onward.

When there is superheat, one can find an approximate solidification time for a finite-thickness slab by using Eq. (4.177) with an enhanced latent heat of fusion, given by

$$L'_f = L_f + c_{p\ell}(T_\infty - T_m) \tag{4.179}$$

The extra sensible heat associated with the superheat has been added to the latent heat, and L'_f replaces L_f in Eq. (4.177). The same replacement can be made in the mold-control solution, Eq. (4.145), and in the solid-control solution, Eq. (4.156). With this approximation the total amount of energy removed from the casting is correct. The solidification times predicted this way are accurate as long as the superheat is not too large. Large amounts of superheat tend to inhibit solid formation until almost all the superheat has been removed.

4.3 CONDUCTION WITH PHASE CHANGE

Table 4.5 Material Properties for Sand and Aluminum

Property	Units	Sand	Solid Al	Liquid Al
Thermal conductivity	kW/mK	8.65×10^{-4}	0.211	0.091
Density	kg/m³	1600	2555	2368
Specific heat	kJ/kgK	1.17	1.19	1.09
Latent heat of fusion	kJ/kg	—	398	—

Example: Aluminum Cast into a Sand Mold

To see how this analysis is used in practice, consider pure aluminum cast into a sand mold. The mold and aluminum slab are considered to be semi-infinite, and the material properties are presented in Table 4.5. The sand mold is initially at 303 K, the melt is poured at 943 K, and the melting temperature is 933 K.

Substituting these data into Eq. (4.176), we get

$$\left(\phi \exp(\phi^2) + 0.0102 \frac{\exp(-0.968\phi^2)}{1 - \text{erf}(1.40\phi)}\right)(\text{erf}(\phi) + 19.9) = 1.063 \quad (4.180)$$

The solution to this equation may be found in many ways. If the calculation is set up on a spreadsheet, then a "goal-seeking" capability can be used to adjust ϕ to satisfy Eq. (4.176). If a program is written to solve the problem, then interval halving, as in the slab program in the Appendix (see pp. 130–131), is a convenient method. In either case, Eq. (A.84) can be used to provide accurate evaluation of the error function. Graphical solution is also a practicable option.

The solution to Eq. (4.180) is $\phi = 0.0423$, so that $\text{erf}(\phi) = 0.0477$. The interface temperature is then computed from Eq. (4.174):

$$T_s = \frac{(303)(1270)(0.0477) + (933)(25300)}{(1270)(0.0477) + 25300} = 931.5 \text{ K} \quad (4.181)$$

The interface velocity, from Eq. (4.165), is

$$\frac{d\delta}{dt} = \frac{(2)(0.0477)^2(69.4 \text{ mm}^2/\text{s})}{\delta} = \frac{0.248 \text{ mm}^2/\text{s}}{\delta} \quad (4.182)$$

and a 10-mm-thick layer will solidify in 202 s.

Note that, because the aluminum is a much better conductor than the sand, the solid is nearly isothermal. The computed surface temperature of 931.5 K is very close to the freezing temperature of the aluminum (933 K), which is imposed at the liquid–solid interface by the boundary condition at the liquid–solid interface. Thus the gradient is extremely small in the solid. A measure of the relative ability to transport heat of the mold and metal is embodied in the ratio of the square roots of the heat diffusivities:

$$\frac{\sqrt{k_s \rho_s c_{ps}}}{\sqrt{k_m \rho_m c_{pm}}} = 19.9 \quad (4.183)$$

The large value of this ratio indicates that the transport of heat in the mold is the rate-controlling step in this solidification process.

Table 4.6 Material Properties for Iron and Copper

Property	Units	Copper	Solid Fe	Liquid Fe
Thermal conductivity	kW/mK	0.398	0.083	0.040
Density	kg/m^3	8970	7210	7022
Specific heat	kJ/kgK	0.38	0.67	0.79
Latent heat of fusion	kJ/kg	—	272	—

Example: Iron Solidifying in a Thick Copper Mold

As a contrasting example, consider solidification of pure iron in a copper mold. The iron is poured at a temperature of 1858 K into a copper mold initially at 303 K. The melting point of iron is 1808 K. Material properties for this case are given in Table 4.6.

Once again substituting these data into Eq. (4.176), we obtain

$$\left(\phi \exp(\phi^2) + 0.0517 \frac{\exp(-1.383\phi^2)}{1 - \mathrm{erf}(1.54\phi)}\right)(\mathrm{erf}(\phi) + 0.543) = 2.09 \quad (4.184)$$

For this case, we have $\phi = 0.772$ and $\mathrm{erf}(\phi) = 0.725$. The interface temperature is

$$T_s = \frac{(303)(36800)(0.725) + (1808)(20000)}{(36800)(0.725) + 20000} = 948\,\mathrm{K} \quad (4.185)$$

The corresponding interface velocity is

$$\frac{d\delta}{dt} = \frac{(2)(0.772)^2(17.2)}{\delta} = \frac{20.5\,\mathrm{mm/s}}{\delta} \quad (4.186)$$

and a 10-mm-thick layer solidifies in 2.44 s.

An interesting result from this analysis is that liquid iron, at a temperature well over the melting point of copper (1346 K), can be poured into a thick copper mold without melting the copper. This is because the copper is a better conductor than the iron, so it can conduct away the heat faster than the iron can supply it.

4.4 SUMMARY

In this chapter, we have examined heat conduction, concentrating on geometries that permit analytical solution. In semi-infinite domains, we have found a similarity solution variable $\zeta = x/\sqrt{4\alpha t}$, and solutions in which the temperature is expressed in terms of $\mathrm{erf}(\zeta)$. We also constructed Fourier series solutions to problems in finite-thickness slabs.

We next considered phase change problems. An additional boundary condition, the Stefan condition, is needed to resolve the position of the moving interface. We then found that, for many geometries, the position δ of the solidification front from an external boundary is proportional to $\sqrt{t}$. Several cases of solidification in semi-infinite domains were considered, and we learned that the solution depends strongly on the relative heat conducting capability of the solidifying material and the mold.

BIBLIOGRAPHY

H. S. Carslaw and J. C. Jaeger. *Conduction of Heat in Solids*. Clarendon Press, Oxford, 2nd edition, 1959.

F. P. Incropera and D. P. DeWitt. *Fundamentals of Heat and Mass Transfer*. Wiley, New York, 4th edition, 1996.

W. M. Rohsenow and H. Y. Choi. *Heat, Mass, and Momentum Transfer*. Prentice-Hall, Englewood Cliffs, NJ, 1961.

EXERCISES

1. **Energy consumption by a furnace.** Many metals processing applications, from melting to heat treatment, involve placing the metal in a furnace at rather high temperature. The energy cost for such processes can be very high. Consider a furnace used to heat steel slabs prior to hot rolling, as sketched below. The ends of the furnace are open, and slabs pass through continuously on a conveyor belt. The interior of the furnace is maintained at 1000°C, while the ambient temperature is 30°C. You are to find the heat loss through the walls and estimate the corresponding energy cost. (Heat loss through the open ends contributes significantly to the cost, but is more difficult to estimate.)

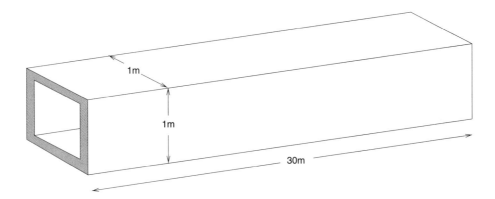

   (a) The furnace wall consists of 100 mm of fire brick ($k = 1$ W/mK), covered by 50 mm of fiberglass insulation ($k = 0.05$ W/mK) and a 5-mm-thick steel shell ($k = 45$ W/mK). Compute the thermal resistance of each component.

   (b) The average heat transfer coefficients on the interior and exterior of the furnace are 30 W/m²K and 5 W/m²K, respectively. Use this and the answer to part (a) to compute the heat flux through the wall.

   (c) If the average cost of electricity is \$0.06/kW h, compute the daily and annual energy cost to operate this furnace.

2. **Insulation on a circular pipe.** Pipes are often insulated to reduce heat loss. In this problem, we use the concept of thermal resistance to determine the effectiveness of the insulating layer, and we find the surprising result that there is an optimal thickness, above which additional insulation actually *increases* the heat loss from the pipe.

(a) Consider a pipe of circular cross section of radius R_i covered with an insulating layer extending to radius R_o. The inside of the pipe is held at temperature T_i, and for now we suppose that the temperature at the surface of the insulation is held at temperature T_o. Consider a length L of the pipe. Write the governing equation and boundary conditions, reducing them appropriately for the geometry.

(b) Solve the differential equation to find the temperature distribution in the insulating layer.

(c) Compute the thermal resistance of the insulating layer.

(d) Now, consider the case in which heat is lost at the surface of the pipe by convection to the ambient temperature T_o with heat transfer coefficient h. Determine the total resistance to heat flow for this system.

(e) Find the value of R_o, at which the resistance to heat flow is minimum.

3 *Temperature along a fin.* When a thin fin is used to transfer heat to a fluid, the steady-state temperature distribution along the fin can be modeled by the following ordinary differential equation and boundary conditions:

$$\frac{d^2T}{dx^2} - C_1 T = 0$$

$$\frac{dT}{dx}(x=0) = 0$$

$$T(x=L) = T_L$$

The constant C_1 depends on the fin thickness, fin conductivity, and surface heat transfer coefficient. The first boundary condition implies that there is no heat loss through the tip of the fin, and the second boundary condition gives the temperature at the fin's base. Solve this equation to find the temperature distribution $T(x)$.

4 *Anisotropic heat conduction.* We most often use the isotropic form of Fourier's law for heat conduction,

$$\mathbf{q} = -k \nabla T$$

where $\mathbf{q}$ is the heat flux vector and ∇T is the temperature gradient; k is the thermal conductivity, a scalar quantity. For *anisotropic* materials the corresponding description is

$$\mathbf{q} = -\mathbf{k} \cdot \nabla T$$

where $\mathbf{k}$ is a thermal conductivity tensor, which is assumed to be symmetric.

(a) Consider a material whose thermal conductivity tensor is

$$\mathbf{k} = \begin{bmatrix} 7.750 & -3.897 & 0 \\ -3.897 & 3.250 & 0 \\ 0 & 0 & 1 \end{bmatrix}$$

Find $\mathbf{q}$ when $\nabla T = \{2, 2, 2\}^T$.

(b) For isotropic materials, **q** and ∇T point in the same direction. Is this always true for anisotropic materials?

(c) Show that the conductivity tensor for an isotropic material is $\mathbf{k} = k\mathbf{I}$.

(d) One particular type of anisotropic material is called *transversely isotropic*. This type of material has one symmetry axis and is isotropic in the plane normal to that axis. A composite material reinforced with aligned fibers is an example: the fiber direction is the symmetry axis and the material is isotropic in the plane perpendicular to the fibers.

Let the thermal conductivity of a transversely isotropic material be k_a along the fiber axis and k_t transverse to the fibers (i.e., in the isotropic plane). Show that the thermal conductivity for this material is

$$\mathbf{k} = (k_a - k_t)\mathbf{pp} + k_t\mathbf{I}$$

where **p** is a unit vector pointing in the fiber direction.

5 ***Thermal contact between bodies at different temperatures.*** In many materials processing problems, two materials at different temperatures are suddenly placed in thermal contact, for example when a hot liquid is quickly poured into a cold mold. Consider the case in which both bodies have constant, but different, thermal properties, and in which convection can be neglected. Also assume that the bodies are in perfect thermal contact, that is, that there is no extra resistance to heat transfer at the interface between them. Consider the case of short-time behavior, so that both bodies can be considered to be semi-infinite. Then the situation is like the sketch below.

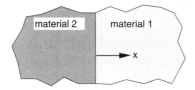

The two materials have different initial temperatures T_{01} and T_{02}. Their thermal properties are ρ_1, c_{p1}, k_1, α_1 and ρ_2, c_{p2}, k_2, α_2, respectively. The interface between them lies at $x = 0$. Call the temperature in the first material $T_1(x, t)$, and the temperature in the second material $T_2(x, t)$. The problem can be formulated by writing an energy equation for each body:

$$\frac{\partial T_1}{\partial t} = \alpha_1 \frac{\partial^2 T_1}{\partial x^2} \quad \text{for} \quad x > 0$$

$$\frac{\partial T_2}{\partial t} = \alpha_2 \frac{\partial^2 T_2}{\partial x^2} \quad \text{for} \quad x < 0$$

The initial conditions are

$$T_1(x, t = 0) = T_{01}$$
$$T_2(x, t = 0) = T_{02}$$

One pair of boundary conditions is

$$T_1(x = \infty, t) = T_{01}$$
$$T_2(x = -\infty, t) = T_{02}$$

We also need two boundary conditions at the interface. The two temperatures must match,

$$T_1(0, t) = T_2(0, t)$$

and the heat fluxes must match,

$$-k_1 \frac{\partial T_1}{\partial x}\bigg|_{x=0} = -k_2 \frac{\partial T_2}{\partial x}\bigg|_{x=0}$$

Suppose that the interface temperature is constant in time, say with value T_{surf}. We know that the semi-infinite body solution is a solution to both energy equations, so that the temperature solution must be

$$T_1 = A_1 + B_1 \operatorname{erf}\left(\frac{x}{\sqrt{4\alpha_1 t}}\right)$$

$$T_2 = A_2 + B_2 \operatorname{erf}\left(\frac{-x}{\sqrt{4\alpha_2 t}}\right)$$

If these equations can be made to satisfy all the initial and boundary conditions (by adjusting the constants A_1, A_2, B_1, and B_2), then they are the solution to the problem.

(a) Solve for the constants A_1, A_2, B_1, and B_2 and write out the temperature solutions. Note: $\operatorname{erf}(-x) = -\operatorname{erf}(x)$.

(b) Carefully sketch the solution for temperature as a function of x at several different times.

(c) Find an expression for the temperature at the interface, T_{surf}. Show that the combination of material properties ($k\rho c_p$) governs the value of the interface temperature in relation to the two initial temperatures. The combined property ($k\rho c_p$) is called the *heat diffusivity*, not to be confused with the thermal diffusivity $\alpha = k/\rho c_p$.

(d) Calculate the interface temperature when polycarbonate at 300°C is placed in contact with steel at 25°C. Material properties are given in the table.

Property Units	k J/m s K	ρ kg/m^3	c_p J/kg K
Polycarbonate	0.192	1200	1250
Steel	43.2	7810	472

6 *Analysis of the Jominy end-quench test.* A common test for the response of steels to heat treatment is the Jominy test. A cylindrical steel rod, initially at 750°C, is quenched from one end by a controlled water stream at 25°C. The heat transfer coefficient for such a configuration will be a strong function of temperature. However, for simplicity of analysis, we will assume that it is a constant, $h = 2 \times 10^{-2}$ W/mm^2 K. We will also assume that the sides and end opposite the quench are perfectly insulated, allowing us to treat the rod as if it were semi-infinite. Then Eq. (4.58) applies, and we can use it to analyze the heat treatment process.

EXERCISES

A typical Jominy curve, also known as a *hardenability curve*, is shown below. The hardness of the steel is plotted as a function of distance from the quenched end. Notice that there is a scale along the top that correlates cooling rate at 700°C with distance from the quenched end. We now show where such a scale comes from by examining the temperature and cooling rate at 6 mm from the quenched end.

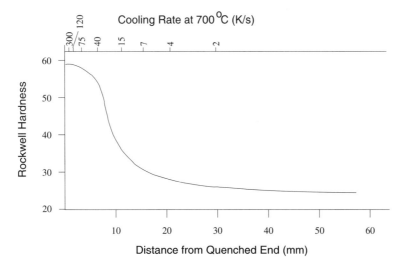

(a) Substitute the appropriate parameters into Eq. (4.58), using $\alpha = 6$ mm^2/s and $k = 30 \times 10^{-3}$ W/mm K, to develop a transcendental equation for the time t when this location reaches 700°C.

(b) Solve the equation for t. A graphical solution is an easy way to do this.

(c) By differentiating the expression for the temperature, or numerically differentiating the graphical solution of the previous part of this exercise, find the cooling rate at 700°C at this location.

7 *Aluminum quenching.* Many aluminum alloys are heat treated by precipitation hardening. In this process, a high temperature heat treatment (~500°C) is first performed to return elements segregated during solidification into solid solution. The part is then water quenched to retain the solid solution, and then reheated to an intermediate temperature to control the precipitation of second phases. In this problem, we explore the quenching process.

If the solid solution is to be retained during quenching, the part must cool rapidly enough to bypass the precipitation reaction. The time delay for the precipitation reaction to begin depends on temperature and is often presented as in the following figure, called a *transformation diagram*.

Consider now a 50-mm-thick plate of aluminum alloy, quenched from 500°C into water at 25°C. Assume that the heat transfer coefficient is constant, $h = 7.6 \times 10^3$ W/m^2 K, and that the thermal properties are also constant: $k_\ell = 190$ W/m K, $\rho_\ell = 2750$ kg/m^3, and $c_p = 1$ kJ/kg.

(a) Compute the Biot number for the process as described. Remember that the plate thickness is defined as $2H$.

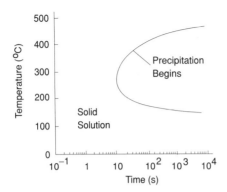

(b) Determine whether the precipitation reaction will be avoided in the plate during quenching. (Hint: Use Fig. 4.7.)

8. **Product solutions.** One can sometimes construct solutions to two-dimensional heat conduction problems as products of one-dimensional solutions. The requirements are:
   - boundaries must be lines of constant coordinate,
   - the initial condition can be expressed as a product $T = f_1(x) f_2(y)$, and
   - the boundary conditions are the same, and independent of position.

   Consider the 90° corner illustrated below, initially at temperature T_0.

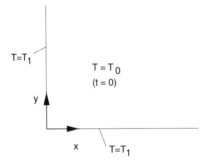

   (a) Write the governing equation and boundary conditions, assuming that there is no velocity, no internal heat generation, and that all properties are constant.

   (b) Find a solution as a product of one-dimensional error function solutions:
   $$T(x, y, t) = C_1 + C_2 \, \text{erf}\left(\frac{x}{\sqrt{4\alpha t}}\right) \text{erf}\left(\frac{y}{\sqrt{4\alpha t}}\right)$$
   Use the boundary and initial conditions to evaluate C_1 and C_2.

   (c) Sketch the isotherms for some $t > 0$. How far from the corner do you need to be before the isotherms are nearly parallel to the wall?

9. **Thermocouple response.** A thermocouple is made by welding two dissimilar metals into a small spherical bead. The potential difference between the

two legs of the thermocouple is a measure of the average temperature in the bead. Suppose that you have a chromel-alumel thermocouple, with the following material properties: $\rho = 2700$ kg/m³, $k = 100$ W/m K, $c_p = 400$ J/kg K. The thermocouple bead has a radius of 1 mm. It is to be plunged suddenly from room temperature into boiling water, and the heat transfer coefficient is 10^4 W/m K.

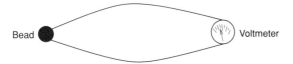

(a) Show that for these data the thermocouple can be considered as a lumped mass; that is, the internal temperature is constant.

(b) Develop an equation that describes the temperature measured by the thermocouple during the experiment; then solve the equation and plot the result.

(c) Suppose that during an experiment the thermocouple is moved through water where the temperature varies with position. The temperature reported by the thermocouple will not necessarily reflect the exact local temperature if the temperature varies faster in time than the typical response time of the thermocouple (see the previous part of this problem as an example). Derive an expression that could be used to reconstruct the real temperature value from the recorded temperature. What difficulties might arise if you tried to use this expression in practice, for example on a real signal that contained noise?

10 *Transient temperatures in microwave heating.* A large slab with thickness $2H$ is heated in a microwave oven. The oven supplies a constant volumetric heat source $\dot{R}$. Assume that the slab has an initial temperature T_0, and its outer surfaces are held at T_0 throughout the heating process. We wish to know the steady-state temperature distribution and how long it takes to achieve it.

(a) Write the governing equation, boundary conditions, and initial conditions. Assume that temperature is a function of x and t only (with x being the direction across the thickness of the slab and $x = 0$ denoting the midplane) and that the properties of the material are constant.

(b) Solve for the steady-state temperature distribution; call it $T_s(x)$.

(c) Guess that the transient solution can be written as the sum of the steady solution and a second, unsteady part $T_u(x, t)$:

$$T(x, t) = T_s(x) + T_u(x, t)$$

Substitute this into the equation from part (a) and derive a governing equation, boundary conditions and initial conditions for $T_u(x, t)$. These should be the same as the transient slab heating problem without any heat source.

(d) Work out the solution to this problem. You may use any part of the solution for transient conduction in a slab without rederiving it.

(e) A polymeric sheet that is 3 mm thick is heated in a microwave oven that provides a heat source of 7×10^6 W/m^3. The polymer has $k = 0.2$ J/m s K, $\rho = 1200$ kg/m^3, and $c_p = 1250$ J/kg K. The initial and boundary temperatures are both 75°C. Calculate the maximum temperature at the midplane, and the time for the midplane temperature to reach a value 90% of the way between the initial and steady-state values.

11 *Casting with interface control.* Section 4.3.1 introduced two simplified treatments of heat transfer in metal casting processes, one in which the mold provided the major resistance to heat transfer, and a second in which the solidifying metal layer itself provided the major resistance. However, when a liquid metal is cast in a metal mold, a thin air gap quickly forms at the interface. This gap results from expansion of the mold as it heats and from contraction of the solid portion of the casting as it cools. The thickness of this gap is difficult to predict, and one commonly relies on experimental data for mold surface temperatures and rates of solidification. This is modeled by using a heat transfer coefficient h,

$$-q_x = h(T_{\text{solid}} - T_{\text{mold}})$$

where all quantities are measured at the mold surface. Interface resistance tends to be important during the initial stages of solidification, when neither the metal nor the mold has changed temperature very much. When the interface resistance is large, the casting is thin, or both, interface resistance can dominate throughout the entire solidification process.

Consider the case in which the interface resistance dominates the heat removed from the mold. In this case, one can assume that the solid temperature is very close to T_m and the mold temperature is very close to T_0.

(a) Derive a formula for $\delta(t)$, the solidified thickness as a function of time, in a slab-shaped casting, when the metal is poured with no superheat. Assume that the heat transfer coefficient h is known. Suggestion: relate the heat flux leaving the surface of the mold to the heat required to grow the frozen layer.

(b) Using A to represent the mold surface area and V to represent the volume of the casting, derive an expression for the solidification time of a casting of general shape. Follow the approach used to derive Chvorinov's rule, namely that every increment of surface area is equivalent to a flat surface. How does your result compare to Chvorinov's rule?

12 *Iron casting.* Consider an iron casting that is poured with no superheat into a sand mold at 303 K. Using the properties in Tables 4.5 and 4.6, compute the solidification time for a 40-mm-thick slab (20 mm half-thickness). Compute first assuming that the mold control formula applies, and then use the more complete general form given in Section 4.3.2. Comment on the error introduced by the mold control approximation in this case.

13 *Aluminum casting.* Use the simplified mold control solution to analyze the problem of aluminum cast in a sand mold given on p. 119. Use both the general result from Section 4.3.2 and the mold control solution to find the solidification time for a slab that is 20 mm thick. Comment on the error introduced by the mold control approximation in this case.

14 *Iron casting with a copper chill.* Copper chills are sometimes used to control the solidification pattern in a sand casting. Consider the casting of a 15-mm-thick iron slab illustrated below, where there is a thick Cu chill on one side and sand on the other. If the iron is poured at its melting point, the solidification from both sides can be treated by using the semi-infinite solutions we have developed in this chapter. Choose an appropriate solution for the solidification from each side, and determine the location of the last point to freeze in the slab. Use the data from Tables 4.5 and 4.6 for the material properties.

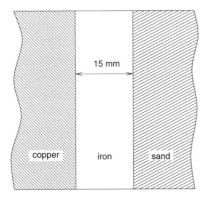

15 *Atomized droplet solidification.* There are many applications for using fine particles of metals. An important example is called *powder metallurgy*, in which solid particles are compacted into a complex shape and then sintered at temperatures just below the melting point to form a dense solid part.

The particulate material is often prepared by atomization, in which a liquid metal is sprayed from an orifice to form fine droplets, which then freeze in the surrounding atmosphere, usually argon or air. In this problem, we determine the amount of time it takes for such a droplet to freeze. Consider a nickel droplet 100 μm in diameter, solidifying in argon at 25°C. The heat transfer coefficient between the droplet and the gas is 45 W/m K. Assume that the nickel has the following thermal properties: $\rho_\ell = 7.9 \times 10^3$ kg/m^3, $k_\ell = 25$ W/m K (estimated), $c_{p\ell} = 620$ J/kg K, $L_f = 292$ kJ/kg, and $T_f = 1453$°C.

(a) Compute the Biot number for the heat transfer from the droplet to the argon gas. State the implications of this calculation for the subsequent analysis of the solidification time.

(b) Compute the approximate solidification time for the droplet.

16 *Freezing of polyethylene.* Low-density crystalline polyethylene is injected into a cold steel mold at 25°C, forming a sheet that is 6 mm thick. The polymer is injected at 120°C, and center must cool below 50°C before the part can be ejected without distortion. Compute the residence time required for the part in the mold before ejection to meet this requirement. Assume that there is perfect thermal contact between the mold and the part.

Property	Polymer	Steel Mold
Thermal conductivity (W/m K)	0.26	37.5
Density (kg/m^3)	920	7730
Specific heat (J/kg K)	1670	610

Appendix

```
      program biot_slab
c
c  Program to compute the series solution for cooling of a slab
c  Input parameter is the Biot number
c
c  Evaluations are done at several values of Biot Number and time
c  as enumerated in the following data sets
c
      implicit double precision (a-h,o-z)
      parameter (nb=5,ntime=51)
      dimension biots(nb), el(50)
      character filnam*15
      data biots / 0.01, 0.1, 1.0, 10., 100./
      data pi,eps1,eps2 / 3.1416, 1.e-4, 1.e-4 /
c
      do 100 n=1,nb
       Bi = biots(n)
       write(filnam,20)Bi
20     format('biot_',e6.1,'.dat')
       open(21,file=filnam)
       do 90 nx=0,2
        x = 0.5*nx
c
c  Find the first MAXTERM roots of the transcendental equation for
c this Bi
c
        MAXTERM = 50
        do 40 m=1,MAXTERM
         elmin = (m-1)*pi + eps1
         elmax = 0.5*(2*m-1)*pi - eps1
         call elroot(elmin,elmax,Bi,eps one,el(m))
40      continue
c
        do 60 nt=1,ntime
         time = 10.**(-3. + 5.5*(nt-1)/ntime)
         sum = 0.0
c
c  Sum the series taking no more than MAXTERM terms
```

APPENDIX

```fortran
c
        do 50 m=1,MAXTERM
          c_m = 2.*sin(el(m))/(el(m) + sin(el(m))*cos(el(m)))
          term = c_m*exp(-time*el(m)**2)*cos(el(m)*x)
          sum = sum + term
          contrib = abs(term/sum)
          if(contrib.lt.eps2)then
           write(21,*)time,sum
           go to 60
          endif
50      continue
60       continue
         write(21,*)'&'
90      continue
        close(21)
100     continue
        end
c
        subroutine elroot(elmin,elmax,Bi,eps1,el)
c
c       Subroutine to find roots to the equation L*tan(L)=Bi
c
        implicit double precision (a-h,o-z)
        elval(x) = x*tan(x)-Bi
c
10      fleft = elval(elmin)
        fright = elval(elmax)
        elmid = (elmin+elmax)*0.5
        fmid = elval(elmid)
        if(abs(fmid).lt.eps1)then
         el = (elmin+elmax)*0.5
         return
        elseif(fleft*fmid.le.0.0)then
         elmax = elmid
         if(abs(elmax-elmin).lt.eps1)then
          el = (elmin+elmax)*0.5
          return
         endif
         go to 10
        else
         elmin = elmid
         if(abs(elmax-elmin).lt.eps1)then
           el = (elmin+elmax)*0.5
           return
          endif
          go to 10
        endif
        end
```

CHAPTER FIVE

Isothermal Newtonian Fluid Flow

In this chapter we turn our attention to problems involving fluid flow. Such problems are solved by using the mass and momentum balance equations derived in Chapter 2. We start by analyzing Newtonian fluids in simple geometries, such as flow between parallel plates, or flow in a tube. We use scaling to determine whether inertial or viscous effects dominate the problem, and we concentrate initially on viscous-dominated flows. These simple-geometry viscous results are useful building blocks for process modeling, especially in polymer processing where the fluid viscosity is very large. Several examples illustrate how this is done. We show how to formulate problems with free or moving boundaries, a situation that is common in materials processing, and we apply these ideas to some simple models of mold filling. Finally, we consider cases in which the inertial terms are dominant, and we provide basic tools for treating these problems.

5.1 NEWTONIAN FLOW IN A THIN CHANNEL

Consider flow in a channel whose thickness is much smaller than its width and length. A great many polymer processing operations involve this type of geometry, for reasons that can be understood by using the results from Chapter 4. Most plastic parts are intended to be inexpensive, which requires that they be produced in a short time, and cooling the polymer melt is the slowest part of the process because polymers are relatively poor conductors of heat. We saw in Chapter 4 that the cooling time for a low-conductivity slab, with half-thickness H, is approximately H^2/α, where α is the thermal diffusivity. For polymers a typical value of α is 1×10^{-7} m²/s, so the total thickness that corresponds to a production time of 10 s is ~2 mm. Note also that the cycle time increases as the *square* of the thickness. Because of this thermal and economic limit, molded plastic parts are rarely thicker than ~4 mm.

With this motivation, let us consider flow in a thin channel defined by two parallel flat plates. A no-slip boundary condition applies at the surfaces of the plates, meaning that the velocity of the fluid and the wall are identical. Fluid flow may be generated in one of two ways: *drag flow*, in which one boundary moves and drags some of the fluid along with it, and *pressure flow*, in which the channel walls are stationary and the flow is driven by a pressure difference between the entrance and the exit. It is also

5.1 NEWTONIAN FLOW IN A THIN CHANNEL

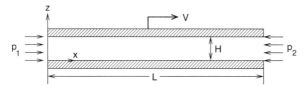

Figure 5.1: Two-dimensional domain for analyzing combined pressure and drag flow.

possible to have a combined flow, in which there is both a pressure gradient as well as boundary motion. We now analyze the combined flow problem for a Newtonian fluid. The goal is to relate the flow rate Q through the channel to the imposed velocity of the boundary and the pressure drop along the length, and determine the external force necessary to sustain the motion of the upper plate.

5.1.1 BASIC SOLUTION

To find the solution for this combined pressure and drag flow problem, we begin with a number of simplifying assumptions:

- the fluid is Newtonian, with constant density and viscosity;
- the flow is steady and two dimensional in the x–z plane;
- there is no slip on the walls of the channel; and
- the body forces are negligible.

The geometry is shown in Fig. 5.1.

Mass Balance Equation

Under the assumptions of two-dimensional flow and constant density, the mass balance equation is

$$\frac{\partial v_x}{\partial x} + \frac{\partial v_z}{\partial z} = 0 \tag{5.1}$$

We now introduce dimensionless velocity and position variables:

$$\begin{aligned} x^* &= \frac{x}{L} \\ z^* &= \frac{z}{H} \\ v_x^* &= \frac{v_x}{V} \\ v_z^* &= \frac{v_z}{V_z} \\ p^* &= \frac{p - p_2}{\Delta p_c} \end{aligned} \tag{5.2}$$

where V_z is an unknown characteristic velocity in the z direction, and Δp_c is a characteristic pressure difference, also to be determined. Substituting these scales into the mass balance yields

$$\left[\frac{V}{L}\right]\frac{\partial v_x^*}{\partial x^*} + \left[\frac{V_z}{H}\right]\frac{\partial v_z^*}{\partial z^*} = 0 \tag{5.3}$$

Because the two terms must balance and the dimensionless derivatives must be $\mathcal{O}(1)$, we can equate the two sets of dimensional parameters to find the unknown characteristic value:

$$V_z = \left[\frac{H}{L}\right]V \tag{5.4}$$

Because V_z is the scale for v_z, this tells us that the velocity in the z direction is smaller than the x-direction velocity by a factor of H/L. Because $H/L \ll 1$, we expect that motion in the z direction will be unimportant. The fact that the gap is narrow will turn out to have some other useful implications later on.

Momentum Balance Equation

We now examine the momentum equation, assuming that the body forces are negligible but retaining all of the other terms for steady, two-dimensional flow. The x-direction Navier–Stokes equation becomes

$$\rho\left(v_x\frac{\partial v_x}{\partial x} + v_z\frac{\partial v_x}{\partial z}\right) = -\frac{\partial p}{\partial x} + \mu\left(\frac{\partial^2 v_x}{\partial x^2} + \frac{\partial^2 v_x}{\partial z^2}\right) \tag{5.5}$$

Introducing the dimensionless variables defined above yields

$$\left[\frac{\rho V^2}{L}\right]v_x^*\frac{\partial v_x^*}{\partial x^*} + \left[\frac{\rho V V_z}{H}\right]v_z^*\frac{\partial v_x^*}{\partial z^*}$$
$$= -\left[\frac{\Delta p_c}{L}\right]\frac{\partial p^*}{\partial x^*} + \left[\frac{\mu V}{L^2}\right]\frac{\partial^2 v_x^*}{\partial x^{*2}} + \left[\frac{\mu V}{H^2}\right]\frac{\partial^2 v_x^*}{\partial z^{*2}} \tag{5.6}$$

We anticipate that the viscous stress associated with $\partial^2 v_x/\partial z^2$ will be important, and we multiply through by $H^2/(\mu V)$. We also substitute the derived relation $V_z = (H/L)V$. This gives

$$\left\{\frac{\rho V H}{\mu}\frac{H}{L}\right\}\left(v_x^*\frac{\partial v_x^*}{\partial x^*} + v_z^*\frac{\partial v_x^*}{\partial z^*}\right) = -\left\{\frac{H^2\Delta p_c}{\mu V L}\right\}\frac{\partial p^*}{\partial x^*} + \left\{\frac{H^2}{L^2}\right\}\frac{\partial^2 v_x^*}{\partial x^{*2}} + \frac{\partial^2 v_x^*}{\partial z^{*2}}$$
(5.7)

The dimensionless parameter $(\rho V H/\mu)$ is the *Reynolds number*, Re. Usually we think of Re as representing the ratio of inertial effects to viscous forces. However, scaling shows us that in this particular problem the ratio of inertial to viscous forces is the Reynolds number multiplied by the aspect ratio: Re (H/L).

Typical material properties and process parameters for a polymer melt flow are

$$\begin{aligned}\rho &= 10^3 \text{ kg/m}^3\\ \mu &= 10^4 \text{ N s/m}^2\\ V &= 10^{-1} \text{ m/s}\\ H &= 10^{-3} \text{ m}\\ L &= 10^{-1} \text{ m}\end{aligned} \tag{5.8}$$

5.1 NEWTONIAN FLOW IN A THIN CHANNEL

Using these values we find that $\text{Re}(H/L) = 10^{-7}$, which certainly makes the inertial terms negligible. We would have expected this if we had simply computed the traditional Reynolds number, which is 10^{-5} for these values. However, if the fluid were a lubricating oil with a viscosity of 10^{-1} N s/m^2, the Reynolds number would equal one, but the inertial terms would still be negligible because $\text{Re}(H/L) = 10^{-2}$. This demonstrates the importance of using scaling analysis to derive meaningful dimensionless groups, rather than relying on traditional definitions.

Returning to Eq. (5.7), we now drop the inertial terms and neglect the term multiplied by H^2/L^2 ($= 10^{-4}$). The x-direction momentum equation is now

$$0 = -\left\{\frac{H^2 \Delta p_c}{\mu V L}\right\} \frac{\partial p^*}{\partial x^*} + \frac{\partial^2 v_x^*}{\partial z^{*2}} \tag{5.9}$$

Recognizing that these two terms must balance and that the dimensionless derivatives are $\mathcal{O}(1)$, we find the characteristic pressure difference to be

$$\Delta p_c = \frac{V \mu L}{H^2} \tag{5.10}$$

and thus we have simplified the x-direction momentum equation to

$$\frac{\partial^2 v_x^*}{\partial z^{*2}} = \frac{\partial p^*}{\partial x^*} \tag{5.11}$$

In dimensionless form, the boundary conditions become

$$v_x^* = 0 \quad \text{at } z^* = 0$$
$$v_x^* = 1 \quad \text{at } z^* = 1 \tag{5.12}$$

For the moment, let us assume that p^* is independent of z^*; we shall prove this shortly. The flow conditions do not vary along the x direction, so the velocity gradient should not vary either. Thus, we assume that $\partial p/\partial x$ is constant and recognize that its value must be determined by p_1, p_2, and L. In dimensional form we have

$$\frac{\partial p}{\partial x} = \frac{p_2 - p_1}{L} = \frac{-\Delta p}{L} \tag{5.13}$$

Here we have defined Δp as the pressure *drop*, so that Δp will be positive when the flow is in the positive x direction (i.e., when $p_1 > p_2$). It is now convenient to define a dimensionless pressure gradient G^* as

$$G^* = -\frac{\partial p^*}{\partial x^*} = \frac{\Delta p H^2}{2 \mu L V} \tag{5.14}$$

The minus sign makes G^* positive when the pressure gradient is driving the flow in the positive x direction. Now Eq. (5.11) reads

$$\frac{d^2 v_x^*}{dz^{*2}} = -G^* \tag{5.15}$$

The partial derivative has been replaced by an ordinary derivative, because G^* does not depend on z^*. To solve this equation we integrate twice:

$$\frac{dv_x^*}{dz^*} = -G^*z^* + c_1 \tag{5.16}$$

$$v_x^* = -\frac{G^*z^{*2}}{2} + c_1 z^* + c_2 \tag{5.17}$$

Using the boundary conditions from Eq. (5.12) to evaluate c_1 and c_2, we find the solution:

$$v_x^* = \underbrace{z^*}_{\text{drag flow}} + \underbrace{G^*(z^* - z^{*2})}_{\text{pressure flow}} \tag{5.18}$$

This solution displays a number of interesting features. First, the velocity field is the sum of the solution for drag flow alone (a linear velocity profile) and the solution for pressure flow alone (a parabolic velocity profile). This linear superposition is a result of the linearity of Eq. (5.11), which comes about because we have a linear (Newtonian) constitutive relation, and because the nonlinear (inertial) terms in the equation of motion were shown to be negligible.

The solution for combined pressure and drag flow is parameterized by the dimensionless pressure gradient G^*. Graphs of the velocity solution are shown in Fig. 5.2 for several values of G^*. When $G^* = 0$ we have pure drag flow, and the velocity profile is linear. When G^* is positive the drag and pressure flows are in the same direction, and v_x is positive everywhere across the gap. When G^* is negative the pressure gradient and drag flows are opposed and, depending on the value of G^*, the fluid can move in the negative x direction in some parts of the channel.

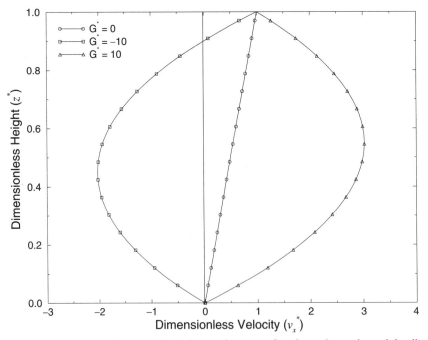

Figure 5.2: Velocity field for combined drag and pressure flow for various values of the dimensionless pressure gradient G^*.

5.1 NEWTONIAN FLOW IN A THIN CHANNEL

Pressure Scaling and z Momentum Balance

Let us now return to our assumption that p^* is not a function of z^*. To examine the validity of this assumption, we must consider the z-momentum equation. By now the substitution of dimensionless variables should be a familiar exercise, and we skip some steps to arrive at the dimensionless form:

$$\left\{\frac{\rho V H}{\mu}\frac{H}{L}\right\}\left(v_x^*\frac{\partial v_z^*}{\partial x^*}+v_z^*\frac{\partial v_z^*}{\partial z^*}\right)=-\left\{\frac{\Delta p_c L}{\mu V}\right\}\frac{\partial p^*}{\partial z^*}+\left\{\frac{H^2}{L^2}\right\}\frac{\partial^2 v_z^*}{\partial x^{*2}}+\frac{\partial^2 v_z^*}{\partial z^{*2}} \quad (5.19)$$

Note that the same Reynolds number and aspect ratio terms appear in this form as in the x-direction momentum equation, Eq. (5.7). We therefore drop the same terms as before, and have

$$\frac{\partial^2 v_z^*}{\partial z^{*2}}=\left\{\frac{\Delta p_c L}{\mu V}\right\}\frac{\partial p^*}{\partial z^*} \quad (5.20)$$

If we now substitute Eq. (5.10) for Δp_c, we obtain an unexpected result:

$$\frac{\partial^2 v_z^*}{\partial z^{*2}}=\left\{\frac{L^2}{H^2}\right\}\frac{\partial p^*}{\partial z^*} \quad (5.21)$$

How do we interpret this result? If the dimensionless derivatives are all $\mathcal{O}(1)$, what is the multiplier L^2/H^2 doing here? The answer to this puzzle lies in the choice of the pressure scale. Earlier we chose Δp_c based on the pressure gradient in the x direction, that is, $\Delta p_c = V\mu L/H^2$. Equation (5.21) is telling us that this pressure scale is not meaningful in the z direction. One way to interpret Eq. (5.21) is that $\partial p^*/\partial z^*$ is $\mathcal{O}(H^2/L^2)$ when we nondimensionalize pressure as in Eq. (5.2), and thus $\partial p^*/\partial z^* \ll \partial p^*/\partial x^*$.

Another way to reach the same conclusion is to treat pressure gradient, which is a vector, in the same way as we treat velocity, where the different components may scale differently. We can then define an unknown characteristic pressure difference Δp_z for the transverse direction and rescale the z-direction momentum equation by using a new dimensionless pressure $\hat{p}$:

$$\hat{p}=(p-p_2)/\Delta p_z \quad (5.22)$$

The result is analogous to Eq. (5.20):

$$\frac{\partial^2 v_z^*}{\partial z^{*2}}=\left\{\frac{\Delta p_z L}{\mu V}\right\}\frac{\partial \hat{p}}{\partial z^*} \quad (5.23)$$

We can require that $(\Delta p_z L/\mu V) = \mathcal{O}(1)$, because we now expect this new scaling to give $\partial \hat{p}/\partial z^* = \mathcal{O}(1)$. Using this to find Δp_z and comparing the result with Eq. (5.21) shows that

$$\Delta p_z = \left\{\frac{H^2}{L^2}\right\}\Delta p_c \quad (5.24)$$

More importantly, it is now clear that

$$\frac{\partial p}{\partial z} \sim \left\{\frac{H^2}{L^2}\right\}\frac{\partial p}{\partial x} \quad (5.25)$$

In other words, in a narrow gap the transverse pressure gradient is negligible in comparison to the axial pressure gradient. This justifies the assumption that p^* depends only on x^*. Finally, if we return to Eq. (5.21) and set $\partial p^*/\partial z^* = 0$, we see that

$$\frac{\partial^2 v_z}{\partial z^{*2}} = 0 \tag{5.26}$$

so that $v_z^* = c_1 + c_2 z^*$. Because $v_z^* = 0$ at $z^* = 0$ and $z^* = 1$, v_z^* must equal zero everywhere.

The Lubrication Approximation and Fully Developed Flow

The phenomena involved in this problem arise frequently enough that they have been given a special name. Whenever a viscous fluid is confined to a narrow gap ($H/L \ll 1$) and the inertial terms are negligible, we expect that the transverse pressure gradient will be negligible in comparison with the axial pressure gradient, and that the shear stresses across the gap direction will be the dominant viscous terms. This combination of phenomena is known as the *lubrication approximation*. The lubrication approximation is regularly used to model oil flow in journal bearings and in other types of hydrodynamic lubrication problems. It is also the basis for modeling non-Newtonian flow in injection molding of polymers, and in a variety of other polymer melt flows.

Let us examine this limiting case a bit further. In the asymptotic limit where $H/L \ll 1$, we found $v_z^* \equiv 0$. Using this in the mass balance equation, we find

$$\frac{\partial v_x^*}{\partial x^*} = 0 \tag{5.27}$$

and v_x^* is independent of x^*. We do not expect Eq. (5.27) to apply near the entrance or exit of a channel, because the velocity profile $v_x(z)$ will be changing with x in those regions. In entrance and exit regions, the proper length scale for x is H, not L, and v_x and v_z can be comparable in magnitude. Our analysis applies away from these regions, and we sometimes refer to the condition in Eq. (5.27) as describing a "fully developed" flow, meaning that entrance effects are negligible.

An alternative approach is to *assume* that the flow is fully developed, that is, that the velocity profile and pressure gradient are independent of x ($\partial v_x/\partial x = 0$ and $\partial p/\partial x = $ constant). In the case of two-dimensional fully developed flow, the mass balance equation reduces to

$$\frac{\partial v_z}{\partial z} = 0 \tag{5.28}$$

Together with either one of the two boundary conditions $v_z = 0$ at $z = 0$ and $z = H$, this implies that $v_z \equiv 0$. The inertial terms then vanish in the x-momentum equation, and we obtain the identical result. We may assume that the flow is fully developed and then show that the inertial terms vanish, or we may make assumptions about the material properties which make the inertial terms negligible, and then show that the flow is fully developed. Neither approach is really better than the other, and the conclusions are the same.

Note that scaling analysis never tells us that any particular term is identically zero. However, it is not necessary for a term to equal zero for us to ignore it; we

5.1 NEWTONIAN FLOW IN A THIN CHANNEL

need to show only that the term is small compared with other terms in the governing equations. This is an important point, for it allows us to use solutions like Eq. (5.14) when the flow geometry is slightly different from parallel plates.

Volume Flow Rate and Applied Force

Now let us return to the original objectives, which were to compute the volumetric flow rate Q and the force F required to move the upper plate. We also return to the dimensional form. The volumetric flow rate is evaluated by integrating v_x over the cross-sectional area. We take the width of the channel in the y direction to be W, so that

$$Q = \int_0^W \int_0^H v_x \, dz \, dy \qquad (5.29)$$

Carrying out the integrals we find

$$Q = \frac{VWH}{2} + \frac{WH^3 \Delta p}{12\mu L} \qquad (5.30)$$

Like the velocity distribution, the flow rate is the sum of a drag flow term and a pressure flow term.

The force on the upper plate is computed from the shear stress,

$$F = WL\,\tau_{zx}|_{z=H} \qquad (5.31)$$

For a Newtonian fluid we know that

$$\tau_{zx} = 2\mu D_{xz} = \mu \left(\frac{\partial v_x}{\partial z} + \frac{\partial v_z}{\partial x} \right) \qquad (5.32)$$

but $v_z = 0$ here, so in this particular flow we have

$$\tau_{zx} = \mu \frac{\partial v_x}{\partial z} \qquad (5.33)$$

This type of flow, in which the velocity is unidirectional and varies only in the direction perpendicular to the flow, is called a *simple shear flow*, and the velocity gradient dv_x/dz is called the *shear rate*.

Let us consider the special case in which $\Delta p = 0$ and only drag flow is present. Then the force is given simply by

$$\frac{F}{WL} = \frac{\mu V}{H} \qquad (5.34)$$

This is the basis of a common method to measure the viscosity. Given the geometry of the apparatus, L and W are known. If we were to drive the upper plate at a prescribed velocity V we could create any desired shear rate V/H. We could measure the force required to sustain this motion, and rearrange Eq. (5.34) to compute the viscosity:

$$\mu = \frac{F}{WL} \frac{H}{V} \qquad (5.35)$$

It is not practical to build such an apparatus with flat parallel plates because it is difficult to seal the edges and, after a moderate amount of motion, the plates no longer lie on top of one another. However, same this type of flow can be created easily by confining the fluid between two concentric cylinders. If the gap H between the cylinders is much smaller than their radius, then the curvature will have little effect on the flow. The relative motion is created by rotating one of the cylinders, and instead of measuring the force one measures the torque required to rotate the cylinder. Such an apparatus is called a *Couette viscometer*, and the drag flow between concentric cylinders is called a *Couette flow*. The Couette viscometer is analyzed in detail in one of the exercises.

5.1.2 APPLICATION: MELT PUMPING IN A SINGLE SCREW EXTRUDER

There are many applications for this basic solution to flow between parallel plates. One important example is the modeling of single screw extruders for polymers. As shown in Fig. 5.3, a single screw extruder consists of a long, helical screw fitted into a cylindrical barrel. The barrel is heated from the outside, and the screw is rotated by an electric motor. Pellets of solid polymer enter the barrel from a hopper and are compacted, conveyed down the barrel, melted, and pumped out the end of the extruder. In the simplest operation, profile extrusion, a die is attached at the end of the barrel. The die provides a shape to the extrudate, which is cooled and solidified immediately after leaving the die. Profile extrusion is used to make rods, tubes, pipes, thick sheets, and channels. Extruders are the primary device for melting thermoplastic polymers, and they are used in many other processes, including wire coating, blown film extrusion, and film casting.

In this example we will develop a simple model of the melt pumping capacity of an extruder. We will assume that all of the polymer is melted, and that the melt can be modeled as a Newtonian fluid of constant viscosity. These assumptions are not especially realistic, but they will give us a simple model that shows some important qualitative behaviors. We will also assume that the process is operating at steady state, that the density of the fluid is constant, there is no slip at the walls, and that the inertial and body force terms in the momentum balance equation can be neglected, because of the high viscosity of the polymer melt. (We could follow the procedure of the previous section to provide a rigorous basis for these assumptions, which are quite realistic.) The screw has total axial length of L_{screw}, and the entrance and exit pressures p_{in} and p_{out} are known. The main goal is to find the volume flow rate Q of fluid pumped by the screw.

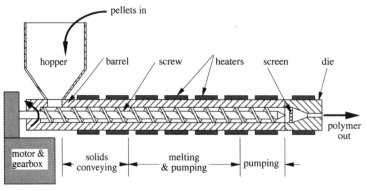

Figure 5.3: A single screw extruder for polymers.

5.1 NEWTONIAN FLOW IN A THIN CHANNEL

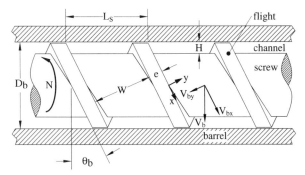

Figure 5.4: Actual geometry and nomenclature for a single screw extruder; N is the rotation rate of the screw relative to the barrel, and V_b is the velocity of the barrel surface relative to the screw.

The geometry of the screw and the nomenclature are shown in Fig. 5.4. The raised portion of the screw is called the *flight*, and the space in between the flights that is filled with polymer is called the *channel*. The barrel diameter is D_b, and this is essentially the same as the diameter of the screw at the top of the flights. The flight width is e, the channel width is W (both measured perpendicular to the flight), and the channel depth is H. The axial repeat length of the flight, called the *lead* of the screw, is L_s. Many screws have square pitch, which means $L_s = D$. The screw is rotated at a speed N. It is customary to measure N in revolutions per unit time (e.g., RPM), so the angular velocity of the screw is $2\pi N$.

At the barrel surface the helix angle θ_b of the screw is given by

$$\tan \theta_b = \frac{L_s}{\pi D} \tag{5.36}$$

The helix angle actually varies with radius from the axis of the screw, but in this example we will analyze only shallow channels, $H \ll D$, so we take the helix angle to be θ_b at any point in the channel. A little work with the geometry shows that

$$W = L_s \cos \theta_b - e \tag{5.37}$$

which reduces the number of independent parameters of the screw geometry.

In analyzing this problem we first change our point of view to a coordinate frame in which the screw is fixed and the barrel rotates. This coordinate frame is rotating with respect to the laboratory frame, and ordinarily we would have to account for this in the momentum balance equation by introducing the accelerations associated with the rotary motion of the coordinate system (the centripetal and Coriolis accelerations). However, we already know that all inertial terms are negligible, so the centripetal and Coriolis terms can also be neglected. In these screw-fixed coordinates, the barrel in Fig. 5.4 appears to move downward in front. The tangential velocity of the barrel surface (accounting for the definition of N) is

$$V_b = \Omega r = (2\pi N)\left(\frac{D_b}{2}\right) = \pi N D_b \tag{5.38}$$

Next we transform from the actual, helical geometry of Fig. 5.4 to an approximate geometry, in which the top, bottom, and side surfaces of the channel are planar.

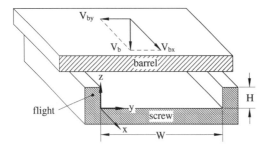

Figure 5.5: Approximate rectangular geometry for the shallow-channel model of a single screw extruder. The x and y axes and the dimensions H and W correspond to the actual geometry shown in Fig. 5.4.

It may be helpful to imagine removing the screw from the barrel, packing the channel with clay, and then unwrapping the long strip of clay from the barrel and laying the clay strip flat. The result is a long, rectangular strip with thickness H and width W. This new, approximate geometry is shown in Fig. 5.5. Some distortion is required to unwrap the channel and lay it flat, as the upper and lower surfaces of the channel have different lengths in the helical geometry, but this distortion is small if $H \ll D$, as we have assumed.

In this new, approximate geometry we adopt Cartesian coordinates, with x as the down-channel direction, y as the cross-channel direction, and z as the direction of the channel height. These axes are shown on both the actual geometry, Fig. 5.4, and in the approximate geometry, Fig. 5.5. The total length of this unwrapped channel, measured in the x direction, is $L_{\text{screw}}/\sin\theta_b$.

In this new rectangular geometry the relative motion of the barrel has both down-channel and cross-channel components, whose magnitudes are

$$V_{bx} = V_b \cos\theta_b = \pi N D_b \cos\theta_b \tag{5.39}$$
$$V_{by} = V_b \sin\theta_b = \pi N D_b \sin\theta_b \tag{5.40}$$

These are also shown in Figs. 5.4 and 5.5. Note that the barrel moves across the channel in the negative y direction, but we have defined V_{by} so that its value will be positive.

With all of these geometric definitions and transformations in hand, we can now turn to the fluid mechanics problem. If we also have $H \ll W$ (consistent with the assumption that the channel is shallow), then the fluid mechanics problem in the x direction is exactly the combined pressure and drag flow problem we just solved. We can use Eq. (5.30), and simply replace V by V_{bx}, L by $L_{\text{screw}}/\sin\theta_b$, and Δp by $(p_{\text{in}} - p_{\text{out}})$. For an extruder, or any other pump, we normally expect p_{out} to be greater than p_{in}, so we find the flow rate in the extruder to be

$$Q = \frac{V_{bx}WH}{2} - \frac{WH^3(p_{\text{out}} - p_{\text{in}})}{12\mu L} \tag{5.41}$$

or

$$Q = \frac{\pi N D_b W H \cos\theta_b}{2} - \frac{WH^3 \sin\theta_b(p_{\text{out}} - p_{\text{in}})}{12\mu L_{\text{screw}}} \tag{5.42}$$

The first term in Eq. (5.42) is the drag flow, which is controlled by the screw speed N. The second term is the pressure flow, which is controlled by the exit pressure p_{out}.

Figure 5.6 shows how the flow rate varies with pressure difference across between the inlet and outlet of the screw, for a fixed screw speed N. The maximum flow rate

5.2 OTHER SLOW NEWTONIAN FLOWS

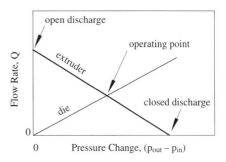

Figure 5.6: Flow rate vs. pressure change for a single screw extruder, for the isothermal Newtonian model of Eq. (5.42). The relationship for the die attached to the end of the extruder is also shown.

occurs for an open discharge, where the outlet of the extruder is simply open to the atmosphere and $p_{out} = p_{in}$. For a pump to be useful it must raise the pressure of the fluid as well as have a positive flow rate Q, so useful pumping always involves $p_{out} > p_{in}$. At the other limit, a single screw extruder can actually be operated with the outlet sealed, so that $Q = 0$. This gives the maximum possible pressure rise, but of course nothing is being pumped.

The other aspects of our channel flow solution also apply here. For example, the down-channel velocity profile $v_x(z)$ will have a shape like those shown in Fig. 5.2. The shape with negative G^* is the most likely, as the pressure gradient in an extruder opposes the flow.

If a die is attached to the end of the extruder, it resists flow and provides a pressure drop as the polymer passes through it. The flow rate through the die equals the flow exiting the extruder, and the pressure rise across the extruder equals the pressure drop across the die. For isothermal Newtonian fluids, the die has a pressure drop–flow rate curve like that shown in Fig. 5.6. The actual operating point of the extruder–die combination is the intersection of the extruder and die curves, where the flow rates and pressure changes are both equal.

The cross-channel (y-direction) flow can also be analyzed the same way, noting that $\partial p/\partial y$ is unknown, but the flow rate in the y direction is zero. This leads to a single velocity profile $v_y(z)$, regardless of the flow rate passing through the extruder. This cross-channel part of the flow is important both to melting the solid polymer and to mixing in the liquid polymer.

This analysis of a single screw extruder is quite simplified and can be expected to give only rough estimates of the flow rate or pressure drop. The main uses of this simple analysis are to provide some insight into the way that extruders work and to give some guidance in choosing the geometric parameters of the screw. To create a model that gives useful quantitative predictions, one must include the following: thermal effects, especially the melting of the solid granules; frictional conveying of the granules before they melt; and non-Newtonian behavior of the molten polymer. We will treat the thermal phenomena in later parts of this book. For more details on extruder modeling, see Tadmor and Gogos (1979) and Rauwendaal (1986).

5.2 OTHER SLOW NEWTONIAN FLOWS

Analytical solutions for Newtonian fluid flow at low Reynolds number have been found for a variety of other geometries. Deriving these results provides good practice in solving Newtonian flow problems. Here we provide only the solutions, leaving the derivations as exercises. Some of these problems also have analytical solutions for a power-law fluid, and these are given in Section 6.3.

5.2.1 PRESSURE FLOW BETWEEN PARALLEL PLATES

When the channel flow analyzed in the previous section is driven exclusively by a pressure gradient, with no drag flow component, then the solution is conveniently developed in a coordinate system whose origin lies halfway between the plates. This takes advantage of the symmetry of the flow with respect to the midplane. Let h be the distance from the midplane to one plate ($2h = H$ from the previous section), L be the length of the plates, W the width of the channel, and Δp the pressure drop. The velocity distribution is

$$v_x = \frac{\Delta p}{2\mu L}(h^2 - z^2) \tag{5.43}$$

This is a parabola that is symmetric about $z = 0$. The shear stress varies linearly across the gap,

$$\tau_{zx} = -\frac{\Delta p}{L}z \tag{5.44}$$

and the volume flow rate is

$$Q = \frac{2Wh^3 \Delta p}{3\mu L} \tag{5.45}$$

This is the same as the pressure-flow component of Eq. (5.30) when we account for the fact that $H = 2h$. This solution applies when $H \ll L$ and $H \ll W$.

5.2.2 AXIAL PRESSURE FLOW IN A TUBE

Consider the fluid flow in a tube of radius R and length L, with $R \ll L$, under the action of a driving pressure difference $\Delta p = p_{in} - p_{out}$. We use cylindrical coordinates, and we find by scaling that p depends only on the axial coordinate z and that v_z depends only on the radial coordinate r. The pressure is linear in z, and the velocity profile is parabolic:

$$v_z = \frac{\Delta p}{4\mu L}(R^2 - r^2) \tag{5.46}$$

The shear stress varies linearly with radius,

$$\tau_{rz} = -\frac{\Delta p \, r}{2L} \tag{5.47}$$

and the volume flow rate Q is given by

$$Q = \frac{\pi R^4 \Delta p}{8\mu L} \tag{5.48}$$

This solution is often referred to as Poiseuille flow.

5.2.3 AXIAL PRESSURE FLOW BETWEEN CONCENTRIC CYLINDERS

This is also a flow in a tube, but now a second cylinder with outer radius κR is inserted into the original tube, whose inner radius is R; κ is a dimensionless constant, and $0 < \kappa < 1$. The fluid is confined between the two cylinders. If both the inner and outer surfaces are stationary and the flow is driven by an axial pressure gradient $-\partial p/\partial z = \Delta p/L$, then the velocity profile, shear stress distribution, and flow rate are

$$v_z = \frac{\Delta p R^2}{4\mu L}\left[1 - \left(\frac{r^2}{R^2}\right) + \left(\frac{1-\kappa^2}{\ln(1/\kappa)}\right)\ln\left(\frac{r}{R}\right)\right] \tag{5.49}$$

$$\tau_{rz} = -\frac{\Delta p R}{2L}\left[\left(\frac{r}{R}\right) - \left(\frac{1-\kappa^2}{2\ln(1/\kappa)}\right)\left(\frac{R}{r}\right)\right] \tag{5.50}$$

$$Q = \frac{\pi R^4 \Delta p}{8\mu L}\left[(1-\kappa^4) - \frac{(1-\kappa^2)^2}{\ln(1/\kappa)}\right] \tag{5.51}$$

5.2.4 AXIAL DRAG FLOW BETWEEN CONCENTRIC CYLINDERS

As in the previous problem, the fluid is confined between two concentric cylinders of radii κR and R, respectively, with $0 < \kappa < 1$. In this case there is no axial pressure gradient, but the inner cylinder translates in the z direction with velocity V to create a drag flow. An example application in which this flow appears is wire coating. The solution is

$$v_z = V\frac{\ln(r/R)}{\ln \kappa} \tag{5.52}$$

$$\tau_{rz} = \frac{\mu V}{r \ln \kappa} \tag{5.53}$$

$$Q = \frac{\pi R^2 V}{2 \ln \kappa}[2\kappa^2 \ln(1/\kappa) - 1 + \kappa^2] \tag{5.54}$$

5.2.5 TANGENTIAL DRAG FLOW BETWEEN CONCENTRIC CYLINDERS

This flow has the same geometry as the two previous problems, but now all the motion is in the tangential (θ) direction. The inner cylinder rotates with an angular velocity Ω, and the outer cylinder is stationary. We assume that there is motion only in the θ direction, although in a real situation there will be end effects. The velocity and stress fields are

$$v_\theta = \Omega r \frac{1 - (R^2/r^2)}{1 - (1/\kappa^2)} \tag{5.55}$$

$$\tau_{r\theta} = \frac{\mu \Omega R^2}{r^2[1 - (1/\kappa^2)]} \tag{5.56}$$

This is the Couette flow geometry discussed on p. 140. When the gap between the cylinders becomes small compared to their radius (κ is close to one) then the velocity field is nearly linear across the gap.

5.2.6 RADIAL PRESSURE FLOW BETWEEN PARALLEL DISKS

This axisymmetric flow occurs in a thin gap of height $2h$ between two flat, parallel disks. Fluid is injected at the center of the disks at a volume flow rate Q and flows radially outward, exiting at the outer radius R_0. This flow is a prototype for injection molding. The solution requires that the gap be narrow, $h \ll R_0$, and is valid only when $r \gg h$. We choose the origin such that $z = 0$ lies at the midplane between the upper and lower surfaces. All of the flow is in the radial direction, but v_r depends on both r and z:

$$v_r = \frac{1}{r} \frac{3Q}{8\pi h} \left(1 - \frac{z^2}{h^2}\right) \tag{5.57}$$

If we take $p = p_0$ at $r = R_0$, the pressure distribution is

$$p - p_0 = \frac{3\mu Q}{4\pi h^3} \ln\left(\frac{R_0}{r}\right) \tag{5.58}$$

Note that in this solution the pressure becomes infinite as $r \to 0$. However, the underlying approximations are not valid near $r = 0$, so this does not present any logical inconsistency. When applying this solution one must choose a finite inner radius R_i, preferably with $R_i \gg h$, at which the inlet pressure is p_i. Then p_i will be finite, and the overall flow rate is related to the pressure drop $\Delta p = p_i - p_0$ by

$$Q = \frac{4\pi h^3 \Delta p}{3\mu} \frac{1}{\ln(R_0/R_i)} \tag{5.59}$$

5.2.7 TORSIONAL DRAG FLOW BETWEEN PARALLEL DISKS

A very different flow between parallel disks is created by holding the lower disk stationary and rotating the upper disk with angular velocity Ω. This geometry is used in some instruments that measure viscosity. Here it is convenient to call the total gap between the disks H and to choose $z = 0$ on the lower disk. The azimuthal velocity is linear in z, and it is proportional to the local tangential velocity of the upper disk:

$$v_\theta = \frac{\Omega r z}{H} \tag{5.60}$$

In fact the flow looks locally like parallel plate drag flow. The shear stress is constant in z but increases with radius,

$$\tau_{z\theta} = \frac{\mu \Omega r}{H} \tag{5.61}$$

and the shear rate is $\Omega r/H$. This flow geometry is mainly useful for measuring the viscosities of Newtonian fluids and other materials with a linear response, because the shear rate is not constant within the fluid.

5.2.8 MODELING OTHER FLOW GEOMETRIES

The preceding solutions can also be used as building blocks to create approximate models for flows in more complex geometries. As a simple example, suppose that we

5.2 OTHER SLOW NEWTONIAN FLOWS

want to relate pressure drop and flow rate for a tube that has radius R_1 over length L_1, and then reduces to radius R_2 for length L_2. The total flow rate Q is the same for both tubes, whereas the total pressure drop Δp is the sum of the pressure drops for the two tubes. Applying Eq. (5.48) once for each tube, the total pressure drop is predicted to be

$$\Delta p = \frac{8\mu Q L_1}{\pi R_1^4} + \frac{8\mu Q L_2}{\pi R_2^4} \qquad (5.62)$$

Note that the tube radius is raised to the fourth power, so if R_1 is more than three times larger than R_2, then the contribution of the larger-diameter tube to Δp is probably negligible.

Equation (5.62) is *not* an exact solution to the problem, because it omits the details of the flow at the junction of the tubes, where the radius changes. This full problem can be solved numerically, or one can make experimental measurements of Δp versus Q and compare them to Eq. (5.62). The results by either method show that the real solution has a somewhat higher pressure drop than would be calculated by Eq. (5.62):

$$\Delta p_{\text{actual}} = \frac{8\mu Q L_1}{\pi R_1^4} + \frac{8\mu Q L_2}{\pi R_2^4} + \Delta p_e \qquad (5.63)$$

This "extra" pressure drop Δp_e is called the *entrance loss*. Of course the entrance loss is not extra at all. Rather, it is the part of the pressure drop not accounted for by our simple model.

If the Reynolds number is small, then the entrance pressure loss will scale like the pressure drop for flow in a tube, except that there is no tube length associated with the entrance effect. That is, we expect

$$\Delta p_e = \frac{8\mu Q \ell_e}{\pi R_2^4} \qquad (5.64)$$

Here ℓ_e is called the entrance length, a quantity that depends on the ratio (R_1/R_2) and on other details of the entrance geometry, such as the angle of the entrance. One can think of ℓ_e as the amount by which we must increase the length of the smaller tube in Eq. (5.62) to correct for the entrance effect. For Newtonian fluids the entrance length is of the order of R_2, so the entrance correction is a small part of the total pressure drop if the smaller tube is long compared to its radius.

5.2.9 APPLICATION: FLOW IN AN EXTRUSION DIE

To illustrate the use of the results from this section, we now show an example calculation for the flow in an extrusion die. This particular die is used with a laboratory-sized extruder to produce small strips. The geometry of the flow passage within the die is shown in Fig. 5.7. The die inlet is fed by an extruder, and the polymer exits the die into the atmosphere. Here we take the polymer melt to be a Newtonian fluid with a viscosity of 600 N s/m², and calculate the flow rate through the die when the inlet pressure is 35 MPa.

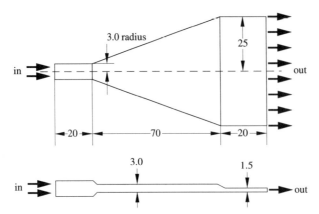

Figure 5.7: Geometry of the flow passage in a small strip extrusion die. All dimensions are in millimeters.

To model this die we approximate the geometry as a series of segments drawn from the repertoire of the preceding section. The first 20 mm of the die can be modeled as a tube, and the last 20 mm as a strip. We will call the pressure drops across these two segments Δp_1 and Δp_3, respectively. Let us handle these two segments and then return to the intermediate segment.

For the first segment we use Eq. (5.48) for pressure-driven flow in a tube, and we write the pressure drop as

$$\Delta p_1 = \left(\frac{8\mu L}{\pi R^4}\right) Q = K_1 Q \tag{5.65}$$

Here K_1 is simply a convenient name for the constant of proportionality between flow rate and pressure drop. Note that Q is the flow rate for the entire die, because all of the flow must pass through each segment. Using $\mu = 600$ N s/m^2, $L = 0.020$ m, and $R = 0.003$ m, we find $K_1 = 3.77 \times 10^{11}$ Pa/(m^3/s).

For the third segment we use Eq. (5.30) for flow between parallel plates, with $V = 0$ because both the upper and lower walls are stationary. In this segment the pressure drop is

$$\Delta p_3 = \left(\frac{12\mu L}{W H^3}\right) Q = K_3 Q \tag{5.66}$$

With $L = 0.020$ m, $W = 0.050$ m, and $H = 0.0015$ m, we find $K_3 = 8.53 \times 10^{11}$ Pa/(m^3/s).

Now we return to the transition segment that connects these two. This is a thin channel, but its width increases gradually in the flow direction. We can model this by using the solution for radial flow between parallel disks of Section 5.2.6. We model our transition segment as a sector of the disk, with included angle θ, inner radius R_i, outer radius R_o, and total thickness $2h$. Equation (5.59) applies for an entire disk, for which the included angle θ equals 2π and the flow rate is Q_{whole}. We rewrite Eq. (5.59) as

$$\Delta p_2 = \left(\frac{3\mu \ln(R_o/R_i)}{2h^3}\right)\left[\frac{Q_{\text{whole}}}{2\pi}\right] \tag{5.67}$$

Note that $[Q_{\text{whole}}/2\pi]$ is the flow rate per total included angle, so we can replace it by $[Q/\theta]$ when modeling a segment of the disk. We then have

$$\Delta p_2 = \left(\frac{3\mu \ln(R_o/R_i)}{2h^3\theta}\right) Q = K_2 Q \tag{5.68}$$

Some judgment (or guesswork) is required to match R_o and R_i to the real geometry, and a reasonable set of values is $R_i = 0.010$ m and $R_o = 0.0815$ m. It is easy to find $h = 0.0015$ m (note: h is the half-height of the gap) and $\theta = 0.609$ rad. With the use of Eq. (5.68), this gives $K_2 = 9.16 \times 10^{11} \text{Pa}/(\text{m}^3/\text{s})$.

Now we combine these results to model the overall problem. The total pressure drop is the sum of the drops from each segment,

$$\begin{aligned}\Delta p &= \Delta p_1 + \Delta p_2 + \Delta p_3 \\ &= K_1 Q + K_2 Q + K_3 Q\end{aligned} \tag{5.69}$$

so that the overall flow rate is

$$Q = \frac{\Delta p}{K_1 + K_2 + K_3} \tag{5.70}$$

Using $\Delta p = 35 \times 10^6$ Pa and the Ks we have found, we note that the overall flow rate is $Q = 1.63 \times 10^{-5}$ m^3/s. With this result in hand one may readily compute the exit velocity of the strip, the shear rates at various points inside the die, and other detailed results.

5.3 FREE SURFACES AND MOVING BOUNDARIES

So far we have dealt with flows that occur within well defined, solid boundaries. However, a great many materials processing problems involve flows in which the position of some part of the boundary is not known a priori. Figure 5.8 shows three examples. In polymer extrusion the polymer swells and stretches as it exits the die, and the cross section of the extrudate will be different from the cross section of the die. In mold filling, a liquid polymer or metal moves progressively across the mold, its surface taking on many different shapes. In float zone crystal growth, surface tension holds a band of liquid metal in place between two solid pieces, a feed rod and the growing crystal.

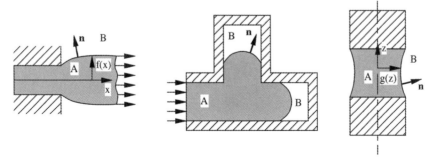

Figure 5.8: Examples of problems with free and moving boundaries. From left to right: polymer exiting an extrusion die, mold filling, and float zone crystal growth.

Problems in which the surface position is unknown, but does not vary with time, are called *free surface* or free boundary problems. The extrusion and crystal growth processes in Fig. 5.8 are examples. When the surface position changes with time, as in the mold-filling process in Fig. 5.8, we have a *moving boundary problem*. In both cases the position and shape of the surface must be determined as part of the solution. Despite the difference in terminology, there is very little difference in how free and moving boundary problems are formulated, so we treat them together.

Because free and moving boundary problems can involve a good deal of geometric complexity, few problems have analytical solutions. In this section we will first provide a general formulation of the problem and examine the role of surface tension through scaling. We then show a simple analysis of mold filling, treating both the global behavior and the local behavior near the flow front.

5.3.1 GENERAL FORMULATION

We will call a free or moving boundary problem *one sided* if we are solving for flow on only one side of the surface, and *two sided* if we solve the flow on both sides. All of the examples in Fig. 5.8 are one-sided problems; fluid B is air, which we assume to be readily displaced, and we care only about the motion of fluid A and the position of the interface. An example of a two-sided problem is polymer coextrusion, in which two polymers are extruded simultaneously through the same die. There we might wish to predict where the interface between the polymers will be in the final cross section. Another example is the mixing of polymer blends, in which the objective is to form fine, micrometer-sized droplets of one polymer dispersed in another. The two-sided problem is more general, so we treat it first and then recover the one-sided formulation as a special case.

A first step in formulating any free or moving boundary problem is to choose a mathematical description for the location and shape of the free surface. There is no universal way to do this, and one usually seeks the simplest mathematical form that captures the surface geometry. In some two-dimensional problems the surface can be described by giving its y coordinate as a function of x, $y = f(x)$, and most axisymmetric surfaces can be described by giving the radius as a function of the axial coordinate, $r = g(z)$; see Fig. 5.8. This approach works well when the surface does not fold back on itself in the x or z direction. A variety of other methods are used to represent free and moving boundaries, with most schemes being oriented toward numerical solutions (e.g., Hirt and Nichols, 1981; Kistler and Scriven, 1983; Wang and Lee, 1989; Unverdi and Tryggvason, 1992; Pichelin and Coupez, 1998).

Once the surface shape is specified, we can compute the direction **n** that is normal to the surface at each point and calculate the surface curvature. (See Section A.6 for a review of curvature). We adopt the convention that the normal vector **n** points from fluid A into fluid B. This makes the surface curvature positive when the surface is concave when viewed from the B side.

Two-Sided Problems

The next step is to write the governing equations. A two-sided problem requires a continuity equation and a momentum equation for each fluid. To match these

5.3 FREE SURFACES AND MOVING BOUNDARIES

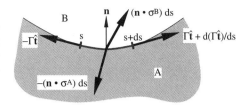

Figure 5.9: A small increment ds of the interface between fluids A and B; **n** is the normal direction and $\hat{\mathbf{t}}$ is the tangential direction. Large arrows indicate the forces acting on this part of the interface. Here curvature κ is positive.

two solutions we need two boundary conditions at the interface between the fluids. The first condition is kinematic: velocity must be continuous across the interface, so at each point on the interface we require

$$\mathbf{v}_A = \mathbf{v}_B \tag{5.71}$$

The second boundary condition is dynamic, and it corresponds to a balance of forces acting on the interface. These forces can come from the stress in fluid A, the stress in fluid B, and from forces transmitted along the surface. Forces are transmitted along the surface as a result of *interfacial tension* or *surface tension*, Γ. From a thermodynamic viewpoint the surface tension Γ is the extra energy per unit area (e.g., joules per square meter) associated with the interface. This arises because atoms of material A or B close to the interface do not have the same pattern of nearest neighbors as atoms in the interior. An equivalent mechanical viewpoint, which we shall use, is that the surface carries a tangential force, with Γ being the force per unit length (e.g., Newtons per meter). The same tensile force acts in all directions in the surface; hence the name *surface tension*.

Returning to the boundary condition, consider a two-dimensional problem as sketched in Fig. 5.9. Now the interface is represented by a plane curve. We will measure position along the surface by using the arc length s, and we consider the forces acting on a small increment ds of the surface. For convenience we consider a unit depth in the direction perpendicular to the page. The surface increment ds has no thickness, so it has no mass. Accordingly, the total force acting on this area must be zero.

The stress in fluid B provides a force of $\mathbf{n} \cdot \boldsymbol{\sigma}^B \, ds$, where we have used Eq. (2.49) to relate total stress $\boldsymbol{\sigma}$ to the force per unit area acting on a surface, and with a unit depth the area is ds. The contribution from fluid A is $-\mathbf{n} \cdot \boldsymbol{\sigma}^A \, ds$, because the unit vector normal to the surface and pointing into fluid A is $-\mathbf{n}$. To describe the surface tension contributions we use a unit vector $\hat{\mathbf{t}}$ that is tangent to the interface, and points to the right in this figure. The contribution of interfacial tension on the left is $-\Gamma\hat{\mathbf{t}}|_s$, the negative sign occurring because the force acts opposite to the direction of $\hat{\mathbf{t}}$. On the right the force is $\Gamma\hat{\mathbf{t}}|_s$, plus the change in this quantity with distance ds along the surface. Adding these contributions together gives a force balance of

$$-\mathbf{n} \cdot \boldsymbol{\sigma}^A \, ds + \mathbf{n} \cdot \boldsymbol{\sigma}^B \, ds - \Gamma\hat{\mathbf{t}}|_s + \left(\Gamma\hat{\mathbf{t}}|_s + \frac{d(\Gamma\hat{\mathbf{t}})}{ds} ds\right) = 0 \tag{5.72}$$

As usual, we cancel a few terms and then divide through by ds. The derivative $d(\Gamma\hat{\mathbf{t}})/ds$ expands as

$$\frac{d(\Gamma\hat{\mathbf{t}})}{ds} = \Gamma\frac{d\hat{\mathbf{t}}}{ds} + \frac{d\Gamma}{ds}\hat{\mathbf{t}} = \Gamma\kappa\mathbf{n} + \frac{d\Gamma}{ds}\hat{\mathbf{t}} \tag{5.73}$$

In the second of these expressions we have used Eqs. A.64 and A.65 to introduce κ, the curvature of the interface. The force balance on the interface is now in a form that can be used as a boundary condition, at least for two-dimensional problems:

$$\mathbf{n} \cdot \boldsymbol{\sigma}^A - \mathbf{n} \cdot \boldsymbol{\sigma}^B = \Gamma \kappa \mathbf{n} + \frac{d\Gamma}{ds}\hat{\mathbf{t}} \qquad (5.74)$$

The left-hand side of this expression represents the force per unit area exerted by the fluids on either side of the interface. In the absence of surface tension ($\Gamma = 0$), these forces must balance. Note, however, that this does not mean that the two *stresses* must be equal. For instance, either fluid may carry tension parallel to the interface, and the boundary condition imposes no restrictions on those values.

The terms on the right-hand side are the contributions from interfacial tension. The first term consists of a scalar multiplying the normal vector $\mathbf{n}$. Thus, if the surface is curved ($\kappa \neq 0$) then interfacial tension provides a net force normal to the interface. This is the most familiar effect of interfacial tension. The second term multiplies the tangent vector $\hat{\mathbf{t}}$ and includes the derivative of Γ with respect to position along the interface. In many problems we can assume that Γ is a constant, so this term will be zero. However, Γ usually depends on temperature, so if the temperature varies along the interface then $d\Gamma/ds$ will be nonzero, this term will be active, and it must be balanced by a shear stress acting on the interface. Flows that arise from this effect are called *Marangoni flows*, and they can be important in welding and crystal growth processes.

If we extend Eq. (5.74) to three-dimensional problems, we get the general dynamic boundary condition at a fluid–fluid interface:

$$\mathbf{n} \cdot \boldsymbol{\sigma}^A - \mathbf{n} \cdot \boldsymbol{\sigma}^B = 2\kappa_m \Gamma \mathbf{n} + \nabla_{\text{surf}}(\Gamma) \qquad (5.75)$$

Here κ_m is the mean curvature of the surface. Note that

$$2\kappa_m = \frac{1}{R_1} + \frac{1}{R_2} \qquad (5.76)$$

where R_1 and R_2 are the principal radii of curvature of the surface (see Section A.6). The symbol ∇_{surf} is the gradient operator in the surface, so $\nabla_{\text{surf}}(\Gamma)$ is a vector that is tangent to the interface and points in the direction in which Γ increases most rapidly.

If the position of the interface were known, we would now have enough equations and boundary conditions to solve the problem. However, we do not know where the interface is, so we need one more equation. For a *free surface problem* the surface itself must be stationary, but this does not restrict the fluid from moving tangentially to the interface. Thus, we restrict only the normal velocity at the interface and require

$$\mathbf{v} \cdot \mathbf{n} = 0 \qquad (5.77)$$

on the interface. For a *moving boundary problem* the normal velocity will be the velocity of the interface itself, so we use

$$\mathbf{v}_{\text{interface}} = (\mathbf{v} \cdot \mathbf{n})\mathbf{n} \qquad (5.78)$$

to find the interface velocity; this governs its motion in time.

5.3 FREE SURFACES AND MOVING BOUNDARIES

To summarize, the governing equations for a free surface problem include the continuity and momentum balance equations for fluids A and B, whereas the boundary conditions are Eqs. (5.71), (5.75), and (5.77) on the interface, plus conditions on the external boundaries. Note that the position and shape of the free surface, whether $y = f(x)$, $r = g(z)$, or any other description, is *unknown*, and must be found as part of the solution. A moving boundary problem has exactly the same governing equations, except that the interface position is a function of time and we use Eq. (5.78) in place of Eq. (5.77). For a moving boundary problem the initial position of the interface must be given as one of the initial conditions, and the solution marches forward in time from that point.

One-Sided Problems

For a one-sided problem we discard the continuity and momentum equations in fluid B. One less boundary condition is needed at the interface and, because we are no longer solving for $\mathbf{v}_B$, we discard Eq. (5.71). Usually a problem is one sided because fluid B has negligible viscosity. In this case the only stress in B is due to its pressure, $\sigma^B = -p_B \mathbf{I}$. If we also separate σ^A into a pressure and an extra stress, then the dynamic boundary condition from Eq. (5.75) becomes

$$(p_A - p_B)\mathbf{n} - \mathbf{n} \cdot \tau^A = -2\kappa_m \Gamma \mathbf{n} - \nabla_{\text{surf}}(\Gamma) \tag{5.79}$$

The governing equations for a one-sided problem are the continuity and momentum equations in fluid A, and the free-surface boundary conditions are Eq. (5.79) and either Eq. (5.77) or Eq. (5.78), depending on whether the problem is steady or transient.

To gain more insight into the boundary condition, Eq. (5.79), consider a two-dimensional problem as illustrated in Fig. 5.9 and assume that Γ is constant, so the Marangoni term is zero. Taking the dot product of Eq. (5.79) with $\mathbf{n}$, and noting that $\mathbf{n} \cdot \mathbf{n} = 1$, gives the normal component of the interfacial force balance:

$$p_A - p_B - \tau^A : (\mathbf{nn}) = -2\kappa_m \Gamma \tag{5.80}$$

Recall that $\mathbf{n} \cdot \tau^A$ is the traction vector (force per unit area) that is due to extra stress in fluid A, so $\tau^A : \mathbf{nn} = \tau_{ij}^A n_i n_j$ is the component of this vector normal to the surface. The negative sign on the right-hand side comes from our choice of the direction of $\mathbf{n}$ and our sign convention for curvature.

There is also a tangential component to the interfacial force balance, which is found by taking the dot product of Eq. (5.79) with the tangent vector $\hat{\mathbf{t}}$. Noting that $\hat{\mathbf{t}} \cdot \mathbf{n} = 0$, we find that

$$\tau^A : (\mathbf{n}\hat{\mathbf{t}}) = 0 \tag{5.81}$$

The left-hand side here is the tangential component of the extra stress acting on the interface. In other words, the shear stress on the interface must be zero. Note that a two-dimensional problem has only one tangent direction, whereas in a three-dimensional problem Eq. (5.81) holds for *any* choice of $\hat{\mathbf{t}}$.

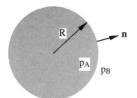

Figure 5.10: A spherical bubble of fluid A in fluid B.

As a simple application, consider a spherical bubble of fluid A immersed in fluid B (Fig. 5.10), where both fluids are at rest and body forces are negligible. The radius of the bubble is R, and in the absence of fluid motion the only stresses come from fluid pressure. Hence, $\tau^A = 0$ and Eq. (5.81) is automatically satisfied. We can then apply Eq. (5.80) to relate the pressures. Following the sign convention for curvature discussed in Section A.6, the principal curvatures are $R_1 = R_2 = -R$. (Curvature is negative when the surface is convex toward fluid B). Substituting this in we find

$$p_A - p_B = \frac{2\Gamma}{R} \tag{5.82}$$

This familiar result tells us that the pressure inside the bubble must be higher than the pressure outside to balance the surface tension.

5.3.2 SCALING OF SURFACE TENSION EFFECTS

We can use scaling to determine the relative importance of surface tension in a free or moving boundary problem. As an example, consider the filling of a long striplike mold, as shown in Fig. 5.11. This is the same geometry as the thin-channel flow at the beginning of this chapter, but we now start with an empty mold and fill it gradually, applying either a known pressure p_1 or a known flow rate Q at the inlet, $x = 0$.

We adopt all of the same assumptions as in Section 5.1. However, because the position of the fluid–air interface changes with time, we cannot assume that the flow is steady. We will add the assumption that the flow is dominated by viscous forces (Re $\ll$ 1), so that the scaling of Section 5.1 still applies. We also assume that the interfacial tension Γ is constant, and that the air can escape freely from the mold, so p_B is constant. We are interested only in the fluid that is filling the cavity, so this is a one-sided problem. We will omit the subscript A, and all variables (except p_B) will refer to fluid A.

We have already scaled the mass and momentum balance equations for this problem in Section 5.1, so our main task is to scale the boundary condition at the fluid–air interface, also called the *flow front*. We can adopt Eq. (5.79) and, because the problem

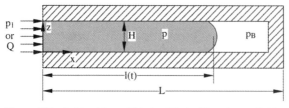

Figure 5.11: A thin strip mold being filled with a viscous fluid.

5.3 FREE SURFACES AND MOVING BOUNDARIES

is two dimensional, the second principal radius of curvature $R_2 \to \infty$. The dimensional form of the boundary condition is then

$$(p - p_B)\mathbf{n} - \mathbf{n} \cdot \boldsymbol{\tau} = -\frac{\Gamma}{R_1}\mathbf{n} \quad (5.83)$$

From previous experience with this problem we know that the largest velocity components will have a magnitude of V, the characteristic velocity in the x direction, whereas the largest velocity gradients will scale like V/H. This means that the magnitude of the largest extra stresses will $\mu V/H$, so we scale the stresses according to

$$\tau^* = \frac{\tau}{\mu V/H} \quad (5.84)$$

We will adopt p_B as a reference pressure and use the characteristic pressure drop Δp_c from Eq. (5.10), so the scaled pressure is

$$p^* = \frac{p - p_B}{\Delta p_c} = \frac{p - p_B}{\mu V L/H^2} \quad (5.85)$$

The radius of curvature R_1 also varies with position, but experimental observations show that the shape of this interface is roughly a semicircle, so we scale the radius of curvature as

$$R_1^* = \frac{R_1}{H} \quad (5.86)$$

Note that $\mathbf{n}$ is a unit vector, so it is already scaled. Substituting the dimensionless variables into the boundary condition, Eq. (5.83) gives

$$\left[\frac{\mu V L}{H^2}\right] p^* \mathbf{n} - \left[\frac{\mu V}{H}\right] \mathbf{n} \cdot \boldsymbol{\tau}^* = -\left[\frac{\Gamma}{H}\right] \frac{1}{R_1^*}\mathbf{n} \quad (5.87)$$

One possibility is to multiply through by $H/\mu V$, in which case we get

$$\left\{\frac{L}{H}\right\} p^* \mathbf{n} - \mathbf{n} \cdot \boldsymbol{\tau}^* = -\left\{\frac{\Gamma}{\mu V}\right\} \frac{1}{R_1^*}\mathbf{n} \quad (5.88)$$

The dimensionless group on the right-hand side is the inverse of a classical dimensionless group, the *capillary number*:

$$\mathrm{Ca} = \frac{\mu V}{\Gamma} \quad (5.89)$$

The capillary number can be regarded as the ratio between viscous forces and surface tension forces. Note the sense of the definition: a *small* Ca indicates *large* surface tension effects, whereas a *large* Ca indicates *small* surface tension effects.

In the present problem {1/Ca} is not the relevant dimensionless group, however, because the left-hand side of Eq. (5.88) has a term $\{L/H\}$ that is much greater than one. The correct scaling for this problem is found by multiplying Eq. (5.88) by H/L, to get

$$p^* \mathbf{n} - \left\{\frac{H}{L}\right\} \mathbf{n} \cdot \boldsymbol{\tau}^* = -\left\{\frac{\Gamma H}{\mu V L}\right\} \frac{1}{R_1^*}\mathbf{n} \quad (5.90)$$

The dimensionless group that describes the importance of surface tension for this problem is $\{\Gamma H/\mu V L\} = (\text{Ca}\, L/H)^{-1}$, and the extra stress term carries a multiplier of $\{H/L\}$. Regardless of the scaling, the tangential component of the boundary condition is

$$\tau : (\mathbf{n}\hat{\mathbf{t}}) = 0 \tag{5.91}$$

That is, the shear stress on the surface is zero. The main issue is the normal component of the boundary condition, which contains the pressure.

For polymer injection molding, we can use the typical material properties given in Eq. (5.8), adding the data that most polymers have an interfacial tension between 30 and 70 mN/m.[1] We then find that $\{\Gamma H/\mu V L\} = 5 \times 10^{-7}$, so we can ignore interfacial tension and the scaling of Eq. (5.90) is correct. If, however, the fluid were a light lubricating oil with $\mu = 10^{-2}$ N s/m^2, $\Gamma = 100$ nM/m, and the remaining parameters unchanged, we would have $\{\Gamma H/\mu V L\} = 1$, and surface tension would play a significant role. As a third possibility, for die casting with a liquid metal we expect to have Re $\gg 1$, so the pressure scaling of Eq. (5.85) are no longer correct, and a different dimensionless group will appear in the scaled boundary condition. This is explored in one of the exercises.

Returning to the parameter values for polymer molding, with $\{\Gamma H/\mu V L\} \ll 1$ and $\{H/L\} \ll 1$, the normal component of the scaled boundary condition reduces to

$$p^* = 0 \tag{5.92}$$

or, in dimensional form,

$$p = p_B \tag{5.93}$$

at the flow front. The scaling analysis has shown us that, when solving the global problem for polymer melts, we can ignore the difference between the fluid pressure and the gas pressure at the flow front.

This is a *global* scaling result. Suppose that we wanted to solve the *local* problem near the flow front, say to calculate the exact shape of the free surface, and we kept the same material parameters. Because Ca $\gg 1$ we can ignore surface tension, almost regardless of the value of L/H. However, Eq. (5.90) seems to suggest that the importance of the extra stress τ near the front depends on the length of the cavity L, which does not make sense; physically there is no way for the fluid near the flow front to "know" how long the cavity is. The parameter $\{H/L\}$ multiplying τ appears in Eq. (5.90) because we used different characteristic values to scale pressure and extra stress. This is meaningful for the global problem, but not for the local behavior near the flow front. In fact, we have something like a boundary layer problem, where the global scaling tells us to ignore a term, but the term is important locally near a boundary. If we rescale locally, using the same characteristic value $(\mu V/H)$ to scale both pressure and extra stress, then the scaled boundary condition becomes

$$p^*\mathbf{n} - \mathbf{n} \cdot \boldsymbol{\tau}^* = -\left\{\frac{1}{\text{Ca}}\right\}\frac{1}{R_1^*}\mathbf{n} \tag{5.94}$$

---

[1] A useful conversion factor for surface tension is 1 mN/m $\equiv 10^{-3}$ N/m = 1 dyn/cm.

5.3 FREE SURFACES AND MOVING BOUNDARIES

Now we see that pressure and extra stress play equal roles in determining the behavior near the flow front (for low Reynolds number problems), and the relative importance of interfacial tension is determined by the capillary number Ca. It is perfectly correct to use Eq. (5.93) as the boundary condition for the global problem, but we need the local scaling of Eq. (5.94) to solve for the behavior near the flow front. We will explore these solutions next.

5.3.3 FILLING A SIMPLE MOLD: GLOBAL ANALYSIS

Finally we are ready to solve the global problem sketched in Fig. 5.11. We will assume that the parameters are typical of polymer injection molding, so that Re $\ll 1$ and Ca $\gg 1$. The assumptions of Section 5.1 will be retained, except that the flow is not steady. However, with Re $\ll 1$ the inertial terms are still negligible, so the scaling and the governing equations of Section 5.1 are still valid. From this point onward there is no particular advantage to working the problem in dimensionless form, and we return to the dimensional variables.

We still have $v_z \ll v_x$ from the global scaling, and the scaled x-direction momentum equation is still Eq. (5.11). All of the solid walls are stationary, so the velocity boundary conditions are

$$v_x = 0 \quad \text{at} \quad z = 0$$
$$v_x = 0 \quad \text{at} \quad z = H \tag{5.95}$$

Let us assume that the inlet boundary condition is a constant pressure. Without loss of generality we can take $p_B = 0$. The boundary condition at the flow front is applied at a distance ℓ from the inlet (see Fig. 5.11), where ℓ is a function of time. (We will be more precise in a moment about exactly how ℓ is measured.) Now the pressure boundary conditions are

$$p = p_1 \quad \text{at} \quad x = 0$$
$$p = 0 \quad \text{at} \quad x = \ell(t) \tag{5.96}$$

and we need to find $\ell(t)$ as part of the solution. We also need a second boundary condition in order to solve for $\ell(t)$, and this should derive from Eq. (5.78). However, we already have enough information to solve the momentum equation, so we will do that first.

Recall that p is independent of z and the pressure gradient $\partial p/\partial x$ is independent of x. The pressure gradient is then $\partial p/\partial x = p_1/\ell(t)$. The x-direction momentum balance equation is the dimensional form of Eq. (5.11):

$$\mu \frac{\partial^2 v_x}{\partial z^2} = \frac{\partial p}{\partial x} \tag{5.97}$$

We integrate this equation twice and apply boundary conditions of Eq. (5.95) to find the velocity solution:

$$v_x(z, t) = \frac{H^2 p_1}{2\mu \ell(t)} \left(\frac{z}{H} - \frac{z^2}{H^2} \right) \tag{5.98}$$

The velocity distribution at any instant is parabolic, with zero velocity at the upper and lower surfaces and a maximum at the midplane. This is readily integrated as in Eq. (5.30) to find the volume flow rate Q,

$$Q(t) = \frac{WH^3 p_1}{12\mu \ell(t)} \tag{5.99}$$

with W as the width of the plates in the y direction. With constant inlet pressure p_1 the flow rate Q is a function of time, because the location of the flow front changes with time. In the early stages of filling ℓ is small, the pressure gradient p_1/ℓ is large, and the flow rate is high. As the mold fills, the same pressure drop occurs over a longer distance ℓ, the gradient is smaller, and the flow rate decreases.

To find $\ell(t)$ we need to introduce another boundary condition, but Eq. (5.78) is not compatible with our velocity solution. It seems reasonable that the flow front should translate down the mold with a constant shape, and experiments bear this out. This means that v_x should be constant on the flow front, and there is no way to match this to the parabolic velocity profile in Eq. (5.98). The local behavior at the flow front must be different from the global solution we just developed. This is the "boundary layer" behavior hinted at by the scaling analysis. Fortunately we can complete the global solution here without knowing the details of the local solution. We will do this now, and return to the local behavior in a moment.

In place of Eq. (5.78) we note that the interface velocity must be the same as the gapwise average velocity $\bar{v}_x$ of for our global velocity solution. This can be developed by making a mass balance on a control volume that has height H and includes the interface in the x direction, and we find

$$\frac{d\ell}{dt} = \bar{v}_x \equiv \frac{1}{H}\int_0^H v_x(z)\,dz \tag{5.100}$$

or, in terms of the volume flow rate,

$$\frac{d\ell}{dt} = \frac{Q(t)}{WH} \tag{5.101}$$

If we do a formal time integration of this equation with $\ell = 0$ at $t = 0$, we find

$$WH\ell(t) = \int_0^t Q(t')\,dt' \tag{5.102}$$

The right-hand side of the equation is the total volume of fluid delivered to the cavity, so we now have a precise definition of ℓ; it is the position of a flat flow front that corresponds to the volume of fluid currently in the cavity.

To actually find $\ell(t)$ we use Eq. (5.99) to replace Q in Eq. (5.101), and we get

$$\frac{d\ell}{dt} = \frac{H^2 p_1}{12\mu \ell(t)} \tag{5.103}$$

This is an ordinary differential equation for $\ell(t)$, and all of the other parameters are known. We separate the variables,

$$\ell\,d\ell = \frac{H^2 p_1}{12\mu}\,dt \tag{5.104}$$

5.3 FREE SURFACES AND MOVING BOUNDARIES

and integrate, giving

$$\frac{\ell^2}{2} = \frac{H^2 p_1}{12\mu} t + c_1 \qquad (5.105)$$

The constant c_1 must be determined from an initial condition, and we set $t = 0$ when $\ell = 0$, corresponding to the time the fluid first enters the mold. The final solution for the filled length of the cavity is then

$$\ell(t) = \sqrt{\frac{H^2 p_1 t}{6\mu}} \qquad (5.106)$$

Thus, the filled length grows like the square root of time, because the driving pressure is constant but the viscous resistance to flow increases as the cavity fills. The time to fill the entire cavity (i.e., to make $\ell = L$) is

$$t_{\text{fill}} = \frac{6\mu L^2}{H^2 p_1} \qquad (5.107)$$

From this result we see that, when filling under constant pressure, the time to fill a long cavity is not proportional to the total flow length, but grows much more rapidly than L. The dependency is L^2 here. Actual injection-molding processes do not scale quite so simply, because of non-Newtonian effects and because of an interplay between heat transfer and temperature-dependent viscosity. Still, this simple analysis provides useful insight into the general nature of mold-filling flows.

This analysis can be extended readily to the case in which the flow rate is controlled and the injection pressure p_1 is an unknown function of time. This is actually more representative of injection-molding practice. If Q is constant then Eq. (5.101) tells us that $d\ell/dt$ is constant,

$$\frac{d\ell}{dt} = \frac{Q}{WH} \qquad (5.108)$$

and we integrate this to find

$$\ell(t) = \frac{Q}{WH} t = Vt \qquad (5.109)$$

where V is the average velocity. The filling time is simply

$$t_{\text{fill}} = \frac{WHL}{Q} \qquad (5.110)$$

Combining Eq. (5.109) with Eq. (5.99) gives the injection pressure in terms of the known parameters,

$$p_1(t) = \left(\frac{12\mu Q}{W^2 H^4}\right) t \qquad (5.111)$$

The pressure rises linearly in time, reaching a maximum value of

$$p_{1\,\text{max}} = \frac{12\mu L Q}{W H^3} \qquad (5.112)$$

at the instant of complete filling.

Similar global mold-filling solutions can be developed for other simple geometries, like axial flow in a tube using Eq. (5.48), and radial flow between disks using Eq. (5.59). This analysis is possible because, in these simple geometries, the time-dependent position of the flow front can be described by a single variable. More complex geometries can be modeled by approximating them as a series of simple shapes, as was done for the extrusion die in Section 5.2.9. However, if more than two or three simple geometrical units are needed to model the mold shape, then the algebra will be too cumbersome for useful hand solutions. One can write a computer program to handle the details, and indeed this type of analysis was the basis for the earliest injection mold-filling software. However, the preferred numerical treatment at this time is to use the generalized Hele–Shaw formulation, which applies to any type of thin cavity and which does not require the user to guess the shape of the flow front. We will present the Hele–Shaw formulation in Section 6.6, after we discuss non-Newtonian fluids.

5.3.4 LOCAL ANALYSIS NEAR THE FLOW FRONT: THE FOUNTAIN FLOW

Now let us return to the local behavior near the flow front. We need to match the global solution, with its parabolic velocity profile, to a local solution in which the x velocity on the flow front equals $\bar{v}_x$. For a parabolic velocity profile the average velocity $\bar{v}_x$ equals exactly two-thirds of the center line velocity, so fluid entering the mold near the center line travels faster than the flow front and catches up with it. As this fluid approaches the flow front, it must slow down until, upon reaching the flow front, its x velocity has decreased to $\bar{v}_x$. Thus, near the flow front there is a region where $\partial v_x/\partial x$ is negative. The continuity equation then tells us that in this region $\partial v_z/\partial z$ must be positive, so there must be z-direction motion to satisfy the mass balance. We will not rework the scaling analysis in detail, but for this local problem the characteristic length is H and the characteristic velocity is $V = \bar{v}_x$ for both the x and z directions.

The streamlines in the vicinity of the front are sketched in Fig. 5.12. The local flow near the flow front is referred to as a *fountain flow*, because the material near the center line catches up to the flow front and the "fountains" out toward the walls. This behavior is confined to a region whose width in the x direction is about H.

The fountain flow problem does not have an analytical solution, but the problem can be solved numerically (Coyle et al., 1987). The classical approach is to pose the problem in a coordinate system that moves with the flow front (i.e., with velocity $\bar{v}_x$ relative to the laboratory frame). In this coordinate frame the mold walls appear to have a velocity $v_x = -\bar{v}_x$, and the flow front appears to be stationary. The advantage

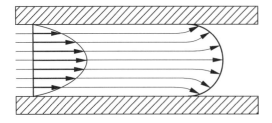

Figure 5.12: Streamlines for the fountain flow, viewed in the fixed (laboratory) frame.

5.4 FLOWS WITH SIGNIFICANT INERTIA

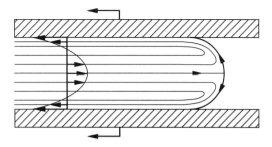

Figure 5.13: Streamlines for the fountain flow, viewed in a frame that moves with the flow front. In this frame the flow front is stationary and the mold walls appear to move to the left.

is that the flow is now steady, with a free surface rather than a moving boundary. The governing equations are the continuity equation and the momentum balance equations for the x and z directions. On the flow front we use Eqs. (5.94) and (5.77). On the left-hand side, the inlet boundary condition is the parabolic distribution of Eq. (5.98) with $\bar{v}_x$ subtracted off,

$$v_x = 6\bar{v}_x \left(\frac{z}{H} - \frac{z^2}{H^2} - \frac{1}{6} \right) \tag{5.113}$$

$$v_z = 0 \tag{5.114}$$

and, consistent with this, we use $v_x = -\bar{v}_x$ and $v_z = 0$ on the upper and lower walls. This provides a well-posed free boundary problem.

The streamlines in the moving frame have the character sketched in Fig. 5.13. This picture is often presented to describe the fountain flow. Because the flow is steady in the moving frame, the streamlines here are also particle pathlines.[2] Thus, Fig. 5.13 shows that the material that is deposited on the walls of the mold arrived there by traveling up the midplane of the mold, catching up to the flow front, and then rolling out onto the walls through the action of the fountain flow. However, in the laboratory frame the x-direction velocity is always positive, and no material every moves backward toward the injection point.

An interesting application of the fountain effect is co-injection molding. In this process one injects a first polymer to partially fill the mold, and then completes the filling process by injecting a second polymer. Because of the fountain effect, the first material ends up on the outside of the part, and the second material ends up in the core. This is sketched in Fig. 5.14. This process provides a way to use recycled thermoplastics as the core of a part, with virgin polymer on the outside to retain good appearance.

5.4 FLOWS WITH SIGNIFICANT INERTIA

The flow solutions we have seen so far are for problems in which the inertia is negligible. Generally this requires that the Reynolds number be small, $\mathrm{Re} \ll 1$, where for example $\mathrm{Re} = \rho V H/\mu$ for fluid in a channel of height H, or $\rho V D/\mu$ for flow in a tube of diameter D. Flows in which inertia is completely negligible are called *creeping flows* or *Stokes flows*.

---

[2] The streamlines in Fig. 5.12 are *not* particle pathlines, because the flow is unsteady in the laboratory frame.

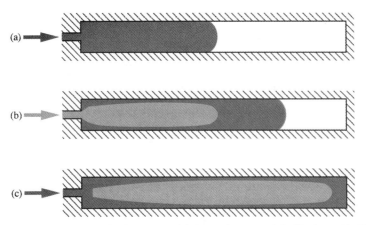

Figure 5.14: The co-injection process. (a) One polymer partially fills the mold. (b) Then a second polymer is injected. Because of the fountain effect the first polymer coats the mold walls. (c) Sometimes the cycle finishes by injecting some of the first polymer, so that only that material appears on the outside of the part.

What about more general flows where the Reynolds number is not small? What types of phenomena occur, and how can we model these flows? This section provides an introduction to modeling flows with significant inertia. We begin by examining models that ignore viscosity entirely, and then we look at more detailed models for laminar and turbulent flows.

5.4.1 INVISCID FLOW MODELS

If inertia is very important, $Re \gg 1$, then one way to formulate a model is to assume that inertial effects completely dominate the momentum balance, and ignore the viscous effects entirely. The appropriate momentum balance equation is then the Euler equation, Eq. (2.101), or one of its special forms. If the fluid is experiencing a steady potential flow, or if we have a good idea what the streamlines look like, then one of the forms of Bernoulli's equation is appropriate as a momentum balance. We illustrate this type of analysis with some examples.

Pouring Liquid Metal from a Ladle

Inertia is usually quite important in flows encountered in metal casting, because of the low kinematic viscosity of liquid metals. In many applications, the flow can be effectively modeled by using Bernoulli's equation. For example, consider the filling of a casting from a ladle, as depicted schematically in Fig. 5.15. In a typical cast house, a relatively large ladle is filled with molten alloy from either a melting or holding furnace. The ladle is then transported to another part of the facility where individual molds are filled. Sometimes the melt is poured by tilting the ladle, and in other cases it is discharged through an orifice at the bottom, as shown in Fig. 5.15.

The orifice is usually much smaller than the ladle itself, and many molds are filled from a single ladle. We can determine the time required to fill a casting by using Bernoulli's equation. We will use the steady form along a streamline, Eq. (2.115), even though the flow is unsteady, and we will justify this simplification later. Choose

5.4 FLOWS WITH SIGNIFICANT INERTIA

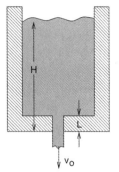

Figure 5.15: A ladle discharging liquid metal.

point s at the top surface of the liquid metal and point o at the orifice. Bernoulli's equation then says that

$$p_s + \frac{1}{2}\rho v_s^2 + \rho g h_s = p_o + \frac{1}{2}\rho v_o^2 + \rho g h_o \tag{5.115}$$

Because the top and the orifice are both open to the air, $p_s = p_o = p_{\text{atm}}$. Also, we anticipate that the cross-sectional area of the ladle will be much larger than that of the orifice, so that $v_s^2 \ll v_o^2$, and we can neglect the term containing v_s^2. Substituting these results into Eq. (5.115) and rearranging terms gives

$$v_o = \sqrt{2gH} \tag{5.116}$$

where $H = h_s - h_o$ is the instantaneous height of the metal in the ladle (see Fig. 5.15).

This expression can now be used to determine the time required to fill a casting of volume V_c. Denote the cross-sectional area of the ladle by A_ℓ and that of the orifice by A_o. The volumetric flow rate from the orifice is simply $A_o v_o$, and this flow rate must match the rate of volume decrease in the ladle, so that

$$-A_\ell \frac{dH}{dt} = A_o v_o = A_o \sqrt{2gH} \tag{5.117}$$

Note that dH/dt is negative when v_o is positive. We first rearrange this equation to separate the variables H and t,

$$-\frac{A_\ell}{A_o}\sqrt{\frac{1}{2g}}\frac{dH}{\sqrt{H}} = dt \tag{5.118}$$

and then integrate it to find

$$-\frac{A_\ell}{A_o}\sqrt{\frac{2H}{g}} = t + c_1 \tag{5.119}$$

We choose $t = 0$ at the start of filling, when $H = H_0$. Using this initial condition to evaluate c_1, we can write the relationship between H and t as

$$t = \sqrt{\frac{2}{g}}\frac{A_\ell}{A_o}(\sqrt{H_0} - \sqrt{H}) \tag{5.120}$$

The casting will be full when the height of metal in the ladle has dropped enough to deliver a volume $V_c = A_\ell(H_0 - H_f)$. This can be solved for H_f:

$$H_f = H_0 - V_c/A_\ell \tag{5.121}$$

Using this result along with Eq. (5.120) we can find t_f, the time to fill the casting:

$$t_f = \sqrt{\frac{2}{g}} \frac{A_\ell}{A_o}(\sqrt{H_0} - \sqrt{H_f}) \tag{5.122}$$

To complete the problem we must justify the use of Eq. (5.115), which ignored the unsteady term $\rho \int_1^2 \partial v/\partial t \, ds$ from Bernoulli's equation, Eq. (2.114). When the orifice is first opened, a short startup time is required for the fluid to accelerate. If the orifice has length L then this time is of the order of $2L/\sqrt{2gH_0}$, as shown in one of the exercises. This is approximately the time required to deliver a volume of fluid that just fills the orifice. It is unlikely that one would want to pour such a small amount of liquid metal, so the startup transient should be insignificant.

We must also ensure that the unsteady term is not important during the pouring period, when the fluid velocity is decreasing as a result of decreasing height in the ladle. From Eq. (5.122) the time over which H decreases will scale as $(A_\ell/A_o)\sqrt{2H_o/g}$. We can neglect the transient term if this time scale is large compared with the transient time scale. This will be true if

$$\frac{A_\ell}{A_o} \gg \frac{L}{H_o} \tag{5.123}$$

This requirement is easily satisfied, because we expect the left-hand side to be much greater than unity and the right-hand side to be much smaller than unity. Thus, it is appropriate to drop the unsteady term from Bernoulli's equation for this problem.

Sprue and Runner Design for Metal Casting

Another application of Bernoulli's equation in casting relates to the sizing of sprues and runners in metal casting. A layout for a typical casting mold is shown in Fig. 5.16. Metal is poured from a ladle into a pouring basin, after which it flows through a sprue (or downsprue), into a runner, and then finally into the casting through one or more ingates. Goals for designing the sprue and runners are to fill the casting completely before it freezes, to avoid picking up sand or other impurities from the mold, and to avoid trapping air in the liquid metal.

A heat transfer analysis as developed in Section 4.3 can be used to estimate the solidification time at the thinnest cross section. The fill time should be substantially

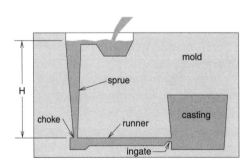

Figure 5.16: Layout of a typical casting mold, and the terminology for various parts.

5.4 FLOWS WITH SIGNIFICANT INERTIA

smaller than this time. Together with the volume of the casting, this sets the minimum flow rate at which the mold should be filled.

Next the cross-sectional area of the runners is chosen to keep the velocity of the liquid metal below some critical value, usually ~1 m/s. This avoids strongly turbulent flow, which might erode sand or other foreign matter away from the mold surface, trapping it in the casting. It also reduces the risk of folding surface oxides into the liquid during filling. Making the runner area large keeps the flow velocity down, but at the expense of pouring more metal that does not become part of the casting.

Flow rate into the casting is controlled by setting the area of the choke, which is the small cross section at the bottom of the sprue. The same Bernoulli analysis we did for the ladle can be applied here, showing that in the initial stages of mold filling the velocity at the choke is $v = \sqrt{2gH}$. Later in the filling stage the velocity will be lower, as the hydrostatic pressure from the metal in the casting rises, offsetting the pressure head from the sprue and pouring basin.

The sprue is tapered so that it will always be full as the metal is poured. The fluid accelerates as it travels down the sprue, because it is falling in a gravitational field. At steady state the flow rates at the top and bottom of the sprue are equal, so the stream must contract as it accelerates. If the sides of the sprue were vertical, then the stream would pull away from the walls. Then, when the stream of metal hit the bottom of the sprue and starts into the runner it would splash, possibly mixing air into the metal, which can lead to porosity in the casting. Thus, it is important to taper the sprue, to keep it full, and to avoid aspirating air.

A simple analysis can be done to determine an appropriate shape for the sprue. As a preliminary step, let us analyze a stream of molten metal that is falling freely in the atmosphere, unconstrained by a sprue or any other channel. This situation is illustrated in Fig. 5.17. The liquid metal exits a container (perhaps the pouring basin

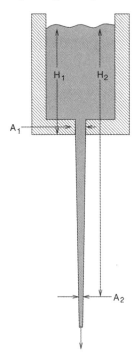

Figure 5.17: A stream of liquid metal falling freely in the atmosphere.

of a casting mold) a distance H_1 below the free surface, though a passage of area A_1. At some later point the distance below the free surface is H_2 and the cross-sectional area is A_2.

As in the previous example, we use Bernoulli's equation for steady flow, realizing that the unsteady term is likely to be small in almost all situations of interest. Following the analysis for the ladle we have

$$v_1 = \sqrt{2gH_1}, \qquad v_2 = \sqrt{2gH_2} \tag{5.124}$$

Note that the pressure is atmospheric all along the freely falling stream, $p_1 = p_2 = p_{atm}$. At steady state, conservation of mass requires that

$$A_1 v_1 = A_2 v_2 \tag{5.125}$$

Combining this with the previous equation to eliminate the velocities, we find that

$$\frac{A_2}{A_1} = \sqrt{\frac{H_1}{H_2}} \tag{5.126}$$

H_2 can be taken at any point along the stream, so this equation describes the shape of a freely falling stream when viscous effects are negligible. If the stream has a circular cross section (which will naturally result from the action of surface tension), then the diameter varies according to

$$\frac{D_2}{D_1} = \left(\frac{H_1}{H_2}\right)^{1/4} \tag{5.127}$$

Returning to the casting mold, we find that the sprue diameter should taper down more rapidly than would a free-falling stream. This ensures that the pressure increases as one moves down the sprue, and it guarantees that the liquid metal always occupies the full cross section. One could design a sprue with curved walls following Eq. (5.127), but the one-fourth power is a weak dependence, and it is much simpler to make a sprue with a straight taper such as that in Fig. 5.16. The actual taper is usually chosen by using empirical rules that are very conservative in comparison to Eq. (5.127).

5.4.2 DETAILED SOLUTIONS: LAMINAR FLOWS

Inviscid models usually capture the large-scale features of a flow very well, but they do not accurately represent the flow near solid boundaries, and they can leave out important effects such as instabilities, flow separation, and turbulence. To find a complete, detailed solution for fluid flow one must include the viscous terms, even when they are very small. Our ability to solve these detailed models depends strongly on whether the flow is laminar or turbulent. Laminar flows can be modeled quite accurately, whereas creating an accurate model of a turbulent flow is very difficult.

Writing a model for a laminar flow with significant inertia is straightforward: use the full mass balance and momentum balance equations, impose the same types of boundary conditions we have used so far, simplify the equations, and solve them. However, compared to creeping flows, the solutions have some differences.

5.4 FLOWS WITH SIGNIFICANT INERTIA

These differences come from the inertial terms in the momentum balance equation. For Newtonian creeping flows with constant viscosity, the momentum equation reduces to

$$0 = -\nabla p + \mu \nabla^2 \mathbf{v} + \rho \mathbf{b} \tag{5.128}$$

However, with significant inertia we must start with

$$\rho \left(\frac{\partial \mathbf{v}}{\partial t} + \mathbf{v} \cdot \nabla \mathbf{v} \right) = -\nabla p + \mu \nabla^2 \mathbf{v} + \rho \mathbf{b} \tag{5.129}$$

Time does not appear in Eq. (5.128), so a creeping flow can only be unsteady if the boundary conditions change with time. However, the unsteady inertial term $\rho(\partial \mathbf{v}/\partial t)$ in Eq. (5.129) can lead to time-dependent solutions, even when the boundary conditions are steady. In addition, the creeping flow form in Eq. (5.128) is linear in the velocity $\mathbf{v}$, but the advection term $\rho(\mathbf{v} \cdot \nabla \mathbf{v})$ makes the general form in Eq. (5.129) nonlinear. Unsteadiness and nonlinearity can introduce many interesting phenomena, which we now explore.

Unsteady Solutions

As an example of an unsteady solution, consider fluid near a plane wall, where initially the fluid and the wall are at rest. The wall surface lies in the x–y plane. The wall is very large in the x and y directions, so we will treat it as being infinite in extent. Similarly, any other walls are far away from this one, so the fluid extends infinitely in the positive z direction. At $t = 0$ the wall begins to translate in the x direction with velocity V. The pressure is uniform, and we wish to find the fluid velocity as a function of position and time.

We will assume that v_y and v_z are zero, so we only need to find v_x as a function of z and t. The continuity equation is satisfied identically, and the x-direction momentum equation simplifies to

$$\rho \frac{\partial v_x}{\partial t} = \mu \frac{\partial^2 v_x}{\partial z^2} \tag{5.130}$$

The initial condition is

$$v_x = 0 \quad \text{at } t = 0 \tag{5.131}$$

and for $t > 0$ the boundary conditions are

$$v_x = V \quad \text{at } z = 0 \tag{5.132}$$
$$v_x = 0 \quad \text{at } z = \infty \tag{5.133}$$

This problem has no natural length or time scale, so we immediately suspect that it has a similarity solution. In fact, these equations and boundary conditions describe a problem that is mathematically identical to the semi-infinite body heat conduction problem in Section 4.2.1. To see this, compare Eqs. (5.130)–(5.133) with Eqs. (4.36)–(4.39). Because the two problems have the same governing equation and the same boundary conditions, they have the same solution. Instead of repeating the solution, we can simply rewrite Eq. (4.55), replacing the variables with their current meanings.

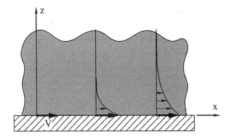

Figure 5.18: Velocity profiles at $t = 0$ and two later times for a wall that is set into motion at $t = 0$.

The solution to our fluid flow problem is then

$$\frac{v_x}{V} = \text{erfc}\left(\frac{z}{\sqrt{4\nu t}}\right) \tag{5.134}$$

The velocity profiles at several different times are sketched in Fig. 5.18. Note that a finite time is required for the motion of the boundary to be felt inside the fluid. Also, comparing Eq. (5.134) to (4.55), note that kinematic viscosity $\nu = \mu/\rho$ plays the same role in this problem that thermal diffusivity $\alpha = k/\rho c_p$ plays in the thermal problem.

Developing Flows and Boundary Layers

When we analyzed the channel flow in Section 5.1, we assumed that the flow was fully developed, which required that the channel be long and/or the Reynolds number be small. This gave a parabolic velocity profile for the pressure-driven flow. In contrast, the Bernoulli analysis of inviscid flow gives a flat velocity profile in a straight channel, with fluid slipping along the walls. For flows with significant inertia, the real situation is somewhere between these two extremes, and the velocity profile depends on the distance from the channel entrance.

Suppose we have a steady, pressure-driven flow in a channel where the Reynolds number is greater than one. The situation near the entrance to the channel is sketched in Fig. 5.19. At the very entrance to the channel the velocity profile is flat. Just inside the channel the no-slip condition on the walls is felt, but only in a thin boundary layer near the wall. As we move down the channel the boundary layer thickness grows, until eventually the boundary layers from the two walls meet in the middle. From this point onwards the flow is fully developed, and has the parabolic velocity profile given in Eq. (5.43). The distance from the entrance to the channel to the point where the flow is fully developed is called the *entrance length*, L_e in Fig. 5.19. At distances less than L_e from the entrance we have a *developing flow*.

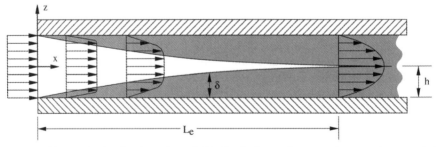

Figure 5.19: Developing flow in a channel. Shading indicates the approximate region occupied by the boundary layer.

5.4 FLOWS WITH SIGNIFICANT INERTIA

The analytical solution of this developing flow problem requires a good deal of mathematics, but we can extract most of the important information by scaling. We define the dimensionless variables and scale the continuity equation just as in Section 5.1, but this time we use the channel half-height h (see Fig. 5.19) as the characteristic z dimension. The x-direction momentum balance equation is still

$$\rho\left(v_x \frac{\partial v_x}{\partial x} + v_z \frac{\partial v_x}{\partial z}\right) = -\frac{\partial p}{\partial x} + \mu\left(\frac{\partial^2 v_x}{\partial x^2} + \frac{\partial^2 v_x}{\partial z^2}\right) \tag{5.135}$$

but when scaling this equation we now assume that the inertial terms on the left-hand side will be important. This leads to a characteristic pressure difference of

$$\Delta p_c = \rho V^2 \tag{5.136}$$

and the scaled equation becomes

$$v_x^* \frac{\partial v_x^*}{\partial x^*} + v_z^* \frac{\partial v_x^*}{\partial z^*} = -\frac{\partial p^*}{\partial x^*} + \left\{\frac{\mu L}{\rho V h^2}\right\} \left(\left\{\frac{h^2}{L^2}\right\} \frac{\partial^2 v_x^*}{\partial x^{*2}} + \frac{\partial^2 v_x^*}{\partial z^{*2}}\right) \tag{5.137}$$

We can immediately drop the viscous term multiplied by $\{h^2/L^2\}$. However, the remaining viscous term cannot be dropped, even when $\{\mu L/\rho V h^2\} \ll 1$, because it contains the highest-order derivative of v_x with respect to z. Recall that a small dimensionless group multiplying the highest-order derivative in a governing equation is the signal that a boundary layer is present. If $\{\mu L/\rho V h^2\} \ll 1$, which means that the Reynolds number is large and/or the channel is not very long, then scaling tells us to expect a boundary layer near the wall, where the viscous stresses will be important.

Scaling can also tell us how thick the boundary layer is and can give an estimate of the entrance length. Following the procedure of Section 3.3, we define a local dimensionless z coordinate within the boundary layer,

$$\eta = \frac{z}{\delta} \tag{5.138}$$

where $\delta(x)$ is the local thickness of the boundary layer (see Fig. 5.19). We then rescale by using this new variable, and we find the dimensionless momentum balance equation to be

$$v_x^* \frac{\partial v_x^*}{\partial x^*} + v_z^* \frac{\partial v_x^*}{\partial \eta} = -\frac{\partial p^*}{\partial x^*} + \left\{\frac{\mu L}{\rho V \delta^2}\right\} \left(\frac{\partial^2 v_x^*}{\partial \eta^2}\right) \tag{5.139}$$

Within the boundary layer the viscous term must be equal in importance to the inertial and pressure gradient terms, so we set its multiplier equal to one. This gives the boundary layer thickness as

$$\delta = \sqrt{\frac{\mu L}{\rho V}} \tag{5.140}$$

In fact, this local scaling analysis works at *any* distance L from the entrance of the channel, provided that L is less than the entrance length. Thus, we can replace L by

x and find that the boundary layer thickness grows like the square root of distance from the entrance,

$$\delta = \sqrt{\frac{\mu x}{\rho V}} \tag{5.141}$$

In dimensionless form this is

$$\frac{\delta}{h} = \sqrt{\frac{x/h}{\text{Re}}} \tag{5.142}$$

with $\text{Re} = \rho V h/\mu$.

The entrance length L_e is simply the value of x at which δ equals h, where the boundary layers from the two sides completely fill the channel. Using Eq. (5.142), we find this is

$$\frac{L_e}{h} = \text{Re} \tag{5.143}$$

Note that this is a scaling estimate, not an exact result, for L_e. If the total channel length is longer than L_e, then the flow will eventually become fully developed as in Fig. 5.19. If the total channel length is less than L_e, then the boundary layers will only occupy thin regions near the walls, and there will always be a core region where the flow is unaffected by viscosity. In this flow the core region will have a flat velocity profile. Figure 5.20 illustrates this point by showing numerically computed results for the velocity profile near the entrance to a channel, for several values of the Reynolds number.

Separated Flows

Consider the flow around a cylinder at moderate Reynolds number, say $\text{Re} = 50$, where $\text{Re} = \rho V D/\mu$ and D is the cylinder diameter. If we assume that viscosity is

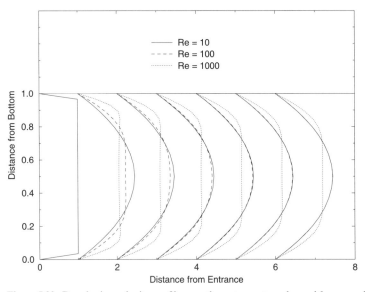

Figure 5.20: Developing velocity profile near the entrance to a channel for several values of the Reynolds number.

5.4 FLOWS WITH SIGNIFICANT INERTIA

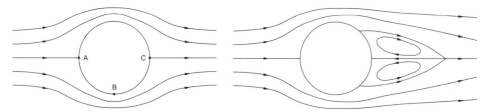

Figure 5.21: Streamlines for flow past a cylinder. Left: For inviscid potential flow the streamlines are symmetric fore and aft and there is no recirculation. Right: The real flow separates, creating large eddies on the downstream side, as a result of small viscous effects.

negligible, then this is a potential flow (because the far-field flow is irrotational; see Section 2.3.6). Streamlines for the potential flow solution are sketched on the left in Fig. 5.21. With no viscosity, the potential flow solution requires the fluid to slip over the surface of the cylinder. We can also argue, based on the fact that the velocity field satisfies Eq. (2.163), that there can never be any recirculation (i.e., no closed streamlines) in any potential flow. For flow around a cylinder the velocity field will have fore-and-aft symmetry, and the pressure field will be similarly symmetric. In particular, the pressure will decrease as we move from point A to point B in Fig. 5.21, because the fluid velocity is increasing, whereas from point B to point C the pressure will increase.

The real situation, in which viscosity plays a role, is sketched on the right in Fig. 5.21. First, the fluid does not actually slip over the surface of the cylinder; instead there is a thin boundary layer near the surface where the fluids is moving slowly. Second, the flow will become *separated*, as shown in Fig. 5.21. Vortices develop immediately behind the body, and close to the back surface the fluid flows "upstream." Flow separation is intimately associated with an adverse pressure gradient, that is, pressure increasing as one moves downstream. For the cylinder, the pressure increase from B to C (in the potential flow solution) causes the flow to separate. The boundary layer feels this pressure gradient, and close to the solid surface the fluid flows backward in response. Flow separation can occur on any body that is blunt on its downstream side.

For a potential flow, the only force than acts on the cylinder is fluid pressure. Because the pressure distribution has fore-and-aft symmetry in this problem, the potential flow solution says there is no net force (no drag) on the cylinder. In the real, separated flow we lose this symmetry, and the pressure does not increase nearly as much from point B to point C as the potential flow solution suggests. This is the primary source of drag on the cylinder. Even though viscous effects are small, they can trigger flow separation, which has a major effect on the flow.

One avoids flow separation by creating smooth flow passages, particularly where the cross section of the passage is expanding. For example, in metal casting runners we prefer to avoid separated flows and large eddies, which may trap air in the metal or draw in loose material from the mold.

Secondary Flows and Instabilities

Inertial effects can also lead to flow patterns that are not obviously related to the boundary conditions that drive the flow. Consider tangential drag flow between concentric cylinders. The fluid is contained between an outer cylinder of radius R

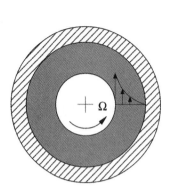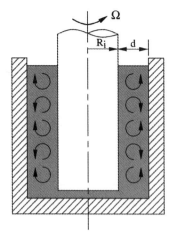

Figure 5.22: Taylor vortex instability for tangential flow between concentric cylinders. Left: Base flow (top view). Right: At sufficiently high rotation speeds, toroidal vortices form in the gap between the inner and outer cylinders (side view).

and an inner cylinder of radius $R_i = \kappa R$ that is rotated at angular velocity Ω. The gap between the two cylinders is d. The solution for this problem when inertia is negligible, sketched on the left in Fig. 5.22, is given in Section 5.2.5. The only motion is in the azimuthal (θ) direction, with a high velocity close to the inner cylinder and a low velocity close to the outer cylinder.

As the inertial terms become more important, this flow becomes unstable. The first instability to appear is a regular pattern of toroidal (doughnut-shaped) vortices, sketched on the right in Fig. 5.22, called *Taylor vortices*. Each vortex is about as tall as the gap height d, vortices are stacked to fill the entire gap, and each successive vortex rotates in the opposite direction of its neighbors. The onset of this instability occurs when a dimensionless group known as the Taylor number reaches a critical value:

$$\text{Ta} \equiv \frac{\Omega R_i d}{\nu} \sqrt{\frac{d}{R_i}} \geq 41.3 \tag{5.144}$$

The Taylor vortex flow in Fig. 5.22 and the separated flow sketched in Fig. 5.21 are steady, but if inertia is increased further then these flows will also become time dependent. The Taylor vortices will become wavy in shape when viewed from the side, and the waves will slowly rotate around the cylinder. For the separated flow around a cylinder, beyond $\text{Re} = 60$ the vortex on the top will detach from the body and move downstream, then the vortex on the bottom will detach and follow it while the top vortex reforms, then the top vortex will detach again, and so forth. This type of vortex shedding produces an oscillating force on the cylinder, which is what causes power lines to hum or sway on windy days.

When inertia is significant, we cannot assume that a flow solution will be steady just because the geometry and the boundary conditions are steady. Mathematically the unsteadiness enters through the transient term $\rho(\partial \mathbf{v}/\partial t)$ in the momentum balance.

In many materials processing operations an instability or secondary flow is highly undesirable, and the main concern is how to avoid it. In such cases, a criterion like

5.4 FLOWS WITH SIGNIFICANT INERTIA

Eq. (5.144) is all that one needs to know. Stability criteria of this type can be predicted quite accurately by using linear stability analysis. We provide an example of this type of analysis in a different context in Section 8.3. The Taylor vortex flow is particularly significant in this connection, because it was the first fluid flow problem in which analysis accurately predicted the onset of an instability.

Linear stability analysis will reveal the point at which the flow first becomes unstable, and it can identify the fastest-growing instability, but it cannot describe the full details of the unstable flow. In principle these are available by simply solving the full governing equations, perhaps numerically, but in practice computing the solutions may be quite difficult.

5.4.3 DETAILED SOLUTIONS: TURBULENT FLOWS (ADVANCED)

We have seen that significant inertia can lead to time-dependent instabilities in laminar flows, beginning around $Re = 100$. If the Reynolds number is increased further, these flows become turbulent. For example, the flow behind a cylinder becomes turbulent once Re exceeds approximately 5000, while flow in a smooth pipe is usually turbulent above $Re = 3000$.

Turbulent flow is characterized by irregular, time-dependent variations in velocity, called *eddies*. There are corresponding irregular, time-dependent variations in pressure. Turbulent eddy motion is inherently three dimensional, even when the overall flow is two dimensional. The largest eddies are comparable in size to the length scale of the flow, say the pipe diameter in a pipe flow or the cylinder diameter for flow over a cylinder. These large-scale features are largely independent of Reynolds number, but as the Reynolds number increases the amount of motion at small scales increases.

There are almost no turbulent flow problems that can be solved analytically, so here we introduce the main approaches for numerical calculation. The reader should understand at the outset that there is no universal model for turbulence. The basic choices are to use an enormous amount of computing power, a large number of empirical fitting factors, or both.

Direct Numerical Simulation

In principle, turbulent flows obey the same mass and momentum balance equations that govern all other fluid flow problems, so in principle turbulent flows should be completely predictable. We should be able to buy software that solves the Navier–Stokes equations, set up our problem, and solve it. This approach is called *direct numerical simulation*, or DNS. From a scientific point of view, DNS is very appealing: the physics of the problem is contained in the Navier–Stokes equations, with no extra assumptions and no fitting factors. However, in practice DNS leads to extremely expensive computations, and only "easy" problems with simple geometries can be modeled with this approach.

One reason that DNS is expensive is that all DNS problems are time dependent. Even when the average flow is steady, the turbulent eddies are time dependent, and this must be captured by the simulation. To find a time-average velocity distribution one must run a transient simulation for a finite time period, and then time-average the results.

A second reason that DNS is expensive is the wide range of length scales involved. Let D represent a characteristic length scale for the largest eddies in the flow, for example, the pipe diameter for flow in a pipe. A rather lengthy scaling argument (Richardson, 1989) shows that the average length scale of the eddies, called the *Taylor microscale*, is

$$\lambda_T = \frac{D}{\mathrm{Re}^{1/2}} \tag{5.145}$$

where $\mathrm{Re} = \rho V D/\mu$, with V the characteristic velocity. There is also a *minimum* eddy size, because eventually the eddies become small enough that they are damped out by viscosity. This length is called the *Kolmogorov microscale*, and it is given by

$$\lambda_K = \frac{D}{\mathrm{Re}^{1/4}} \tag{5.146}$$

For a pipe flow at $\mathrm{Re} = 10^4$, the Taylor microscale is $D \times 10^{-2}$ and the Kolmogorov microscale is $D \times 10^{-3}$. The grid for a DNS solution must resolve the Kolmogorov microscale, so an intractable number of nodes is required, even for a small portion of the pipe. The fine grid scale also forces the use of a small time step. As the Reynolds number increases, both the grid size and the time step size must decrease, and the time required to run the calculation increases rapidly.

At this book's writing, fluid dynamics calculations are limited to approximately 10 million nodes and 1 million time steps. This restricts DNS solutions to fairly small regions and to modest Reynolds numbers, say $\mathrm{Re} \leq 10^4$. Given these limitations, DNS is a tool for turbulence researchers, rather than a tool for practicing engineers. However, if computing power continues to increase and computing costs decrease in the way they have in the past two decades, then DNS could become a practical tool in modeling materials processing.

Reynolds-Averaged Navier–Stokes Equations

If it is too expensive to calculate the details of every little eddy in a turbulent flow, then perhaps we can find a way to calculate only the average velocity. Calculations of this type are called RANS models, because they are based on the *Reynolds-averaged Navier–Stokes equations*. Virtually all commercial software for simulating turbulent fluid flow use some type of RANS scheme.

The virtue of RANS is that the solution is relatively cheap and quick. The grid can be relatively coarse, with just enough nodes to capture the large-scale features of the flow. Also, if the problem is macroscopically steady then the solution is computed without any time stepping. The problem with RANS is that it requires one to make guesses about some of the fine-scale details of the flow, guesses that are reflected in approximate formulas and empirical constants. Unfortunately, the formulas and constants that work for one flow may not work for another.

To see where the guessing comes in, let us examine the governing equations. We are interested in the average velocity at every point in the flow, and we can define this as a time average:

$$\bar{\mathbf{v}}(\mathbf{x}, t_0) = \frac{1}{\Delta t} \int_{t_0 - \Delta t/2}^{t_0 + \Delta t/2} \mathbf{v}(\mathbf{x}, t)\, dt \tag{5.147}$$

5.4 FLOWS WITH SIGNIFICANT INERTIA

The average pressure $\bar{p}$ is defined in a similar way. The averaging window Δt should be much longer than the time scale of the turbulent fluctuations, but much shorter than the time over which the large-scale aspects of the flow change. If the flow is steady, then Δt can be very long indeed. In practice Δt is a conceptual tool, and we never actually have to choose its value.

The difference at any place and time between the actual velocity and the average velocity is call the *velocity fluctuation*, $\mathbf{v}'$:

$$\mathbf{v} = \bar{\mathbf{v}} + \mathbf{v}' \tag{5.148}$$

The average of $\bar{\mathbf{v}}$ is just $\bar{\mathbf{v}}$, so we can take the average of both sides of Eq. (5.148) and deduce that the average velocity fluctuation is zero:

$$\overline{\mathbf{v}'} = 0 \tag{5.149}$$

Taking the time average of the continuity equation for a constant-density fluid we find, after some manipulation, that

$$\nabla \cdot \bar{\mathbf{v}} = 0 \tag{5.150}$$

That is, the time-average velocity in a turbulent flow satisfies the same continuity equation as the pointwise velocity. If we perform the same averaging operation on the momentum balance equation for a Newtonian fluid with constant density, we obtain

$$\rho \frac{D\bar{\mathbf{v}}}{Dt} = -\nabla \bar{p} + \nabla \cdot (\bar{\boldsymbol{\tau}} + \boldsymbol{\tau}^{\text{Re}}) + \rho \mathbf{b} \tag{5.151}$$

Here $\bar{\boldsymbol{\tau}}$ is the average viscous extra stress,

$$\bar{\boldsymbol{\tau}} \equiv 2\mu \bar{\mathbf{D}} \tag{5.152}$$

with $\bar{\mathbf{D}}$ being the rate-of-deformation tensor based on the average velocity,

$$\bar{D}_{ij} \equiv \frac{1}{2}\left(\frac{\partial \bar{v}_i}{\partial x_j} + \frac{\partial \bar{v}_j}{\partial x_i}\right) \tag{5.153}$$

Equation (5.151) is identical to the Navier–Stokes equation, Eq. (2.78), with $\bar{\mathbf{v}}$ substituted for $\mathbf{v}$, except for one term. This term is the *Reynolds stress*, which results from the averaging process, and which is given by

$$\boldsymbol{\tau}^{\text{Re}} \equiv -\rho \overline{\mathbf{v}'\mathbf{v}'} \tag{5.154}$$

The Reynolds stress represents the transport of momentum that is due to velocity fluctuations. It arises mathematically because, although the average of $\mathbf{v}'$ is zero, the average of the product $\mathbf{v}'\mathbf{v}'$ is not.

Equations (5.151)–(5.154) involve no approximations, and the average velocity $\bar{\mathbf{v}}$ must satisfy these equations exactly. If we could accurately predict the Reynolds stress, then solving the RANS equations would give us an accurate and reliable model for turbulent flows. However, we do not know the velocity fluctuation $\mathbf{v}'$, and the whole idea of RANS is *not* to know it, so we must approximate the Reynolds stress in terms of the quantities we do know. The art in creating a good RANS model is to find a good approximation for $\boldsymbol{\tau}^{\text{Re}}$.

From this point onward, there are many variations of the RANS approach. The simplest version starts with the assumption that the Reynolds stress should have a form similar to the Newtonian constitutive equation:

$$\tau^{Re} = 2\mu_T \bar{\mathbf{D}} - \frac{2}{3}\rho \bar{k} \mathbf{I} \tag{5.155}$$

In this equation $\bar{k}$ is the average kinetic energy of the turbulent fluctuations,

$$\bar{k} \equiv -\overline{\mathbf{v'v'}}/2 \tag{5.156}$$

and it appears in Eq. (5.155) to give τ^{Re} the proper trace. Here μ_T is called the *turbulent viscosity*, and for a simple shear flow it is sometimes approximated as

$$\mu_T = \rho \ell^2 \left| \frac{\partial \bar{v}_x}{\partial y} \right| \tag{5.157}$$

Equation (5.157) applies when the flow is in the x direction and the average velocity is changing in the y direction. This equation is suitable for channel flows and flows near walls (boundary layers), but it must be modified for more general flows. The quantity ℓ is called the *Prandtl mixing length*.

The mixing length is not a constant. For fully developed flows in pipes and channels, $\ell = \kappa y$ in the fully turbulent region, where the constant κ is usually taken to be 0.41 and y is the distance from the wall. Very close to a wall the turbulent eddies are suppressed, and $\ell = 0$. Several mixing-length models make ℓ a function of distance from the wall that smoothly joins these two limits.

There are many variations on the mixing-length model, as well as other RANS models that do not use Eq. (5.155) and the subsequent approximations (see Tannehill et al., 1997). These other members of the RANS family add new variables to the problem, such as the turbulent kinetic energy $\bar{k}$. A transport equation for $\bar{k}$ (or any other new variable) is derived by using an averaging process similar to the derivation of the RANS equation, and that equation is solved as part of the solution. The $\bar{k} - \varepsilon$ model, which is widely used in commercial software, is derived this way. Other versions set up equations for the components of τ^{Re} and are called *Reynolds stress models*. No matter what extra variables are chosen, their governing equations always contain some new terms that must be approximated. For instance, the transport equation for $\bar{k}$ contains the average of three velocity fluctuations, $\overline{\mathbf{v'v'v'}}$. It is difficult to put a clear physical interpretation on such variables, and still more difficult to develop accurate approximations for them.

Ultimately, any RANS model relies a great deal on the experience and intuition of the modeler. The simplest mixing-length model contains four constants, whose values are often adjusted to make the predictions fit experimental data. When the modeler has some experience with a similar flow, and has a good idea of which Reynolds stress approximation to use and how to adjust the constants, RANS calculations can give useful predictions. However, it is quite unlikely that any RANS code will give an accurate prediction of a flow problem that has never been examined before.

Large Eddy Simulation

An approach that is intermediate between RANS and DNS is *large eddy simulation*, or LES. In LES one still solves for an average velocity, but now the velocity is a *spatial* average. Averaging apertures of various shapes can be used, the simplest being a cube with sides of length Δx. The spatially averaged velocity is then

$$\tilde{\mathbf{v}}(\mathbf{x}, t) = \frac{1}{\Delta x^3} \int_{x_1-\Delta x/2}^{x_1+\Delta x/2} \int_{x_2-\Delta x/2}^{x_2+\Delta x/2} \int_{x_3-\Delta x/2}^{x_3+\Delta x/2} \mathbf{v}(\mathbf{x}, t) \, dx_3 \, dx_2 \, dx_1 \quad (5.158)$$

The spatial averaging window Δx is chosen to be large enough to filter out the small-scale turbulent eddies, but small enough to capture the large eddies directly. Usually Δx is approximately the same as the grid spacing in the numerical solution. This spatially averaged velocity is time dependent, even when the time-average flow is steady. Thus, in LES a steady flow requires a time-dependent simulation.

The averaging process of Eq. (5.158) is applied to the continuity equation and the momentum equation, producing governing equations that contain $\tilde{\mathbf{v}}$. These are called the *filtered Navier–Stokes equations*, because the small-scale turbulent fluctuations have been filtered out by the averaging process. The resulting equations look exactly like Eqs. (5.150) and (5.151), except that the Reynolds stress is replaced by a new term called the *subgrid scale stress*, τ^{SGS}. Like the Reynolds stress, τ^{SGS} represents the transfer of momentum by velocity fluctuations, but only at scales smaller than Δx. To compute solutions, the subgrid scale stress must be approximated in terms of know quantities. One simple approximation that has been used is

$$\tau^{SGS} = 2\mu_T \tilde{\mathbf{D}} \quad (5.159)$$

where $\tilde{\mathbf{D}}$ is the rate of deformation tensor formed from the filtered velocities. The turbulent viscosity μ_T given by

$$\mu_T = \rho (C_s \Delta x)^2 \sqrt{2\tilde{\mathbf{D}} : \tilde{\mathbf{D}}} \quad (5.160)$$

Here C_s is an adjustable constant. Note that the size of the averaging window Δx influences the value of the turbulent viscosity, and thus of τ^{SGS}. An LES calculation looks like a DNS calculation on a grid that is a bit too coarse, with an extra viscosity μ_T added to account for eddies that are smaller than the grid size.

By directly representing the large-scale eddies, which tend to be anisotropic and highly dependent on the particular flow, LES hopes to provide accurate solutions without adjusting a large number of model parameters. By filtering out the small-scale eddies, which tend to be isotropic and largely independent of the particular flow, LES hopes to provide an affordable computation. At this time, LES is probably the most detailed type of turbulent flow calculation that one would attempt on any practical problem.

5.5 SUMMARY

This chapter has examined the flow of Newtonian fluids. The governing equations are the balance equations for mass and for momentum, and the solution to these equations provides a velocity field and a pressure field.

For viscous-dominated flows, analytical solutions can be found for a variety of simple geometries. These solutions can then be used as building blocks, to model flows in more complex geometries. This scheme is not as accurate as solving for the flow in the exact geometry, but it can provide a good deal of insight with a modest computation.

Free surface problems and moving boundary flows arise when a fluid surface or interface is not constrained by a solid wall. These problems use the same governing equations, with additional boundary conditions applied at the interface. The position or motion of the interface is then determined as part of the solution.

When inertia is dominant, one modeling option is to ignore viscosity entirely. Inviscid modeling tools, such as Bernoulli's equation and potential flow solutions, can reveal many of the large-scale details of the flow. However, inviscid models cannot predict effects such as boundary layer formation, viscous drag, flow separation, and inertial instabilities. These effects can be analyzed by solving the complete mass and momentum balance equations. Such solutions are accurate, though perhaps difficult to find, when the flow is laminar. When the flow is turbulent a direct solution is rarely feasible, and an additional level of modeling is introduced to account for the effects of turbulent velocity fluctuations. There is no universal turbulence model, and one must have both experimental data and computing experience with a similar flow to make accurate predictions.

BIBLIOGRAPHY

D. J. Coyle, J. W. Blake, and C. W. Macosko. The kinematics of fountain flow in mold filling. *AIChE J.* 33:1168–1785, 1987.

C. W. Hirt and B. D. Nichols. Volume of fluid (VOF) method for the dynamics of free boundaries. *J. Comp. Phys.* 39:201–225, 1981.

S. F. Kistler and L. E. Scriven. Coating flows. In J. R. A. Pearson and S. M. Richardson, editors, *Computational Analysis of Polymer Processing*. Applied Science, New York, 1983.

E. Pichelin and T. Coupez. Finite element solution of the 3-D mold filling problem for viscous incompressible fluid. *Comput. Meth. Appl. Mech. Eng.* 163:359–371, 1998.

C. Rauwendaal. *Polymer Extrusion*. Carl Hanser Verlag, Munich, 1986.

S. M. Richardson. *Fluid Mechanics*. Hemisphere, New York, 1989.

Z. Tadmor and C. G. Gogos. *Principles of Polymer Processing*. Wiley, New York, 1979.

J. C. Tannehill, D. A. Anderson, and R. H. Pletcher. *Computational Fluid Mechanics and Heat Transfer*. Taylor and Francis, Washington, DC, 2nd edition 1997.

S. O. Unverdi and G. Tryggvason. A front-tracking method for viscous, incompressible, multifluid flows. *J. Comp. Phys.* 100:25–37, 1992.

H. P. Wang and H. S. Lee. Numerical techniques for free and moving boundary problems. In C. L. Tucker III, editor, *Fundamentals of Computer Modeling for Polymer Processing*. Carl Hanser Verlag, Munich, 1989.

EXERCISES

1 *Pressure flow between parallel plates.* Solve for the velocity profile, stress profile, and flow rate for a pressure-driven flow between parallel plates, as given in Section 5.2.1. Assume a constant density, a constant Newtonian viscosity, and a fully developed two-dimensional flow. Take the following steps.

(a) Use the continuity equation to show that $v_z = 0$ for a fully developed flow.

(b) Write the equation of motion in the x (flow) direction, simplify it according to your assumptions, and integrate to find the velocity profile.

(c) Use the Newtonian constitutive equation to find the stress profile.

(d) Integrate the velocity distribution to find the volume flow rate.

2. *Axial pressure flow in a tube.* Solve for the velocity profile, stress profile, and flow rate for a pressure-driven flow in a tube, as given in Section 5.2.2. Assume a constant density, a constant Newtonian viscosity, and a fully developed axisymmetric flow. Take the following steps.

(a) Use the continuity equation to show that $v_r = 0$ for a fully developed flow.

(b) Write the equation of motion in the z (flow) direction, simplify it according to your assumptions, and integrate to find the velocity profile.

(c) Use the Newtonian constitutive equation to find the stress profile.

(d) Integrate the velocity distribution to find the volume flow rate.

3. *Axial pressure flow between concentric cylinders.* Solve for the velocity profile, stress profile, and flow rate for a pressure-driven flow between concentric cylinders, as given in Section 5.2.3. Assume a constant density, a constant Newtonian viscosity, and a fully developed axisymmetric flow. Take the following steps.

(a) Use the continuity equation to show that $v_r = 0$ for a fully developed flow.

(b) Write the equation of motion in the z (flow) direction, simplify it according to your assumptions, and integrate to find the velocity profile.

(c) Use the Newtonian constitutive equation to find the stress profile.

(d) Integrate the velocity distribution to find the volume flow rate.

4. *Axial drag flow between concentric cylinders.* Solve for the velocity profile, stress profile, and flow rate for a drag flow between concentric cylinders, as given in Section 5.2.4. Assume a constant density, a constant Newtonian viscosity, and a fully developed axisymmetric flow. Take the following steps.

(a) Use the continuity equation to show that $v_r = 0$ for a fully developed flow.

(b) Write the equation of motion in the z (flow) direction, simplify it according to your assumptions, and integrate to find the velocity profile.

(c) Use the Newtonian constitutive equation to find the stress profile.

(d) Integrate the velocity distribution to find the volume flow rate.

5. *Tangential drag flow between concentric cylinders (Couette flow).* Solve for the velocity profile and stress profile for a tangential drag flow between concentric cylinders, as given in Section 5.2.5. Assume a constant density, a constant Newtonian viscosity, and an axisymmetric flow. Also assume that $v_z = 0$ and that $\partial/\partial z = 0$; this assumption is reasonable when the gap between the cylinders is small compared to their length, and it means that we are ignoring end effects. Take the following steps.

(a) Use the continuity equation and the assumptions to show that v_θ is a function of r only, and that $v_r = 0$.

(b) Write the equation of motion in the θ (flow) direction, simplify it according to your assumptions, and integrate to find the velocity profile.

(c) Use the Newtonian constitutive equation to find the stress profile.

(d) Write the r-direction equation of motion, and simplify it according to the assumptions of the problem. Do not neglect the inertia terms. What information can be found by solving this equation?

(e) This flow is used in some instruments that measure viscosity. Find the torque T required to rotate the inner cylinder, and show how the viscosity can be computed when both T and Ω have been measured.

(f) The shear rate in this flow is equal to $2D_{r\theta}$. In an instrument that measures viscosity, it is desirable to have a constant shear rate. How does shear rate vary with r in this geometry? How could you build an apparatus with this geometry that would give nearly constant shear rate?

6 **Torsional drag flow between parallel disks.** Solve for the velocity profile and stress profile for a torsional drag flow between parallel disks, as given in Section 5.2.7. Assume a constant density, a constant Newtonian viscosity, and an axisymmetric flow. Also assume that $v_r = 0$. Take the following steps.

(a) Use the continuity equation to show that $v_z = 0$.

(b) Write the equation of motion in the θ (flow) direction, simplify it according to your assumptions, and integrate to find the velocity profile.

(c) Use the Newtonian constitutive equation to find the stress profile.

(d) This flow is used in some instruments that measure viscosity. Find the torque T required to rotate the upper disk, and show how the viscosity can be computed when both T and Ω have been measured.

(e) The shear rate in this flow is equal to $2D_{\theta z}$. In an instrument that measures viscosity, it is desirable to have a constant shear rate. How does shear rate vary with r in this geometry? Could you build an apparatus with this geometry that would give nearly constant shear rate?

7 **Radial flow between parallel disks.** A situation frequently encountered in the injection molding of polymers is a very viscous fluid flowing between parallel disks, moving radially outward from an injection point. This is problem has no analytic solution for the full equations of motion, but there is a good approximate solution when the gap between the plates is very small compared to the radius of the disk ($h \ll R_0$). This is the problem discussed in Section 5.2.6.

Make the usual assumptions: steady flow (with a known volume flow rate Q), incompressible Newtonian fluid, constant viscosity, negligible inertia and body forces, and no-slip boundary conditions. The flow is axisymmetric and pressure equals p_0 at the outer radius R_0. You may also assume that p is a function of r only, that is, that $\partial p/\partial z \ll \partial p/\partial r$. (A detailed justification of this assumption is given in Section 6.6.)

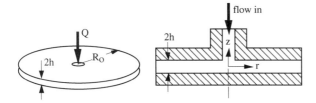

(a) Use the continuity equation and scaling to show that v_z is much smaller than v_r.

(b) Apply scaling to the r-direction equation of motion, and show that the only significant viscous term will be $\mu\,(\partial^2 v_r/\partial z^2)$.

(c) Solve your simplified r-direction momentum equation for v_r, treating $\partial p/\partial r$ as an (unknown) function of r (i.e., just let it ride along in the equations for now).

(d) Integrate the velocity solution to find an expression for the volume flow rate Q. In fact Q is known, so this equation tells you the value of $\partial p/\partial r$. Substitute that back into the velocity solution to get an answer in terms of only known quantities. Verify that this velocity solution satisfies the continuity equation, provided $v_z = 0$.

(e) Integrate your expression for $\partial p/\partial r$ to find the pressure distribution $p(r)$.

(f) According to your solution, what is p at $r = 0$? What is v_r at $r = 0$? Why is your solution poor there? How far would you expect to go out in r until your solution became accurate? What do you think the streamlines really looks like near $r = 0$?

8 **Fluid on a wall.** A thin layer of fluid is coated on a vertical wall. The layer thickness is H, and the fluid has constant density ρ and viscosity μ. The fluid flows down the wall under the action of gravity. The flow is steady, two dimensional, and fully developed.

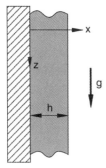

(a) Using the continuity equation, show that v_z is zero.

(b) What are the boundary conditions at the inner and outer surfaces of the fluid layer? How does fluid pressure vary with z?

(c) What are the components of the body force vector **b** for the coordinate system in the sketch?

(d) Write the z-direction equation of motion in terms of stress, simplify it for this problem, and solve for the stress distribution $\tau_{xz}(x)$.

(e) Combine your stress distribution with the constitutive equation for a Newtonian fluid, and solve for the velocity distribution $v_z(x)$.

(f) Sketch the stress and velocity distributions.

9 *Optimal channel depth in a screw extruder.* Consider a screw extruder that is pumping a Newtonian fluid through a die, for which

$$Q = \frac{K}{\mu} \Delta p$$

where K is a constant that depends on the geometry of the die. The screw has a constant channel depth and obeys the shallow-channel model, Eq. (5.42).

(a) Find an expression for the channel depth H that gives maximum flow rate.

(b) Why does a channel that is too deep produces a lower flow rate? Why does a channel that is too shallow produces a lower flow rate?

10 *Extrusion die analysis.* For the extrusion die example in Section 5.2.9, do the following:

(a) Find the speed of the polymer as it exits the die. Assume that downstream of the die the thickness of the extruded sheet equals the gap between the die lips at the exit of the die.

(b) Find the shear rate at the die wall surface a three locations: at the beginning of the second segment, at the end of the second segment, and in the die lip region (the third segment).

(c) Find the shear stress at the die wall surface at the same three locations.

(d) The phenomenon of *melt fracture* is a flow instability that causes extruded polymers to have rough surfaces. Melt fracture is associated with high wall shear stresses, of the order of 10^5 Pa. What is the largest flow rate that can be used with this die without causing melt fracture?

11 *Flow in a wire-coating die.* Plastic insulation is coated onto metal wires in an operation called *wire-coating extrusion*. The polymer is melted and pumped continuously into a die by an extruder. The bare wire is also fed into the back of the die. Inside the die the polymer encircles the wire, and coated wire emerges from the front of the die. Approximately 50 million miles of wire is coated this way each year in the United States. The die looks like the figure below.

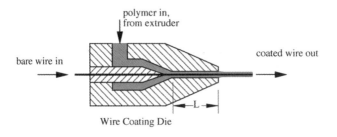

Wire Coating Die

In this problem you are to carry out an analysis of the polymer flow in the last part of the die, over the length marked L in the figure above. Assume that the wire is cylindrical with radius R_i, and that the die surface is also a straight cylinder with radius R_0. The wire speed is V. Treat the polymer as a Newtonian fluid with constant density and viscosity. Assume that inertia and body forces are negligible, and that the flow is axisymmetric and fully developed.

(a) Show that this is the same problem as the axial flow between concentric cylinders in Sections 5.2.3 and 5.2.4. You may use results from these sections without rederiving them.

(b) As the wire and the polymer exit the die there is a rearrangement of the polymer until, some distance downstream, all the polymer is moving at the same speed as the wire and the thickness is constant. Relate the thickness of the coating h at this point to the volume flow rate Q of polymer in the die. (Hint: Use a mass balance on a control volume.)

(c) For the special case in which there is no pressure gradient along the die, write an expression for h as a function of the die geometry and operating parameters. Qualitatively, what effect does increasing the pressure inside the die have on h?

12 *Fiber spinning.* The *fiber-spinning* process is used to make all types of synthetic textile fibers (nylon, polyester, rayon, etc.). In the *melt-spinning* version of the process, molten polymer is extruded through a die called a *spinneret* to create a thin cylinder. Far away from the spinneret, the fiber is wrapped around a drum, which pulls it away at a predetermined take-off speed. The take-off speed is much higher than the extrusion speed, so the filament is stretched considerably in length, and decreases in diameter, between the spinneret and the take-up point. The filament also cools in between these two points, so that it is solid by the take-up point. In the actual process a bundle of many filaments are extruded and stretched together, but for analysis purposes we will consider a single filament.

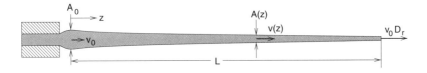

The polymer swells slightly as it exits the die, and one usually starts the analysis from the point of maximum swell. Let z be the distance along the fiber measured from this point, and let $A(z)$ be the cross-sectional area of the filament ($A = \pi R^2$, where R is the radius). The total length of the spinline is L. The area A_0 and speed v_0 at the spinneret are specified, and the speed at the take-up point, $v_o D_r$, is also specified. D_r, called the *draw ratio*, is an important operating parameter of the process. Because the filament is thin, and because it only sees small shear stresses from air drag, one can assume that the velocity v_z is independent of r. We want to solve for $A(z)$ and $v_z(z)$.

The process is operating at steady state. Assume that the polymer has constant density ρ, that it is a Newtonian fluid, and that its viscosity μ is constant up to the take-up. You should also assume that inertial effects are negligible; this is *not* a realistic assumption, because the speeds of commercial spinlines are quite high (up to 3000 m/min), but it makes the analysis tractable.

(a) In Chapter 2, Exercise 5, we found that the continuity equation for this problem is

$$\frac{\partial(\rho A)}{\partial t} + \frac{\partial}{\partial z}(\rho A v_z) = 0$$

Simplify this equation under the current assumptions. Integrate the equation, and show that $A v_z = Q$, where the constant Q is the volume flow rate of polymer.

(b) Use the results of Example 2.2.1 to relate v_r to dv_z/dz. (We can use an ordinary derivative, because v_z does not depend on r or θ.)

(c) Write the components of **D** in cylindrical coordinates and simplify the expressions as much as possible given the current assumptions. You should be able to express all of the nonzero components of **D** in terms of dv_z/dz.

(d) Write expressions for the nonzero components of τ. Using the boundary condition that $\sigma_{rr} = 0$ (ignoring atmospheric pressure), find p inside the fiber. Now express σ_{zz} in terms of μ and dv_z/dz.

(e) Using the integral form of the momentum balance equation, write a z-direction momentum balance for a control volume of length dz that contains the entire cross section of the fiber. (See the sketch of the control volume in Chapter 2, Exercise 5). Use A to describe the cross-sectional area of the fiber, and σ_{zz} to describe the stress. Let τ_d represent the shear stress on the surface from air drag. Initially include all effects in this equation. Then simplify the equation by using the current assumptions.

(f) Integrate your simplified momentum equation and show that $A\sigma_{zz} = F$, where the constant F is the tension along the spinline. For the moment, think of F as being known.

(g) Combine the results from parts (a), (d), and (e) to get a single equation for dv_z/dz that does not contain A. Integrate this equation and apply the boundary condition at $z = 0$, to get a solution for the velocity $v_z(z)$. Return this result to the continuity equation and find the corresponding result for $A(z)$.

(h) In practice F is not controlled directly. Instead, one controls D_r, which in turn determines F. Use your velocity solution together with the boundary condition at $z = L$ to relate F to the known quantities. Substitute this result into the velocity and area equations to give a complete solution.

(i) Make a plot or a careful sketch of the fiber area and velocity versus z for $D_r = 10$.

Surprisingly, fiber-spinning models based on a Newtonian fluid are used quite widely in industry. To make the model useful in the factory we would need to add inertial effects, air drag on the fiber, and heat transfer effects. Note that we included heat transfer in this exercise in a very rough way, by assuming that the fiber would solidify at $z = L$.

13 **Spin coating.** Spin coating is used to apply a thin, uniform layer of a polymeric material onto a flat substrate. The process and nomenclature are explained in Chapter 3, Exercise 6, which you should review now.

In that exercise we assumed that the fluid quickly acquires the angular velocity of the substrate, and found that the momentum balance equation for the radial velocity is

$$-\rho \frac{v_\theta^2}{r} = \mu \frac{\partial^2 v_r}{\partial z^2}$$

(a) Write the boundary conditions at the free surface of the fluid ($z = h$) and at the substrate ($z = 0$).

(b) Substitute $v_\theta = \Omega r$ into the momentum balance equation given above (i.e., assume that the fluid acquires the angular velocity of the spinning substrate) and solve for the radial velocity distribution $v_r(r, z)$. Assume that the fluid layer thickness h is independent of r. Make a sketch showing how v_r varies with r and z.

(c) Use your result for v_r together with the continuity equation to find $v_z(r, z)$. Then use this result to find an expression for dh/dt, the rate of change of the layer thickness. How does dh/dt change with r? You should be able to show that, according to this solution, if the fluid layer has uniform thickness h at any time, then h remains uniform for all later times. This is the "secret" of spin coating: it naturally produces a layer of uniform thickness.

(d) Integrate your result to find an equation for h as a function of time. Use $h = h_0$ at $t = 0$ as the initial condition. The initial thickness, Ω, and time are the main process variables that determine the final thickness. Comment on the sensitivity of the final thickness to these quantities.

14 **Runner balancing.** A Newtonian fluid with constant density ρ and viscosity μ is injected into a tube of length L and radius R. The tube is initially empty, and the fluid is injected at a constant inlet pressure p_{in} at $z = 0$. If inertia is negligible and the tube is slender ($R \ll L$) then one can ignore the details of the flow near the entrance and the flow front, and treat the problem as fully developed flow in a tube. Let $\ell(t)$ represent the filled length of the tube at time t.

(a) Sketch the pressure distribution $p(x)$ at some instant of time during filling. Indicate the locations ℓ and L on your sketch.

(b) Find the instantaneous flow rate into the cavity at time t when the filled length is $\ell(t)$.

(c) Express the flow front velocity $d\ell/dt$ in terms of the flow rate entering the cavity.

(d) Combine the results from parts (a) and (b) to get a differential equation for $\ell(t)$. Solve for ℓ, taking the initial condition as an empty tube at $t = 0$. What is the time required to fill the tube?

(e) A problem that frequently occurs in polymer injection mold design is the need for *runner balancing*. Runners are the passages that convey polymer from the injection unit to the mold cavities. When the mold has more than one cavity, it is important that all cavities begin to fill at the same time, and fill at the same rate. Whenever possible, this is done by making the cavity layout symmetrical, as shown in the first sketch on the left below. However, sometimes a symmetrical layout is not possible, as shown in the sketch on the right below. The runners must then be balanced by adjusting their diameters.

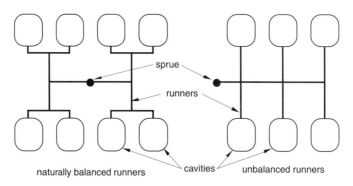

Consider an injection mold that has two cavities and an asymmetric runner layout, as shown in the second sketch below. The two runners have lengths L_1 and L_2. How should their radii R_1 and R_2 be chosen to ensure that the two mold cavities begin to fill at the same time?

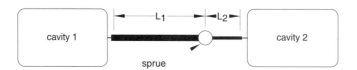

(f) Will your runner design also allow the two cavities to fill at the same rate? If not, what part of the mold design can be used to adjust the cavity fill rates?

15 *Filling of a disk-shaped mold.* A thin disk with a hole in the center is to be make by injection molding. The disk thickness is $2h$ and its outer radius is R_0. The hole in the center has radius R_i, and the polymer enters at this radius, uniformly distributed in θ. This is essentially how a compact disk is made, except that one side of the CD mold has tiny bumps, which form the pits that encode the music or data.

EXERCISES

Here you are to analyze the filling of this type of mold, assuming that the polymer can be modeled as a Newtonian fluid of constant density and viscosity, and that inertial effects and body forces are negligible.

(a) Let $r_f(t)$ represent the position of the flow front at time t. Relate dr_f/dt to the flow rate Q entering the mold. What is the pressure boundary condition at r_f?

(b) Using the solution for flow in a thin disk from Section 5.2.6, relate Q to the injection pressure p_{in}.

(c) Assume that the mold is filled under constant pressure p_{in}. Combine your previous results to get an equation for dr_f/dt in terms of know quantities, and then integrate to find $r_f(t)$. Finally, write an expression for the fill time.

(d) Suppose that a compact disk is being molded, and the viscosity of the polymer is 300 N s/m². Determine the approximate mold dimensions by measuring a compact disk. What injection pressure is required if the mold is filled in 2 s? The mold-clamping force is equal to the integral of pressure over the surface of the mold. What mold-clamping force is required at the instant that the cavity fills?

16 *Importance of surface tension in mold-filling analyses.* Consider two different fluids, a liquid metal and a polymer melt, filling a narrow strip mold with $H = 3$ mm, $L = 150$ mm, and $V = 75$ mm/s. The metal has $\rho = 2360$ kg/m³, $\mu = 0.003$ N s/m², and $\Gamma = 0.7$ N/m, whereas the polymer has $\rho = 1200$ kg/m³, $\mu = 900$ N s/m², and $\Gamma = 0.001$ N/m.

(a) For the polymer, use the results of Section 5.3.2 to determine whether or not surface tension is important. Check both the global and the local effects.

(b) For the metal, redo the scaling analysis. First analyze the equation of motion to determine an appropriate characteristic the pressure difference Δp_c. Then rescale the boundary condition and assess the importance of surface tension. Check both the local and the global importance of surface tension. In place of the capillary number you should get a new dimensionless group, which indicates the importance of inertia relative to surface tension. This is called the *Weber number*.

17 *A Marangoni flow.* Marangoni flows are driven by a temperature-induced variation in surface tension. Consider a layer of liquid metal in a shallow crucible that is withdrawn slowly from a furnace, as in directional solidification. The length of the crucible is L, and the depth of the liquid layer is H.

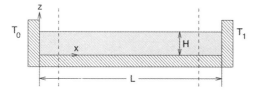

One end of the crucible is hotter than the other, and for a first approximation we will assume that this variation is linear with position:

$$T = T_0 + (T_1 - T_0)x/L$$

The surface tension Γ is a function of temperature, and a good first approximation is that this is also linear:

$$\Gamma = \Gamma_0 - C(T - T_0)$$

Here Γ_0 is the surface tension at T_0 and C is a constant. Here $d\Gamma/dT < 0$ for all liquids, so we expect C to be a positive quantity.

Other assumptions for this problem are that $H \ll L$, density and viscosity are constant, and the flow is steady and two dimensional. You should also assume that the free surface is flat (H is independent of x), as a result of a strong effect of gravity in the z direction.

(a) Apply scaling to the continuity equation, and show that $v_z \ll v_x$, at least away from the ends of the crucible (between the two dashed lines in the sketch).

(b) Write the momentum balance equation in the x direction and simplify it for this problem. Consider only the region between the dashed lines, and assume that the flow is fully developed there.

(c) Using Eq. (5.79), write the x and z components of the boundary condition on the free surface. Note that $\nabla_{\text{surf}}(\Gamma)$ has an x component of $\partial\Gamma/\partial x$ and a z component of zero. Also write the boundary condition at the bottom of the crucible.

(d) Integrate the momentum equation to find a velocity profile and apply the boundary conditions to evaluate the constants.

(e) You need a value for $\partial p/\partial x$, but there are no boundary conditions that give this. What must the value of $\int_0^H v_x \, dz$ be? Use this condition to determine $\partial p/\partial x$ and write an expression for the velocity profile that does not contain $\partial p/\partial x$.

(f) Sketch the velocity profile in the fully developed region and sketch the streamlines for the entire flow.

18 *Filling castings from a ladle.* A foundry makes a series of aluminum castings, each weighing 40 kg. The castings are filled from a ladle that is 0.5 m in diameter, having an orifice at the bottom of diameter 25 mm. The density of liquid aluminum is 2370 kg/m^3. Compute the time that it takes to fill the casting when the ladle is full, with a molten metal depth of 0.5 m. Compare this to the filling time for the same casting when the ladle is nearly empty, with a molten metal depth of 0.1 m. Comment on the validity of using Eq. (5.117) for the velocity at the orifice.

19 *Flow startup with large inertia.* A liquid metal is contained in a tank with a cross-sectional area A_0. A tube with cross-sectional area A and length L is connected to the tank, at a depth h below the surface. The tube is initially full, but its end is closed. At $t = 0$ the end of the tube is opened to the atmosphere. Assume that viscous effects are negligible, and that $A_0 \gg A$.

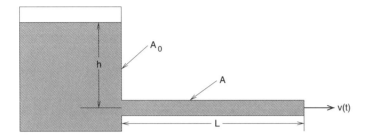

(a) For a streamline that runs from the free surface to the exit, write the unsteady form of Bernoulli's equation. Simplify the equation by using the assumption that $A \ll A_0$. Solve the resulting equation and show that the exit velocity as a function of time is

$$v = \sqrt{2gh}\, \tanh\!\left(\frac{\sqrt{2gh}\, t}{2L}\right)$$

Assume that the elapsed time is short enough that h remains constant.

(b) Suppose that $\rho = 2360$ kg/m^3, $h = 0.5$ m, and $L = 1.0$ m. Find the approximate time required for the fluid in the tube to reach its steady-state velocity.

20 *Deriving the RANS equations (advanced).* The Reynolds-averaged Navier–Stokes equations are a transformation of the Navier–Stokes equations by using a time-averaged velocity $\bar{\mathbf{v}}$ and pressure $\bar{p}$, as defined in Eq. (5.147). The velocity vector is separated into an average component and a deviation, $\mathbf{v}'$:

$$\mathbf{v} = \bar{\mathbf{v}} + \mathbf{v}'$$

The averaging process obeys the following rules:

$$\overline{\bar{\mathbf{u}}\bar{\mathbf{v}}} = \bar{\mathbf{u}}\bar{\mathbf{v}}, \qquad \overline{\frac{\partial \mathbf{v}}{\partial x}} = \frac{\partial \bar{\mathbf{v}}}{\partial x}, \qquad \overline{\mathbf{v}'} = 0$$

Derive the RANS equations by taking the following steps.

(a) Write the continuity equation for a constant-density fluid, and replace $\mathbf{v}$ by $\bar{\mathbf{v}} + \mathbf{v}'$. Take the average of this equation and then use the rules above to simplify it. Show that this leads to Eq. (5.150).

(b) Write the Navier–Stokes equation for a fluid with constant viscosity, and then make the same substitution for velocity and a similar substitution for pressure. Simplify the equation by using the rules above, and show that the result is Eqs. (5.151)–(5.154).

CHAPTER SIX

Non-Newtonian Fluid Flow

Many common fluids, including air, water, honey, and liquid metals, are accurately represented by the Newtonian constitutive equation. However, there are also many materials whose behavior is distinctly non-Newtonian: molten polymers, molten glass, semisolid metals, grease, many types of paint, and foods such as mayonnaise, peanut butter, and melted cheese. Non-Newtonian fluids obey the same mass and momentum balance equations as Newtonian fluids, but they have different constitutive equations for stress. In this chapter we show how to develop constitutive equations for non-Newtonian fluids, and we demonstrate their use in the solving of flow problems. We also develop a simplified set of governing equations for viscous flow in narrow gaps, the generalized Hele–Shaw approximation.

The study of the deformation and flow of materials, particularly of non-Newtonian fluids, is the subject called *rheology*. Rheologists create experimental techniques to measure viscosity and other flow properties, they develop constitutive equations that describe non-Newtonian material behavior, and they relate that material behavior to microscopic structure. A full treatment of rheology is beyond the scope of this book, and readers may wish to consult some of the excellent texts on the subject (Tanner, 1985; Bird et al., 1987; Barnes et al., 1989; Dealy and Wissbrun, 1990; Macosko, 1994).

6.1 NON-NEWTONIAN BEHAVIOR

6.1.1 PURELY VISCOUS FLUIDS

There is only one type of Newtonian fluid behavior, and one Newtonian fluid differs from another only by its value of viscosity. However, there are many different types of non-Newtonian behavior. Assume for the moment that we can subject a material to any type of deformation, and measure the resulting stress. We can group different non-Newtonian behaviors into categories according to how their stress depends on the rate of deformation. If the stress at any time depends only on the rate of deformation *at that time*, that is, if

$$\tau(t) = f(\mathbf{D}(t)) \tag{6.1}$$

then we have a *purely viscous material*. All Newtonian fluids meet this requirement, because they exhibit a linear relationship between $\tau(t)$ and $\mathbf{D}(t)$. If the relationship

6.1 NON-NEWTONIAN BEHAVIOR

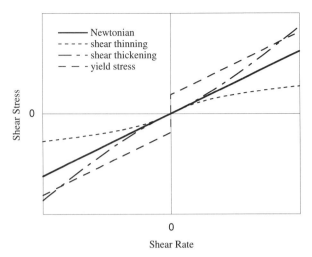

Figure 6.1: The rheological response of different viscous fluids in a simple shear flow, such as the drag flow between parallel plates.

in Eq. (6.1) is nonlinear then we have one type of non-Newtonian model, called a *generalized Newtonian fluid*.

The most common experiments to measure fluid properties use a simple shear flow, such as the drag flow between parallel plates shown in Fig. 2.3. Suppose that we can create this flow at any desired shear rate $\dot{\gamma} = \partial v_x/\partial y$ and measure the resulting shear stress τ_{yx}. Some relationships we might observe between shear stress and shear rate in this flow are sketched in Fig. 6.1. In each case we assume that the fluid is purely viscous, so it obeys Eq. (6.1). The non-Newtonian viscosity η is defined as the ratio of shear stress to shear rate in this experiment:

$$\eta = \frac{\tau_{yx}}{\dot{\gamma}} \tag{6.2}$$

Note that the non-Newtonian viscosity η can vary with the shear rate, whereas the viscosity μ of a Newtonian fluid is independent of shear rate. We will use these two different symbols for viscosity to indicate whether we are dealing with a Newtonian or a non-Newtonian fluid.

Returning to the experiment, if the shear stress versus shear rate curve turns downward as the shear rate is increased, then the fluid is said to be *shear thinning*, or pseudoplastic. This is a very common type of non-Newtonian behavior. For a shear-thinning fluid the viscosity η decreases with increasing shear rate.

If a fluid has the opposite type of behavior, in which its viscosity increases as the shear rate increases, it is called *shear thickening*, or dilatant. This type of behavior is sometimes exhibited by dense suspensions of solid particles in a fluid. In such materials a large fraction of the stress is transmitted by contact between the particles, and higher shear rates lead to increased contacts and higher stresses. However, shear thickening is much less common that shear thinning.

Some fluids will not flow at all unless the stress exceeds a minimum value. Mayonnaise and catsup are familiar materials that act this way. Such fluids are said to possess a *yield stress*, and this is also a type of non-Newtonian fluid behavior. If the stress in a material depends only on the instantaneous rate of deformation (i.e., if it is purely viscous) and it also exhibits a yield stress, then it is said to be *viscoplastic*. Viscoplastic models can be used to represent the deformation of solid metals for some ranges of strain rate and temperature. Figure 6.1 includes a curve for a material with a yield stress. Even if the stress–strain rate curve is linear after yielding, the viscosity η decreases with increasing strain rate. (Recall the definition of η.) Thus, having a yield stress is a particular type of shear-thinning behavior.

6.1.2 HISTORY-DEPENDENT FLUIDS

A second class of non-Newtonian behavior, distinct from the purely viscous fluid, is one when the stress depends on the *history* of the rate of deformation. There are several important subcategories of this type of behavior. The fluid may still be viscous, in the sense that all the work of deformation is converted to heat, but its viscosity may depend on the recent history of deformation. If the viscosity of a fluid decreases over time when it is sheared at a constant rate, the fluid is said to be *thixotropic*. Thixotropic behavior usually indicates that some type of structure within the material is being broken down as the material is sheared. For instance, when a metal alloy is in a semisolid state, the solid phase tends to form long treelike shapes called *dendrites*. If the liquid–solid mixture is sheared at a high enough rate, the dendrites bread up into smaller, rounder particles. This change is accompanied by a reduction in viscosity. If the shearing stops and the material is allowed to sit quietly, the original structure may reform itself, and the viscosity may recover toward its initial value.

Another type of history-dependent non-Newtonian behavior is *viscoelasticity*. A viscoelastic material has a characteristic time constant λ called the *relaxation time*. If the material is deformed over a time scale much longer than λ then it behaves like a viscous fluid, whereas on time scales much shorter than λ the material behaves like an elastic solid. Silly Putty is a viscoelastic material whose relaxation time is a few seconds. If Silly Putty is formed into a ball and thrown onto the floor, it responds elastically and bounces. If it is formed into a cylinder and held at the upper end, then in about 1 min it will sag and flow like a liquid.

In a purely viscous fluid, all the work used to deform the fluid is dissipated by heat, but in a viscoelastic fluid some of the work of deformation may be stored as elastic energy and released at a later time. All polymers are viscoelastic in both the liquid and solid states, though the viscoelastic effects may be subtle or dramatic, depending on the conditions. Inorganic glasses also exhibit viscoelastic behavior.

The description of viscoelastic behavior and the solution of viscoelastic flow problems are extremely challenging subjects, and we will not undertake them in any detail. Fortunately, polymer melts flowing in channels can be quite usefully modeled as purely viscous, but non-Newtonian, fluids. In fact, most fluid flow models in polymer processing use this treatment. In later discussions we will point out where this simplification may be inaccurate, but for now we move ahead with viscous models.

6.2 THE POWER LAW MODEL

One simple way to model either shear-thinning or shear-thickening behavior is to make the shear stress proportional to the shear rate raised to a power. This model is

6.2 THE POWER LAW MODEL

called the *power law fluid*. Most polymer melts can be modeled accurately as power law fluids when the shear rate is high, and the power law fluid has been widely used to model polymer processing flows. In this section we present the power law model without any derivation. We then solve some simple channel flows for a power law fluid, much as we did before for the Newtonian fluid. In a subsequent section we show that the way the power law model is written allows it to be used in any type of flow.

6.2.1 A POWER LAW CONSTITUTIVE EQUATION

One might think to write the relation between extra stress and shear rate for a power law fluid as

$$\tau_{ij} = m(2D_{ij})^n \tag{6.3}$$

but this form cannot handle cases in which $n \neq 1$ and D_{ij} is negative, and it has other deficiencies that we will see later. The actual power law model is

$$\tau_{ij} = 2m\dot{\gamma}^{n-1} D_{ij} \tag{6.4}$$

with $\dot{\gamma}$, called the *strain rate*, given by

$$\dot{\gamma} = \sqrt{2\left(D_{11}^2 + D_{22}^2 + D_{33}^2\right) + (2D_{12})^2 + (2D_{23})^2 + (2D_{31})^2} \tag{6.5}$$

Because $\dot{\gamma} \geq 0$, this form guarantees that each component of τ_{ij} will have the same sign as the corresponding component of D_{ij}. Also, we never have to raise a negative number to a noninteger power. In Eq. (6.4), m and n are material constants; n, called the *power law index*, is dimensionless. If $n < 1$ the fluid is shear thinning, and $n > 1$ gives shear-thickening behavior. The constant m has the rather odd units of N s^n/m^2. The interpretation of m is clearer if we recognize that $(m\dot{\gamma}^{n-1})$ is the viscosity η, which has units of N s/m^2. When $n = 1$ the power law model reduces to the Newtonian fluid, and $m = \mu$.

With Eqs. (6.4) and (6.5) in hand we now examine a drag flow and a pressure flow for the power law fluid. We will treat these flows separately to emphasize the similarities and differences between the power law and Newtonian fluids. In particular, the superposition of solutions we observed earlier will not apply for non-Newtonian fluids.

6.2.2 DRAG FLOW OF A POWER LAW FLUID

The geometry for this case is illustrated in Fig. 6.2. Two long plates separated by a narrow gap contain the fluid, and the top plate is driven at constant speed V in the

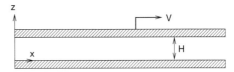

Figure 6.2: Drag flow in a power law fluid.

x direction while the lower plate is held fixed. In order to focus on the physics of the power law fluid, we make the following now-familiar assumptions:

- steady flow,
- two-dimensional flow,
- fully developed flow,
- constant fluid density,
- no body forces,
- negligible inertia, and
- no slip at the walls.

Recall that assuming the flow to be fully developed is equivalent to excluding the entrance and exit regions, regions whose length is only about H. We will do as much of our analysis as possible before invoking a constitutive relation, to help separate and identify the characteristics of the flow and of the material.

As before, fully developed flow implies $\partial v_i/\partial x = 0$, for any component of velocity v_i. The continuity equation, together with the assumption of two-dimensional fully developed flow, leads to $v_z = 0$.

We turn next to the momentum balance equations. Because the fluid is non-Newtonian, we use the general form from Appendix B.5, rather than the Navier–Stokes equation. Consider first the y-direction momentum balance. Because $v_y \equiv 0$, since the flow is two dimensional, the left-hand side vanishes, and we have

$$0 = -\frac{\partial p}{\partial y} + \frac{\partial \tau_{xy}}{\partial x} + \frac{\partial \tau_{yy}}{\partial y} + \frac{\partial \tau_{zy}}{\partial z} \tag{6.6}$$

where we have neglected the body force terms. Invoking the assumptions of two dimensionality and a fully developed flow leaves

$$0 = -\frac{\partial p}{\partial y} + \frac{\partial \tau_{zy}}{\partial z} \tag{6.7}$$

It is certainly true for the power law fluid that τ_{zy} depends on D_{zy}, and we expect this to be the case for virtually all fluids. Writing D_{zy} in terms of the velocity components, we find

$$D_{yz} = \frac{1}{2}\left(\frac{\partial v_z}{\partial y} + \frac{\partial v_y}{\partial z}\right) \tag{6.8}$$

which is exactly zero for this flow, because we have already demonstrated that both velocity components are zero. We conclude that $\partial p/\partial y = 0$. A similar analysis of the z-direction momentum balance equation shows that $\partial p/\partial z = 0$.

Turning to the x-direction momentum balance, the general equation is

$$\rho\left(\frac{\partial v_x}{\partial t} + v_x\frac{\partial v_x}{\partial x} + v_y\frac{\partial v_x}{\partial y} + v_z\frac{\partial v_x}{\partial z}\right) = -\frac{\partial p}{\partial x} + \frac{\partial \tau_{xx}}{\partial x} + \frac{\partial \tau_{yx}}{\partial y} + \frac{\partial \tau_{zx}}{\partial z} + \rho b_x \tag{6.9}$$

The first two terms on the left-hand side are zero, as a result of the assumptions of steady and fully developed flow, respectively, and the next two terms are zero because the y and z velocity components vanish. The assumptions leading to a fully

6.2 THE POWER LAW MODEL

developed, two-dimensional flow also allow us to eliminate $\partial \tau_{xx}/\partial x$ and $\partial \tau_{xy}/\partial y$. We also neglect the body force, which leaves

$$0 = -\frac{\partial p}{\partial x} + \frac{\partial \tau_{zx}}{\partial z} \tag{6.10}$$

Finally, we note that a fully developed flow implies that $\partial p/\partial x$ is independent of x. Therefore, $\partial p/\partial x$ is at most a constant. However, for a pure drag flow the pressure is the same at both ends, so we have $\partial p/\partial x = 0$. The momentum balance is then reduced to

$$0 = \frac{\partial \tau_{zx}}{\partial z} \tag{6.11}$$

We integrate this equation to find

$$\tau_{xz} = c_0 \tag{6.12}$$

where c_0 is a constant of integration. Note that we have reached this point without invoking a constitutive model; Eq. (6.12) is a consequence of the geometry and assumptions, which apply to any highly viscous fluid.

We are now ready to introduce the constitutive relation. The assumption of two-dimensional, fully developed flow has given $v_y = v_z = 0$ and $v_x = f(z)$, so the only nonzero components of **D** are $2D_{xz} = 2D_{zx} = \partial v_x/\partial z$. Using this in Eq. (6.5), we find the strain rate to be

$$\dot{\gamma} = \sqrt{(2D_{xz})^2} = \sqrt{\left(\frac{\partial v_x}{\partial z}\right)^2} = \left|\frac{\partial v_x}{\partial z}\right| \tag{6.13}$$

Then Eq. (6.4) gives the shear stress as

$$\tau_{zx} = m \left|\frac{\partial v_x}{\partial z}\right|^{n-1} \frac{\partial v_x}{\partial z} \tag{6.14}$$

The absolute value in this expression is inconvenient for integration. We can eliminate it in this particular problem by noting that for this flow $\partial v_x/\partial z$ will be positive, provided that V is positive. Otherwise, pushing the fluid in one direction would cause it to flow in the opposite direction, which violates common sense. Thus we have $|\partial v_x/\partial z| = (\partial v_x/\partial z)$, and Eq. (6.14) becomes

$$\tau_{zx} = m \left(\frac{\partial v_x}{\partial z}\right)^n = c_0 \tag{6.15}$$

Manipulating this equation to isolate the derivative of v_x yields

$$\frac{\partial v_x}{\partial z} = \left(\frac{c_0}{m}\right)^{1/n} = c_1 \tag{6.16}$$

where c_1 is a new constant that replaces c_0. Integrating and imposing the boundary conditions that $v_x = 0$ at $z = 0$ and $v_x = V$ at $z = H$ gives us a result we have seen before, a linear velocity profile:

$$v_x = \frac{Vz}{H} \tag{6.17}$$

Surprisingly, the velocity field in this flow is *independent* of the constitutive relation for the fluid. This may seem counterintuitive, at first. However, note that the momentum balance gave a constant shear stress in this flow, without requiring us to say anything about the fluid behavior. If we add only the requirement that the fluid be purely viscous and have uniform properties across the gap, then the shear rate must also be constant. With no-slip boundary conditions at the wall, the linear distribution of Eq. (6.17) must be the velocity profile.

Although the velocity field in drag flow is the same for all isothermal Newtonian and power law fluids, the shear stress is not. The shear rate can be computed directly from the velocity field,

$$2D_{zx} = \frac{\partial v_x}{\partial z} = \frac{V}{H} \tag{6.18}$$

and the shear stress is then given by

$$\tau_{zx} = m \left(\frac{V}{H}\right)^n \tag{6.19}$$

This is a nonlinear relationship between τ_{zx} and V. Note that plotting shear stress versus shear rate on log–log axes produces a straight line with slope n.

When we treated this problem for the Newtonian fluid, we were able to forego the assumption of fully developed flow, and instead demonstrate that the transverse velocity and pressure gradients were negligible by using a scaling argument. This led us to the lubrication approximation. A similar analysis could be performed for the power law fluid, though the details are more tedious. However, this argument cannot be made without using some type of constitutive equation. In particular, there is no scaling basis for eliminating the inertial terms unless one can introduce a characteristic value for the viscosity.

6.2.3 PRESSURE FLOW IN A TUBE

We now consider pressure flow of a power law fluid, but, rather than doing the same geometry as we did for the Newtonian fluid, we consider the flow of a power law fluid in a long circular tube. The pressure flow between parallel plates is slightly easier to work out, and it is given as an exercise at the end of the chapter.

Consider flow in the tube illustrated in Fig. 6.3, driven by a pressure difference between the two ends. Our assumptions are similar to those we used to analyze the drag flow of the power law fluid:

- steady flow,
- axisymmetric flow,
- constant fluid density,
- no body forces,

Figure 6.3: Schematic drawing for flow in a circular tube. We assume that $R/L \ll 1$.

6.2 THE POWER LAW MODEL

- negligible inertia (implies fully developed flow), and
- no slip on the wall.

We begin our analysis once again with the mass balance equation. In cylindrical coordinates this is

$$\frac{1}{r}\frac{\partial}{\partial r}(rv_r) + \frac{1}{r}\frac{\partial v_\theta}{\partial \theta} + \frac{\partial v_z}{\partial z} = 0 \tag{6.20}$$

The assumption of axisymmetry eliminates $\partial v_\theta/\partial \theta$, and the assumption that the flow is fully developed eliminates $\partial v_z/\partial z$. This leaves us with

$$\frac{1}{r}\frac{\partial}{\partial r}(rv_r) = 0 \tag{6.21}$$

which we integrate to find

$$v_r = \frac{c_1}{r} \tag{6.22}$$

where c_1 is a constant. Because v_r is zero on the wall of the tube, we have $c_1 = 0$ and $v_r \equiv 0$. Axisymmetry, together with the condition that $v_\theta = 0$ on the wall, leads to the conclusion that $v_\theta \equiv 0$. The only velocity is in the z direction.

Now consider the z-direction momentum equation. As in the previous example, all of the unsteady and inertial terms vanish (the reader should confirm this by writing out the full equation and canceling the terms), so the momentum balance is

$$0 = -\frac{\partial p}{\partial z} + \frac{1}{r}\frac{\partial}{\partial r}(r\tau_{rz}) + \frac{1}{r}\frac{\partial \tau_{\theta z}}{\partial \theta} + \frac{\partial \tau_{zz}}{\partial z} + \rho b_z \tag{6.23}$$

Once again, the assumption of axisymmetry eliminates $(1/r)(\partial \tau_{\theta z}/\partial \theta)$, and the assumption of a fully developed flow eliminates $\partial \tau_{zz}/\partial z$. The body force b_z is also neglected, leaving

$$0 = -\frac{\partial p}{\partial z} + \frac{1}{r}\frac{\partial}{\partial r}(r\tau_{rz}) \tag{6.24}$$

Following the same procedure as in the previous example, the r- and θ-direction momentum equations reduce to $\partial p/\partial r = 0$ and $\partial p/\partial \theta = 0$, respectively. These results, along with the assumption of a fully developed flow, indicate that $\partial p/\partial z$ is constant. The geometry of the problem shows that

$$\frac{\partial p}{\partial z} = \frac{p_2 - p_1}{L} = -\frac{\Delta p}{L} \tag{6.25}$$

where $\Delta p = p_1 - p_2$. We define Δp to be positive when the flow is in the positive z direction, in which case $\partial p/\partial z$ is negative. Substituting for $\partial p/\partial x$ into Eq. (6.24) yields

$$-\frac{\Delta p}{L} = \frac{1}{r}\frac{\partial}{\partial r}(r\tau_{rz}) \tag{6.26}$$

which we integrate to find

$$\tau_{rz} = -\frac{\Delta p}{2L}r + \frac{c_2}{r} \tag{6.27}$$

The constant c_2 must be zero for the stress to remain finite at $r = 0$, so the shear stress distribution is

$$\tau_{rz} = -\frac{\Delta p}{2L} r \tag{6.28}$$

Note once more that we have not yet assumed any constitutive relation, so Eq. (6.28) gives the shear stress distribution for steady, fully developed axisymmetric tube flow of *any* viscous fluid.

Next we find the strain rate. Referring to Appendix B.3 for the components of **D** in cylindrical coordinates, and using $v_r = v_\theta = 0$ and $v_z = f(r)$, we find that the only nonzero components are $2D_{rz} = 2D_{zr} = \partial v_z/\partial r$. Then Eq. (6.5) gives the strain rate as

$$\dot{\gamma} = \left| \frac{\partial v_z}{\partial r} \right| \tag{6.29}$$

Now using the constitutive equation, Eq. (6.4), for the power law fluid, we find that the shear stress is

$$\tau_{rz} = m \left| \frac{\partial v_z}{\partial r} \right|^{n-1} \frac{\partial v_z}{\partial r} \tag{6.30}$$

To eliminate the absolute value sign, note that v_r should decrease monotonically from its maximum at $r = 0$ to zero on the walls (provided Δp is positive). Therefore $\partial v_z/\partial r < 0$, and $|\partial v_z/\partial r| = -(\partial v_z/\partial r)$. Hence we may rewrite Eq. (6.30) as

$$\begin{aligned}\tau_{rz} &= m \left(-\frac{\partial v_z}{\partial r} \right)^{n-1} \frac{\partial v_z}{\partial r} \\ &= -m \left(-\frac{\partial v_z}{\partial r} \right)^{n} \end{aligned} \tag{6.31}$$

Note that the numerical value of the expression in the parentheses is positive. Equating the expression for τ_{rz} in Eq. (6.31) with the result of the momentum balance, Eq. (6.28), yields

$$-m \left(-\frac{\partial v_z}{\partial r} \right)^n = -\frac{\Delta p}{2L} r \tag{6.32}$$

Rearranging terms and using the notation that $s \equiv 1/n$, we have

$$\frac{\partial v_z}{\partial r} = -\left(\frac{\Delta p}{2mL} \right)^s r^s \tag{6.33}$$

At last we have something that can be integrated! Carrying out the integration, we find

$$v_z = -\frac{1}{s+1} \left(\frac{\Delta p}{2mL} \right)^s r^{s+1} + c_3 \tag{6.34}$$

To this we apply the boundary condition $v_z = 0$ at $r = R$ to evaluate c_3, and we obtain the final velocity distribution:

$$v_z = \frac{1}{s+1} \left(\frac{\Delta p}{2mL} \right)^s (R^{s+1} - r^{s+1}) \tag{6.35}$$

6.2 THE POWER LAW MODEL

The power law fluid reduces to a Newtonian fluid when $s = 1$ and $m = \mu$. Substituting these values into Eq. (6.35), we obtain the familiar Newtonian parabolic profile,

$$v_z = \frac{\Delta p}{4\mu L}(R^2 - r^2) \tag{6.36}$$

which is identical to Eq. (5.46). This reduction to the Newtonian solution is a useful check for power law fluid solutions.

Returning to the power law fluid, we find that the total flow rate passing through the tube is obtained by integrating the velocity over the cross section,

$$Q = \int_0^{2\pi}\int_0^R v_z(r)\, r\, dr\, d\theta$$
$$= \frac{\pi}{s+3}\left(\frac{\Delta p}{2mL}\right)^s R^{s+3} \tag{6.37}$$

The average velocity $\bar{v}_z$ is Q divided by the area,

$$\bar{v}_z = \frac{Q}{\pi R^2}$$
$$= \frac{R^{s+1}}{s+3}\left(\frac{\Delta p}{2mL}\right)^s \tag{6.38}$$

This is actually quite different behavior than the Newtonian solution in Eq. (5.48). Suppose that we have $n = 0.5$, which represents a significant, but not unusual, level of shear thinning. Then $s = 2.0$, and Q is proportional not to Δp, as in the Newtonian case, but to $(\Delta p)^2$. That is, doubling the pressure drop will quadruple the flow rate.

To get a better idea of what is going on here, we normalize v_z by the average velocity $\bar{v}_z$:

$$v_z = \bar{v}_z \frac{s+3}{s+1}\left(1 - \left[\frac{r}{R}\right]^{s+1}\right) \tag{6.39}$$

This provides a convenient way to compare velocity profiles for different values of n. These are shown in Fig. 6.4. The curve for $n = 1$ is the parabola exhibited by a Newtonian fluid. For a shear-thinning fluid, $n < 1$, the higher stress near the walls leads to a lower viscosity, and the lower viscosity requires an increased shear rate $|\partial v_z/\partial r|$ to sustain the higher shear stress. At the center of the tube the shear stress is small, so the viscosity is high there and the shear rate is low. This makes the velocity profile for shear-thinning fluids flattened, compared with the Newtonian profile. As n becomes very small the fluid is almost in a plug flow, with very little deformation in the center of the tube. For shear-thickening fluids, which have $n > 1$, the opposite is true, and the velocity profile becomes more pointed than the Newtonian profile.

Shear-thinning behavior is actually advantageous in many situations, such as profile extrusion of polymers. Inside the die the shear rate is large, so the polymer has a relatively low viscosity, and the pressure drop across the die is not too high. However, once the polymer exits the die it is subjected only to stresses from its own weight, which are small. Under these low stresses the polymer has a high viscosity, which helps it holds its shape as it cools.

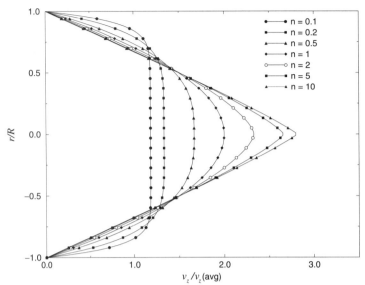

Figure 6.4: Axial flow velocity profiles for power law fluids flowing in a tube. Note that $n = 1$ represents a Newtonian fluid. Fluids with $n < 1$ are shear thinning, and fluids with $n > 1$ are shear thickening. The solutions are scaled so that they all have the same total flow rate.

6.3 POWER LAW SOLUTIONS FOR OTHER SIMPLE GEOMETRIES

There are analytical solutions for the flow of a power law fluid in a variety of simple channel geometries. The corresponding Newtonian solutions are given in Section 5.2. We present the results without derivation. As in the previous section, we use $s \equiv 1/n$ as a convenient shorthand.

6.3.1 PRESSURE FLOW BETWEEN PARALLEL PLATES

Two flat plates are separated by a distance $2h$, with $z = 0$ being the midplane between the plates. A pressure difference Δp is applied over a distance L in the flow (x) direction, and the width of the plates in the y direction is W. This solution requires $h \ll L$ and $h \ll W$. The shear stress distribution is linear, and this is independent of the constitutive equation

$$\tau_{zx} = -\frac{\Delta p}{L} z \qquad (6.40)$$

The scaled velocity distribution has the same shapes as for the tube flow in Fig. 6.4, but the multiplier is different.

$$v_x(z) = \frac{h}{s+1} \left(\frac{h \Delta p}{mL}\right)^s \left(1 - \left|\frac{z}{h}\right|^{s+1}\right) \qquad (6.41)$$

The overall flow rate is

$$Q = \frac{2Wh^2}{s+2} \left(\frac{h \Delta p}{mL}\right)^s \qquad (6.42)$$

As expected, the results for v_x and Q reduce to the Newtonian result when $n \to 1$ and $m \to \mu$.

6.3.2 AXIAL DRAG FLOW BETWEEN CONCENTRIC CYLINDERS

A long cylinder of radius κR is centered inside a cylinder of radius R, and it moves axially with velocity V. There is no pressure gradient. This creates a drag flow, whose velocity is

$$v_z(r) = V \frac{[(r/R)^{1-s} - 1]}{[\kappa^{1-s} - 1]} \tag{6.43}$$

The overall flow rate is

$$Q = \frac{2\pi R^2 V}{(\kappa^{1-s} - 1)} \left[\frac{1 - \kappa^{3-s}}{3-s} - \frac{1 - \kappa^2}{2} \right] \tag{6.44}$$

and the shear stress distribution is

$$\tau_{zr} = -\frac{mR}{r} \left[\frac{V(s-1)}{R(\kappa^{1-s} - 1)} \right]^n \tag{6.45}$$

This solution requires $(1 - \kappa)R \ll L$, where L is the length of the cylinders (i.e., the gap is small compared to the length). There is no restriction on the value of κ.

Like many problems in cylindrical coordinates, these equations do not go smoothly to the Newtonian limit of $n = 1$. Notice that when $s \to 1$, $v_z(r) \to 0$ in Eq. (6.43). The Newtonian solution must be derived separately.

6.3.3 TANGENTIAL DRAG FLOW BETWEEN CONCENTRIC CYLINDERS

Again we have fluid in the annular space between a cylinder of radius κR and a cylinder of radius R, but this time the flow is created by rotating the inner cylinder with angular velocity Ω. The velocity is

$$v_\theta(r) = \Omega r \frac{1 - (R/r)^{2s}}{1 - (1/\kappa^{2s})} \tag{6.46}$$

and the torque required to rotate the inner cylinder is

$$T = 2\pi (\kappa R)^2 m L \left(\frac{2\Omega}{n(1 - \kappa^{2s})} \right)^n \tag{6.47}$$

This corresponds to a shear stress of

$$\tau_{r\theta} = -m \left[\frac{2\Omega}{n(\kappa^{-2s} - 1)} \right]^n \left(\frac{R}{r} \right)^2 \tag{6.48}$$

where the sign applies when Ω is positive. Again we require $(1 - \kappa)R \ll L$. Note that measuring $\tau_{r\theta}$ over a range of values of Ω allows one to measure the power law exponent n. Once n is known, then m can also be deduced. Thus, the Couette viscometer is also useful for non-Newtonian fluids.

6.3.4 RADIAL PRESSURE FLOW BETWEEN PARALLEL DISKS

Two disks are separated by a distance $2h$, with $z = 0$ at the midplane between the disks. Fluid is injected at the center at a volume flow rate Q and flows radially outward.

At some radius R_o the fluid pressure is p_o. The velocity profile has the same shape as the flow between parallel plates, but the magnitude of velocity decreases with radius.

$$v_r(z, r) = \frac{Q}{4\pi r h} \left(\frac{s+2}{s+1}\right) \left(1 - \left|\frac{z}{h}\right|^{s+1}\right) \tag{6.49}$$

The pressure does not change with z, but varies with r according to

$$p(r) - p_o = \frac{mQ^n}{1-n} \left(\frac{s+2}{4\pi h^{s+2}}\right)^n \left[R_o^{1-n} - r^{1-n}\right] \tag{6.50}$$

For $n < 1$ the pressure is well defined as $r \to 0$, though it is more common and physically meaningful to prescribe the injection pressure at some finite-valued inner radius. The Newtonian solution has a $\ln(r)$ pressure dependence, and $p \to \infty$ as $r \to 0$, so choosing a finite radius at which to measure the inlet pressure becomes essential.

This solution requires $h \ll R_o$. If the inlet pressure in measured at a radius R_i then strictly speaking we must also have $h \ll R_i$; however, this restriction is sometimes ignored.

6.4 PRINCIPLES OF NON-NEWTONIAN CONSTITUTIVE EQUATIONS

The previous section introduced Eqs. (6.4) and (6.5) for the power law fluid without much explanation. In this section we show why this particular form is used for the power law fluid, and we provide some general guidelines and tools for developing constitutive equations.

6.4.1 MATERIAL OBJECTIVITY

The physical principles that lead us to the balance equations in Chapter 2 do not provide the same guidance in developing constitutive equations. There are too many different kinds of rheological behavior exhibited by various materials to permit this, and developing a constitutive equation always requires some knowledge about the particular material, either from theory or from experiments.

However, continuum mechanics does provide one very important guideline for constitutive equations, which is known as the *principle of material objectivity* or *material frame indifference*. This principle says that the predicted behavior of a material cannot depend on the coordinate frame we use to do the calculation. The coordinate axes are something that we impose on the problem. We usually choose them for our own convenience, but any choice is legal. When we use a constitutive equation, we want to be sure that the answers it gives us for velocity profiles, stresses, and the like are not going to change if we rework the problem using a different coordinate system. A constitutive equation that satisfies this requirement is said to be *objective*. There is no guarantee that an objective equation will accurately describe the behavior of any given material, but an equation that is not objective is so unreliable and unpredictable that we reject all nonobjective equations.

As an example, we said that Eq. (6.3) was unacceptable because it cannot be used when any of the D_{ij}'s have negative values. We could have patched up the solution

6.4 PRINCIPLES OF NON-NEWTONIAN CONSTITUTIVE EQUATIONS

for that particular problem by writing the power law equation as

$$\tau_{ij} = 2m|2D_{ij}|^{n-1} D_{ij} \qquad (6.51)$$

If we were to use this equation to do the drag flow and pressure flow problems from the previous section, we would get the same answers for stress distribution and velocity profile. However, it is not hard to show that Eq. (6.51) is *not* objective. Using Eq. (6.51) in a multidirectional drag flow (a problem we will do correctly in a moment) gives a force on the upper plate that changes if one rotates the coordinate system. This is not realistic physical behavior, and we would not want to use an equation that has this property.

If we find just one example in which a constitutive equation gives different answers in two coordinate systems, then we have proved that the equation is not objective. But we would much rather know how to be certain that a constitutive equation will be objective. It is neither convenient nor possible to check the equation in all possible flow problems. Instead we need some sort of building blocks that guarantee objectivity. Suppose we wanted to make Eq. (6.51) objective. We could start by writing something like the Newtonian fluid equation, only using η for the viscosity, and letting η depend on the rate of deformation:

$$\tau = 2\eta(\mathbf{D})\mathbf{D} \qquad (6.52)$$

This lets the "direction" of the stress correspond to the direction of the rate of deformation, which is good; but now we need to find some way for the viscosity η to depend on $\mathbf{D}$. This is not easy, because η is a scalar and $\mathbf{D}$ is a tensor. We not only have to match "apples to oranges," we must do so in a way that is objective. That is, the value of η must be independent of the coordinate system in which we express the components of $\mathbf{D}$. How can this be done? We can use the *invariants* of $\mathbf{D}$. Invariants are scalar properties of a vector or tensor that are independent of the reference frame in which they are computed.

6.4.2 COORDINATE TRANSFORMATION AND INVARIANTS

To explain how invariants work, a brief discussion of transformation under coordinate rotation is needed. Suppose that a vector $\mathbf{v} = \{v_x, v_y, v_z\}^T$ is defined in a Cartesian coordinate system, with axes x, y, z. We then choose a new set of Cartesian axes, x', y', z'. These might be rotated relative to the xyz axes by the Euler angles ϕ, θ, and ψ, as illustrated in Fig. 6.5. That is, starting from the xyz axes, one first rotates the axes by ϕ about the z axis, then by θ about the current x axis, and finally by ψ about the new z' axis.

In the new coordinate system, the components of the vector $\mathbf{v}'$ are given by

$$\begin{Bmatrix} v'_x \\ v'_y \\ v'_z \end{Bmatrix} = \begin{bmatrix} \cos\psi \cos\phi - \cos\theta \sin\phi \sin\psi & \cos\psi \sin\phi + \cos\theta \cos\phi \sin\psi & \sin\psi \sin\theta \\ -\sin\psi \cos\phi - \cos\theta \sin\phi \cos\psi & -\sin\psi \sin\phi + \cos\theta \cos\phi \cos\psi & \cos\psi \sin\theta \\ \sin\theta \sin\phi & -\sin\theta \cos\phi & \cos\theta \end{bmatrix}$$
$$\times \begin{Bmatrix} v_x \\ v_y \\ v_z \end{Bmatrix} \qquad (6.53)$$

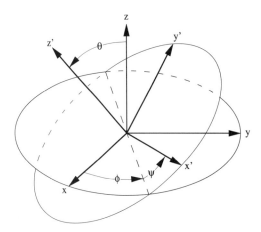

Figure 6.5: Rotation of coordinate axes from the xyz to the $x'y'z'$ system, using the Euler angles ϕ, θ, and ψ.

We can write this in a more compact form as

$$\mathbf{v}' = \mathbf{Q} \cdot \mathbf{v} \quad \text{or} \quad v'_i = Q_{ij} v_j \tag{6.54}$$

Each component Q_{ij} turns out to be the cosine of the angle between a unit vector $\mathbf{e}'_i$ from the new system and a unit vector $\mathbf{e}_j$ from the old system.

If the original set of axes is orthogonal and new set is too, then the transformation matrix $\mathbf{Q}$ will be an *orthogonal matrix*, meaning that its transpose is also its inverse. That is,

$$\mathbf{Q} \cdot \mathbf{Q}^T = \mathbf{Q}^T \cdot \mathbf{Q} = \mathbf{I} \quad \text{or} \quad Q_{ij} Q_{ik} = \delta_{jk} \tag{6.55}$$

for any matrix $\mathbf{Q}$ that represents a transformation between two orthogonal coordinate systems.

Now let us look at invariants. A vector has one invariant: its magnitude. To prove that the magnitude is invariant, we write the magnitude in terms of the vector's components. In the original coordinate system this is

$$|\mathbf{v}|^2 = \mathbf{v} \cdot \mathbf{v} = v_i v_i \tag{6.56}$$

whereas in the rotated coordinate system the magnitude is

$$|\mathbf{v}'|^2 = v'_i v'_i \tag{6.57}$$

We manipulate this latter equation by using the vector transformation rule, Eq. (6.54), and the orthogonality property of $\mathbf{Q}$, Eq. (6.55), to find

$$\begin{aligned}
|\mathbf{v}'|^2 &= (Q_{ij} v_j)(Q_{ik} v_k) \\
&= (Q_{ij} Q_{ik}) v_j v_k \\
&= \delta_{jk} v_j v_k \\
&= v_k v_k
\end{aligned} \tag{6.58}$$

That is, the magnitude of the vector computed by using its components in the primed coordinate system is the same as its magnitude computed by using the components in the original coordinate system. Because the primed system can be rotated by any amount in any direction compared to the original system, the magnitude of the vector will be the same in any rotated coordinate system. We have proved that the magnitude of a vector is an invariant.

A second-rank tensor **A** (such as our stress and rate-of-deformation tensors) transforms according to

$$\mathbf{A}' = \mathbf{Q}\mathbf{A}\mathbf{Q}^T \quad \text{or} \quad A'_{ij} = Q_{im} A_{mk} Q_{jk} \tag{6.59}$$

Using this together with Eq. (6.55), one can show that every second-rank tensor has three invariants. These can be chosen as

$$\begin{aligned} I_A &= \operatorname{tr}\mathbf{A} = A_{ii} \\ II_A &= \operatorname{tr}\mathbf{A}^2 = A_{ik} A_{ki} \\ III_A &= \operatorname{tr}\mathbf{A}^3 = A_{ij} A_{jk} A_{ki} \end{aligned} \tag{6.60}$$

The proof that these quantities are invariant follows the same argument as above, and this is left as an exercise. Some authors use an alternate set of tensor invariants that are combinations of I_A, II_A, and III_A; these are given in Eq. (A.31).

6.4.3 THE GENERALIZED NEWTONIAN FLUID

Now let us return to the creation of a constitutive equation. We choose to adopt the form of Eq. (6.52), and we guarantee objectivity by making the scalar viscosity η a function of the three invariants of the rate-of-deformation tensor:

$$\boldsymbol{\tau} = 2\eta(I_D, II_D, III_D)\mathbf{D} \tag{6.61}$$

However, we can simplify things by noting that the first invariant of **D** is

$$I_D = D_{ii} = \frac{\partial v_i}{\partial x_i} \tag{6.62}$$

which is equal to zero for a constant density fluid [cf. Eq. (2.34)]. Thus we can omit any dependence of the viscosity on the first invariant. Also, the third invariant equals zero for any simple shear flow and viscosity measurements are primarily carried out in simple shear flow, so we *assume* that the viscosity will be independent of the third invariant. This leaves the viscosity as a function of the second invariant. As a matter of convenience, we replace the second invariant by the scalar strain rate $\dot{\gamma}$, which is defined as

$$\dot{\gamma} = \sqrt{2 II_D} = \sqrt{2 D_{ij} D_{ji}} \tag{6.63}$$

Writing out the summation in the last expression and accounting for the symmetry of **D** gives Eq. (6.5). The quantity $\dot{\gamma}$ is called the scalar magnitude of the rate of deformation, or simply the *strain rate*. Note that its value is always greater than or equal to zero.

To see why $\dot{\gamma}$ is useful to describe the rate of deformation, let us compute its value for the simple shear flow analyzed in the previous section. For that flow we had

$v_x = Vz/H$ and $v_y = v_z = 0$. The rate-of-deformation tensor is then

$$\mathbf{D} = \frac{1}{2} \begin{bmatrix} 0 & 0 & V/H \\ 0 & 0 & 0 \\ V/H & 0 & 0 \end{bmatrix} \quad (6.64)$$

and we see that

$$II_D = \frac{1}{2}\left(\frac{V}{H}\right)^2 \quad (6.65)$$

so that

$$\dot{\gamma} = V/H \quad (6.66)$$

That is, in a simple shear flow $\dot{\gamma}$ equals the shear rate, $|\partial v_x/\partial z|$.

With the first and third invariants neglected and the second invariant replaced by $\dot{\gamma}$, our constitutive equation, Eq. (6.61), is now

$$\tau = 2\eta(\dot{\gamma})\mathbf{D} \quad \text{or} \quad \tau_{ij} = 2\eta(\dot{\gamma})\,D_{ij} \quad (6.67)$$

Equation (6.67) is the constitutive equation for a *generalized Newtonian fluid* (GNF). The generalized Newtonian fluid is actually a class of models, as one can choose any function for $\eta(\dot{\gamma})$. The power law fluid is obtained by choosing

$$\eta(\dot{\gamma}) = m\dot{\gamma}^{n-1} \quad (6.68)$$

Equations (6.68) and (6.67) together provide a complete constitutive equation for the power law fluid, one that can be used in any flow (not just simple shear flows), and one that satisfies the principle of material frame indifference. We will see some other generalized Newtonian fluid models in Section 6.5.

6.4.4 EXAMPLE: MULTIDIRECTIONAL DRAG FLOW

To see how the generalized Newtonian fluid constitutive equation works, we consider another example that is slightly more complicated than the strictly one-dimensional problems examined thus far. Suppose that we have a fully developed drag flow between two parallel plates, but now the upper plate moves both in the x direction with velocity V_x and in the y direction with velocity V_y. The geometry is shown in Fig. 6.6.

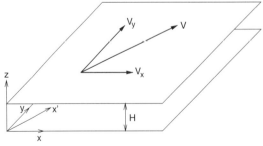

Figure 6.6: Geometry for multidirectional drag flow.

6.4 PRINCIPLES OF NON-NEWTONIAN CONSTITUTIVE EQUATIONS

A little reflection should convince you that this is really the same problem we solved in Section 6.2.2, because we could rotate the coordinate system about the z axis, choosing an x' axis that is lined up with the motion of the upper plate. Then we would have unidirectional drag flow with an upper plate velocity of $V = \sqrt{V_x^2 + V_y^2}$. Solving the problem in the x–y coordinate system should produce the same result, so let us proceed and see what happens.

With $v_z = 0$ and v_x and v_y depending only on z (by the same arguments we used before), the rate-of-deformation components simplify to

$$D_{zx} = \frac{1}{2}\frac{\partial v_x}{\partial z}$$
$$D_{zy} = \frac{1}{2}\frac{\partial v_y}{\partial z} \tag{6.69}$$

These (and their symmetric counterparts D_{xz} and D_{yz}) are the only nonzero components of **D**. Using Eq. (6.63), we see that the scalar rate of deformation is

$$\dot{\gamma} = \sqrt{\left(\frac{\partial v_x}{\partial z}\right)^2 + \left(\frac{\partial v_y}{\partial z}\right)^2} \tag{6.70}$$

The same arguments we used earlier allow us to reduce the x and y momentum equations to

$$\frac{\partial \tau_{zx}}{\partial z} = 0$$
$$\frac{\partial \tau_{zy}}{\partial z} = 0 \tag{6.71}$$

We integrate these in z to find that both shear stress components are constant:

$$\tau_{zx} = c_1$$
$$\tau_{zy} = c_2 \tag{6.72}$$

Combining these results with the constitutive equation, Eq. (6.67), and the expressions for the components of **D** from Eq. (6.69), we have

$$\tau_{zx} = \eta \frac{\partial v_x}{\partial z} = c_1$$
$$\tau_{zy} = \eta \frac{\partial v_y}{\partial z} = c_2 \tag{6.73}$$

Because η depends on the rate of deformation, in general it will be a function of position. However, in this case Eq. (6.72) tells us that τ, the scalar magnitude of the extra stress, is constant, so from Eq. (6.69) the shear rate $\dot{\gamma}$ must be constant, and hence in this problem the viscosity η will also be constant across the gap. With this in hand we can integrate Eqs. (6.73) to find

$$v_x = c_1 z/\eta + c_3$$
$$v_y = c_2 z/\eta + c_4 \tag{6.74}$$

The no-slip boundary conditions allow us to evaluate the constants, and the final velocity solution is

$$v_x = V_x z/H$$
$$v_y = V_y z/H \qquad (6.75)$$

This is probably what one would have guessed: the velocity is simply a combination of a linear drag-flow profile in the x direction, and another linear profile in the y direction.

The results for stress are a bit more interesting. Substituting Eqs. (6.75) into Eq. (6.70), we find that the strain rate is

$$\dot\gamma = \sqrt{\left(\frac{V_x}{H}\right)^2 + \left(\frac{V_y}{H}\right)^2} \qquad (6.76)$$

Next we use this expression for $\dot\gamma$ in Eq. (6.68) to find the viscosity, and use that value in Eqs. (6.73) to calculate the stresses. This gives the stress components as

$$\tau_{zx} = m\left[\left(\frac{V_x}{H}\right)^2 + \left(\frac{V_y}{H}\right)^2\right]^{n-1/2} \frac{V_x}{H}$$

$$\tau_{zy} = m\left[\left(\frac{V_x}{H}\right)^2 + \left(\frac{V_y}{H}\right)^2\right]^{n-1/2} \frac{V_y}{H} \qquad (6.77)$$

Now we see that the x-direction motion influences the y-direction stress τ_{zy}, and the y-direction motion influences τ_{zx}. This occurs because the fluid has only one viscosity at any point, and this viscosity is affected by both the x- and the y-direction deformation. For a shear-thinning fluid, moving the plate in the y direction will actually make an ongoing x-direction motion easier (i.e., the x force needed to move the top plate will decrease), because the extra deformation increases the strain rate and reduces the viscosity. A constitutive equation that is not objective would miss this effect.

As a check on the result, we transform the stress to coordinates x', y', z, where x' is aligned with the top plate velocity. The coordinate transformation rule for stress gives the shear stress in the new system as

$$\tau_{zx'} = \sqrt{\tau_{zx}^2 + \tau_{zy}^2} \qquad (6.78)$$

Substituting Eqs. (6.77) into this gives

$$\tau_{zx'} = m\left[\left(\frac{V_x}{H}\right)^2 + \left(\frac{V_y}{H}\right)^2\right]^{n-1/2}\left[\left(\frac{V_x}{H}\right)^2 + \left(\frac{V_y}{H}\right)^2\right]^{1/2}$$

$$= m\left[\frac{V}{H}\right]^n \qquad (6.79)$$

where $V^2 = V_x^2 + V_y^2$ is the top plate velocity in the x' direction. This is identical to our result for the unidirectional drag flow, Eq. (6.19), so using the full power law constitutive equation has given the same answer in two different coordinate systems.

6.5 MORE NON-NEWTONIAN CONSTITUTIVE EQUATIONS

In addition to the power law model, many other non-Newtonian constitutive equations are in widespread use. Here are a few of the most common ones.

6.5.1 THE CARREAU AND CROSS MODELS

The power law fluid, Eq. (6.68), is one example of a generalized Newtonian fluid, but we can create other generalized Newtonian fluid models by using Eq. (6.67) and letting η be some function of $\dot{\gamma}$ other than a power law.

Many polymer melts follow the power law closely at high strain rates, and this is why the power law is such a popular model. However, as $\dot{\gamma} \to 0$, the power law model with $n < 1$ gives $\eta \to \infty$. Polymer melts do not do this; instead their viscosity becomes constant as the shear rate approaches zero.

Figure 6.7 shows viscosity versus shear rate data for a polymer melt at typical processing temperatures. On this plot, where both η and $\dot{\gamma}$ are on logarithmic scales, the power law model would appear as a straight line with slope $n - 1$. A Newtonian fluid would appear on this plot as a horizontal line, as its viscosity is independent of shear rate. The data tend to follow the power law with shear thinning (negative slope) at high strain rates, but flatten out at a constant viscosity as the shear rate is decreased. This plateau value is called the *zero shear viscosity* and is given the symbol η_0. In this low-strain-rate range, the polymer behaves like a Newtonian fluid.

Several equations have been developed that combine a constant viscosity for low shear rates, a power law at high shear rates, and a smooth transition in between. The most popular models of this type have the form

$$\eta = \frac{\eta_0}{[1 + (\eta_0 \dot{\gamma}/\tau^*)^a]^{(1-n)/a}} \tag{6.80}$$

where the material constants are η_0, n, τ^*, and a. If $a = 2$ then Eq. (6.80) is the *Carreau model*, and if $a = 1 - n$ this is the *Cross model*. Note that η_0 controls the

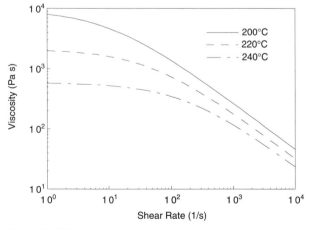

Figure 6.7: Viscosity versus shear rate and temperature for a commercial grade polymethylmethacrylate. Cross–WLF (Williams, Landel, and Ferry) model with parameters from Hieber and Chiang (1992).

Newtonian plateau value at low shear rates, n is the power law index at high shear rates, and τ^* controls the stress level at which the fluid makes the transition from Newtonian to shear thinning behavior. The parameter a determines the shape of this transition, with smaller values of a giving a broader transition.

The Cross and Carreau models are widely used in modeling polymer melt flow, and they can give good quantitative agreement with pressure drops and flow rates in injection molding and extrusion. They normally require a numerical solution to the flow problem, because the complexity of Eq. (6.80) makes analytical solutions intractable. The power law model admits a variety of analytical solutions and is much better that the Newtonian fluid for modeling polymer melts, but it only gives quantitatively accurate predictions if the strain rate is high everywhere in the flow domain.

6.5.2 FLUIDS WITH A YIELD STRESS

An alternate way to write the GNF constitutive equation is to make viscosity a function of τ, the scalar magnitude of the extra stress tensor:

$$\tau = \sqrt{\frac{1}{2}\tau_{ij}\tau_{ij}} \qquad (6.81)$$

Our notation is that $\boldsymbol{\tau}$ (boldface) is the whole tensor, τ (lightface) is the scalar magnitude, while τ_{xz}, τ_{yz}, and so on (lightface with subscripts) are scalar components of $\boldsymbol{\tau}$.

Letting viscosity depend on τ is simply a different way to write the generalized Newtonian model, because Eqs. (6.81), (6.67), and (6.63) tell us that the magnitudes of the extra stress and rate-of-deformation tensors are related by

$$\tau = \eta\dot{\gamma} \qquad (6.82)$$

One model that is conveniently written in terms of τ is the *Bingham plastic* model. The viscosity for the Bingham plastic is

$$\eta = \begin{cases} \infty & \text{if } \tau < \tau_Y \\ \dfrac{\tau_Y}{\dot{\gamma}} + \mu_0 & \text{if } \tau \geq \tau_Y \end{cases} \qquad (6.83)$$

with μ_0 and τ_Y being material constants. This model exhibits a yield stress, like the curve in Fig. 6.1. If the magnitude of the applied stress τ is less than τ_Y, then the material does not flow, and the viscosity is infinite. That is, for $\tau < \tau_Y$ the material acts like a solid. If $\tau \geq \tau_Y$, then the material flows, with higher strain rates requiring higher stresses. In the Bingham model this dependence is linear, and the constant μ_0 has the units of viscosity.

The Bingham model can be used to find analytical solutions in simple geometries, but for many numerical solution methods it is very useful to have a single equation and a smooth curve of shear stress versus shear rate. In those cases the model proposed by Papanastasiou (1987) is useful:

$$\eta = \eta_0 + \frac{\tau_Y(1 - e^{-a\dot{\gamma}})}{\dot{\gamma}} \qquad (6.84)$$

6.5 MORE NON-NEWTONIAN CONSTITUTIVE EQUATIONS

Here a is a constant that has units of time, so that $a\dot{\gamma}$ is dimensionless; a is usually chosen to be large (100 s or larger) to give Bingham-like behavior. The Bingham model does not flow at all for low stresses, but for $\tau \ll \tau_Y$ the Papanastasiou model deforms slowly, with a very high viscosity equal to $\eta_0 + a\tau_Y$. For large strain rates, say $a\dot{\gamma} > 3$, Eq. (6.84) reproduces the Bingham model, and it gives a smooth transition between the two behaviors.

A yield stress is often observed in polymers that are highly filled with small particles, in clays and pastes, and in semisolid metals. Solid metals also exhibit a yield stress, and the yield criterion for the Bingham plastic, $\tau \geq \tau_Y$, is identical to the von Mises yield criterion.

Bingham Fluid in Tube Flow

To see how fluids with a yield stress behave, consider the pressure-driven flow of a Bingham fluid in a long tube. We will use the same assumptions as the power law solution in Section 6.2.3, and the initial treatment of the equation of motion is identical up through Eq. (6.28). Note that none of that derivation makes any reference to the rheological properties of the fluid. This gives the shear stress as

$$\tau_{rz} = -\frac{\Delta p r}{2L} \tag{6.85}$$

Because τ_{rz} and τ_{zr} are the only nonzero components of the extra stress, Eq. (6.81) gives the magnitude of the extra stress as

$$\tau = \frac{\Delta p r}{2L} \tag{6.86}$$

where we have assumed that Δp is positive. The nonzero rate-of-deformation tensor component is

$$D_{rz} = \frac{1}{2}\frac{\partial v_z}{\partial r} \tag{6.87}$$

and, because we expect $\partial v_z/\partial r \leq 0$, the scalar strain rate is

$$\dot{\gamma} = -\frac{\partial v_z}{\partial r} \tag{6.88}$$

From Eq. (6.86) we see that the stress is greatest at the wall ($r = R$) and zero at the center line ($r = 0$). Let us assume for the moment that the shear stress equals the yield stress somewhere in between. The Bingham fluid constitutive equation, Eq. (6.83), has two parts, so we solve the problem in two parts as well. The dividing line is at R_Y, where $\tau(R_Y) = \tau_Y$. The precise location of R_Y is

$$R_Y = \frac{2L\tau_Y}{\Delta p} \tag{6.89}$$

First consider $r \geq R_Y$. For any generalized Newtonian fluid we have

$$\tau_{rz} = 2\eta D_{rz} \tag{6.90}$$

and for $r \geq R_Y$ the viscosity is given by the second part of Eq. (6.83). Combining these equations with Eqs. (6.85) and (6.88), we find

$$-\frac{\Delta p r}{2L} = \left(\frac{\tau_Y}{-\partial v_z/\partial r} + \mu_0\right)\frac{\partial v_z}{\partial r} \qquad (6.91)$$

or, with a little rearranging,

$$\frac{\partial v_z}{\partial r} = -\frac{\Delta p r}{2\mu_0 L} + \frac{\tau_Y}{\mu_0} \qquad (6.92)$$

This is readily integrated to give $v_z(r)$, and we apply a no-slip boundary condition at $r = R$ to get

$$v_z = \frac{\Delta p R^2}{4\mu_0 L}\left[1 - \left(\frac{r}{R}\right)^2\right] - \frac{\tau_Y R}{\mu_0}\left[1 - \left(\frac{r}{R}\right)\right], \quad R_Y \leq r \leq R \qquad (6.93)$$

For $r \leq R_Y$, Eq. (6.83) tells us that the viscosity is infinite, so from Eq. (6.90) we must have $D_{rz} = 0$, or v_z equal to a constant. The constant is found by requiring the velocity to be continuous at $r = R_Y$. Using Eq. (6.93), we find this inner velocity to be

$$v_z = \frac{\Delta p R^2}{4\mu_0 L}\left[1 - \left(\frac{R_Y}{R}\right)\right]^2, \quad 0 \leq r \leq R_Y \qquad (6.94)$$

Figure 6.8 shows the velocity distribution described by Eqs. (6.93) and (6.94). The velocity profile is flat in the central "plug" region, and parabolic outside of that. If the average stress is very high compared with the yield stress ($R_Y \ll R$) then the velocity profile is very close to that of a Newtonian fluid. If the maximum shear stress is only a little bit greater than the yield stress, then the velocity profile is nearly flat everywhere, and all of the shearing is confined to a thin layer near the wall. Note

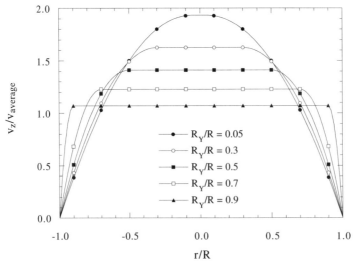

Figure 6.8: Velocity distributions for a Bingham fluid in tube flow. Curves are normalized by the average velocity.

the similarity between the two flattest Bingham velocity profiles in Fig. 6.8 and the power law profiles for $n = 0.1$ and 0.2 in Fig. 6.4.

Integrating the velocity distribution over the cross section gives the flow rate as

$$Q = \frac{\pi R^4 \Delta p}{8\mu_0 L} \left[1 - \frac{4}{3}\left(\frac{R_Y}{R}\right) + \frac{1}{3}\left(\frac{R_Y}{R}\right)^4 \right] \tag{6.95}$$

where R_Y is defined in Eq. (6.89). If the maximum shear stress is less than the yield stress ($R_Y > R$) then the Bingham fluid has $\eta = \infty$ everywhere across the radius of the tube and, with a no-slip boundary condition, there is no flow at all. This is a very different behavior from the power law fluid with small n, because the power law fluid always has some flow rate, however small, when a pressure gradient is applied.

6.5.3 MODELS FOR TEMPERATURE-DEPENDENT VISCOSITY

The models considered so far capture the dependence of viscosity on shear rate or stress, and are quite useful in isothermal flow problems. However, in Chapter 7 we will consider flow problems in which the temperature changes, and we will need a constitutive equation that can represent this effect.

When the temperature of a liquid is increased, its viscosity decreases. If the fluid is shear thinning, then its power law index usually does not change with temperature. Thus, for temperature-dependent models we can retain the general form of the Newtonian or generalized Newtonian constitutive equation, and we can allow the viscosity to depend on temperature.

Let us consider this in the context of the Cross–Carreau model, Eq. (6.80). To make this model temperature dependent we keep n, a, and τ^* constant, and we make η_0 a function of temperature. There are several options for the temperature-dependent function. One widely used choice is an Arrhenius form,

$$\eta_0(T) = \eta^* \exp\left(\frac{E_\eta}{RT}\right) \tag{6.96}$$

Here R is the gas constant (8.31 J/mol K), E_η is the activation energy (a material property), and T must be the *absolute* temperature. Arrhenius models are usually good for low-molecular-weight fluids and for polymers well above their glass transition temperatures. Values for the constants η^* and E_η are typically found by adjusting them to fit experimental data.

For polymers, another widely used model is the WLF equation (Williams, Landel, and Ferry, 1995). This gives the viscosity as

$$\eta_0(T) = \eta^* a_T(T) \quad \text{with} \quad \log_{10} a_T = \frac{-C_1(T - T_0)}{C_2 + (T - T_0)} \tag{6.97}$$

Here T_0 is a reference temperature and C_1, C_2, and η^* are constants. The WLF equation has a theoretical basis and it tends to fit data very well from the glass transition temperature T_g up to $T_g + 100°C$. The original paper proposed this equation as a universal temperature dependence that was based on each material's individual T_g, and when no other data are available one can reasonably estimate material behavior by using the original "universal" constant values: $C_1 = 17.44$ and $C_2 = 51.6$ K. When data are available over a range of temperatures, then it is better to adjust C_1, C_2, and T_0 to fit the data. Figure 6.7 shows viscosity curves at several different temperatures for a commercial grade polymer, using the Cross–WLF model.

For analytical solutions neither of these forms is very convenient, and an exponential model is often used:

$$\eta_0(T) = \eta^* e^{a(T-T_0)} \tag{6.98}$$

This is strictly an empirical equation, and it tends to fit data well over only a modest range of temperatures, say 10–20°C. However, it provides much better fits to data than a linear temperature dependence.

These models are easily adapted to Newtonian and power law fluids. For a Newtonian fluid, make μ a function of temperature by using any one of Eqs. (6.96), (6.97), or (6.98). For a power law fluid, keep n constant and make m temperature dependent by using the same functions.

Recommended practice for numerical solutions is to use a WLF temperature dependence for polymeric materials and an Arrhenius temperature dependence for all other materials. Analytical solutions quickly become intractable for these models, so the exponential model is recommended for analytical solutions. We will use the exponential model in Chapter 7.

6.5.4 VISCOELASTIC FLUIDS (ADVANCED)

The behavior of a viscoelastic fluid combines some aspects of an elastic solid and some aspects of a viscous liquid. Among its material properties, any viscoelastic material has a time constant λ, also called the *relaxation time*. Not surprisingly, this leads to a dimensionless parameter, the *Deborah number*, De,

$$\text{De} = \lambda/t_c \tag{6.99}$$

where t_c is a characteristic time over which the material is deformed. For $\text{De} \gg 1$ (times much shorter than λ) the material behavior is elastic, whereas for $\text{De} \ll 1$ (times much longer than λ) the material behavior is viscous. All polymers melts are viscoelastic, and polymers also exhibit viscoelasticity in the solid state, though with much longer time constants. Other viscoelastic fluids include inorganic glasses and many foods.

Viscoelastic fluids, particularly polymers, exhibit a number of distinctive flow characteristics.

1. They exhibit shear thinning. In a steady simple shear flow, their viscosity decreases as the shear rate increases. This is the *only* viscoelastic behavior that is captured by the generalized Newtonian fluids models.
2. They exhibit extension thickening. In elongational flows, such as fiber spinning, the stress in a viscoelastic fluid increases as the strain increases. This is the property that makes melted cheese "stringy" when you try to lift a slice of pizza away from the whole pie. It is also important to the stability of fiber-spinning processes.
3. They exhibit normal stresses in shear flow. In a simple shear flow, say $v_x = \dot{\gamma} y$, a Newtonian fluid develops only the shear stress τ_{xy}. Polymeric fluids also develop

6.5 MORE NON-NEWTONIAN CONSTITUTIVE EQUATIONS

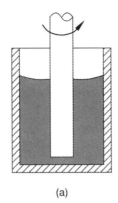
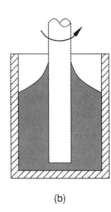

Figure 6.9: Rod climbing. (a) When a rod is rotated in a Newtonian fluid the free surface is slightly lower in the center as a result of fluid inertia. (b) In a viscoelastic fluid, normal stress effects cause the fluid to rise dramatically in the center. This can be seen with rubber cement, which is a solution of a polymer in an organic solvent.

(a) (b)

nonzero normal stresses, and the stress differences $(\tau_{xx} - \tau_{yy})$ and $(\tau_{yy} - \tau_{zz})$ are nonzero. This effect provides an extra tension along the streamlines, and it causes phenomena such as the rod-climbing effect sketched in Fig. 6.9.

4. They exhibit elastic recovery. When a rubber band is stretched and released, it returns to its original shape. A viscoelastic material that is deformed and then released returns part way to its original shape. An important example of this is *die swell*. Figure 6.10 shows a Newtonian and a viscoelastic fluid being extruded from a circular die. As the fluids exit the die, the diameter of the Newtonian fluid increases slightly, but the diameter of the viscoelastic fluid increases dramatically. This is due to elastic recovery of the straining that occurs inside the die.

5. They exhibit elastic instabilities. At high extrusion speeds, polymers exhibit an unstable flow behavior called *melt fracture*. Above a critical wall shear stress, the extruded polymer no longer has a smooth surface. Instead, the flow becomes irregular and the extrudate surface has a rough texture that is called *sharkskin*. This behavior is thought to be related to wall slip at high stress, a violation of the classical no-slip boundary condition. Melt fracture limits the processing speed in many polymer extrusion operations.

Because viscoelastic materials have time-dependent behavior, their constitutive equations contain time derivatives or integrals. The constitutive equations most often

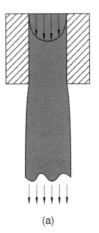
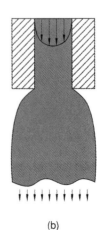

Figure 6.10: Die swell. (a) A Newtonian fluid exiting a die increases diameter slightly. (b) For a viscoelastic fluid, elastic recovery causes a large increase in diameter.

(a) (b)

used for numerical simulation have the form

$$\mathbf{Y} \cdot \boldsymbol{\tau} + \lambda \overset{\square}{\boldsymbol{\tau}} = 2\eta_0 \mathbf{D} \tag{6.100}$$

Here $\mathbf{Y}$ is a tensor function of the extra stress $\boldsymbol{\tau}$, and $\overset{\square}{\boldsymbol{\tau}}$ is a compact notation for the Gordon–Schowalter convected derivative of stress:

$$\overset{\square}{\boldsymbol{\tau}} \equiv \frac{\partial \boldsymbol{\tau}}{\partial t} + \mathbf{v} \cdot \nabla \boldsymbol{\tau} - (\mathbf{L} - \xi \mathbf{D}) \cdot \boldsymbol{\tau} - \boldsymbol{\tau} \cdot (\mathbf{L}^T - \xi \mathbf{D})^T \tag{6.101}$$

Recall that $\mathbf{L}$ is the velocity gradient tensor, $L_{ij} = \partial v_i / \partial x_j$; $\overset{\square}{\boldsymbol{\tau}}$ is a general way to take an objective time derivative of a tensor.

When $\mathbf{Y} = \mathbf{I}$ (so that $\mathbf{Y} \cdot \boldsymbol{\tau} = \boldsymbol{\tau}$) and $\xi = 0$, Eq. (6.100) is called the *upper convected Maxwell model*. This model has only two material properties, a relaxation time λ and a viscosity η_0. For a two-dimensional problem, the xy component of Eq. (6.100) then expands to

$$\tau_{xy} + \lambda \left(\frac{\partial \tau_{xy}}{\partial t} + v_x \frac{\partial \tau_{xy}}{\partial x} + v_y \frac{\partial \tau_{xy}}{\partial y} - \frac{\partial v_x}{\partial x} \tau_{xy} - \frac{\partial v_x}{\partial y} \tau_{yy} - \tau_{xx} \frac{\partial v_y}{\partial x} - \tau_{xy} \frac{\partial v_y}{\partial y} \right)$$
$$= \eta_0 \left(\frac{\partial v_x}{\partial y} + \frac{\partial v_y}{\partial x} \right) \tag{6.102}$$

This gives a better idea of the complexity of the differential viscoelastic constitutive equations.

The upper convected Maxwell model exhibits many of the qualitative behaviors of viscoelastic fluids, including extension thickening, normal stresses in shear, and elastic recovery. However, it does not exhibit shear thinning. To get a reasonable match to polymer behavior, one must introduce some additional nonlinearities by altering the function $\mathbf{Y}$. Two models that are widely used in viscoelastic flow simulation are the *Giesekus model*, which has

$$\mathbf{Y} = \mathbf{I} + \frac{\alpha \lambda}{\eta_0} \boldsymbol{\tau} \tag{6.103}$$

and the *PTT* (Phan-Thien and Tanner) model, which uses

$$\mathbf{Y} = \exp\left(\frac{\varepsilon \lambda}{\eta_0} \mathrm{tr}\, \boldsymbol{\tau} \right) \mathbf{I} \tag{6.104}$$

Each of these models adds another dimensionless parameter, α or ε, that controls the nonlinearity, so both models have four parameters: λ, η_0, ξ, and either α or ε.

In addition, to get a quantitative match with the real flow behavior, one must use what is called a *multimode model*. Here the extra stress is taken to be the sum of a Newtonian term plus a number of "mode stresses," $\boldsymbol{\tau}_i$:

$$\boldsymbol{\tau} = 2\eta_s \mathbf{D} + \sum_{i=1}^{N} \boldsymbol{\tau}_i \tag{6.105}$$

Each $\boldsymbol{\tau}_i$ is given by an equation such as Eq. (6.100), typically using either the Giesekus or the PTT expression for $\mathbf{Y}$. Usually four to eight modes are used, with each successive λ_i being 10 times larger than the previous one.

The form of these constitutive equations makes the viscoelastic flow problem more complex than for Newtonian fluids. The only unknowns in the Navier–Stokes equation are pressure and velocity, because we could combine the equation of motion and the Newtonian constitutive equation to eliminate τ algebraically. The same simplification is possible for any generalized Newtonian fluid. However, we cannot eliminate τ algebraically for a viscoelastic fluid. Instead, in a viscoelastic flow problem we have velocity, pressure, *and* stress as unknowns. For a single-mode model this adds three unknowns for a two-dimensional problem, and six unknowns for a three-dimensional problem. For a multimode model these numbers are multiplied by the numbers of modes. Thus, for a two-dimensional problem with a five-mode constitutive equation, there are 18 different variables to solve for.

Solving viscoelastic flow problems numerically is a very challenging task, not just because of the extra unknowns, but also because common numerical schemes often fail to converge on a meaningful solution. A recent review of the best numerical schemes is given by Baaijens (1998), and the chapter by Keunings (1998) provides useful background. Some recent papers have shown good agreement between computation and experiments for viscoelastic fluids (Baaijens et al., 1997; Debbaut et al., 1997), and they provide a good idea of what is required to characterize a viscoelastic material and solve a flow problem numerically. Readers who need to solve a viscoelastic flow problem are urged to consult these references, as well as the rheology books listed in the bibliography.

6.6 THE GENERALIZED HELE–SHAW APPROXIMATION

6.6.1 INTRODUCTION

Many polymer processing problems involve the flow of a viscous fluid in a narrow gap. In these circumstances one can apply a very useful simplification of the balance equations, called the *generalized Hele–Shaw approximation*. The essence of this approximation is that the flow is always locally like pressure-driven flow between parallel plates. Other than requiring a low Reynolds number and a narrow gap geometry, the Hele–Shaw approximation is quite general. It applies to geometries in which the gap thickness varies with position (provided it varies slowly), and to geometries in which the surfaces of the gap are curved (provided their radius of curvature is large compared to the gap height). It applies to all generalized Newtonian fluids, including the power law fluid and the Cross–Carreau model, and it applies to nonisothermal flows, including problems in which fluid solidifies on the walls.

The generalized Hele–Shaw approximation is important in materials processing, because it is the standard model used to simulate polymer injection molding (Wang and Lee, 1989; Kennedy, 1995). It can also model extrusion dies for sheets and similar products, and it has been extended to treat compression molding and flow in twin screw extruders. The formulation for non-Newtonian and nonisothermal flows is from Hieber and Shen (1980), and we follow their development here. We like the model not only for its utility, but because it is a good example of how scaling can be used to develop a simple but effective model.

Historically, Hele–Shaw (1899) proposed the idea of pumping a viscous fluid through a narrow gap between two sheets of glass as a way of visualizing two-dimensional flows of ideal (inviscid) fluids. A cross section of the shape is cut out of some impermeable sheetlike material and placed between the glass plates, whose edges are then sealed. A viscous Newtonian fluid, with streaks of dye, then enters along the upstream edge of the cell, and exits on the downstream end. The dye streaks reveal the streamline pattern as the fluid passes around the complex shape. Schlichting (1968) shows some nice photographs of flows in a Hele–Shaw apparatus. A curious paradox is that the viscosity-dominated flow of the Hele–Shaw cell gives exactly the same streamlines as an inviscid potential flow around the actual object. This can be proved rigorously (e.g., Schlichting, 1968), though the original paper does not contain this level of justification.

In this section we first present the final equations and discuss how they can be applied to mold filling simulations. We then work through the complete derivation, doing the scaling as well as the other mathematical manipulations that produce the Hele–Shaw equations.

6.6.2 EQUATIONS AND APPLICATION

Consider flow in a narrow gap geometry such as that shown in Fig. 6.11. For the moment we restrict our attention to flat shapes, taking z as the thickness direction and x and y as the in-plane directions. The midplane lies at $z = 0$ and the upper and lower surfaces of the gap are at $z = \pm h$. We assume that the flow is symmetric about $z = 0$. We will later use L as a length scale for the in-plane dimensions of the mold.

The Hele–Shaw approximation states that the pressure does not vary significantly in the z direction, and that the velocity in the z direction is negligible compared with the in-plane velocities. The primary variable is the pressure, and one finds $p(x, y)$ by solving the partial differential equation:

$$\frac{\partial}{\partial x}\left(S\frac{\partial p}{\partial x}\right) + \frac{\partial}{\partial y}\left(S\frac{\partial p}{\partial y}\right) = 0 \qquad (6.106)$$

The quantity $S(x, y)$, called the *flow conductance*, is defined as

$$S(x, y) = \int_0^h \frac{z^2\, dz}{\eta(x, y, z)} \qquad (6.107)$$

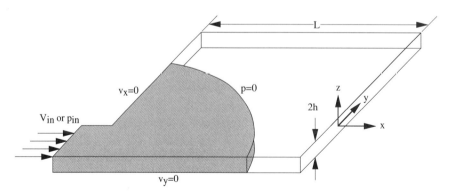

Figure 6.11: Example narrow gap geometry as analyzed by the Hele–Shaw approximation.

6.6 THE GENERALIZED HELE–SHAW APPROXIMATION

where η is the local viscosity. Some additional details are needed to calculate S for a non-Newtonian fluid; these are discussed later, starting with Eq. (6.152).

Once the pressure solution has been found, the in-plane velocities are readily calculated. The gapwise average velocities, defined formally in Eqs. (6.135) and (6.136), are given by

$$\bar{v}_x = -\frac{S}{h}\frac{\partial p}{\partial x}, \quad \bar{v}_y = -\frac{S}{h}\frac{\partial p}{\partial y} \tag{6.108}$$

The full velocity distributions can also be calculated, using

$$v_x(z) = -\frac{\partial p}{\partial x}\int_z^h \frac{z'dz'}{\eta(z')}, \quad v_y(z) = -\frac{\partial p}{\partial y}\int_z^h \frac{z'dz'}{\eta(z')} \tag{6.109}$$

where z' is a dummy variable of integration. Note that the two velocity profiles $v_x(z)$ and $v_y(z)$ have the same shape, differing only by the factors $\partial p/\partial x$ and $\partial p/\partial y$. This tells us that, although the magnitude of velocity varies with z, its direction is independent of z. Thus, the streamlines of any Hele–Shaw flow are the same on any plane of $z = $ constant. It then makes sense to look at the problem from the top (along the z axis) and focus on what is happening in the x–y plane.

Equation (6.106) requires boundary conditions on all boundaries in the x–y plane. On the flow front we use $p = 0$, consistent with Eq. (5.93) and the surrounding discussion. Where the fluid touches a solid boundary in the x–y plane (at the edges of the mold), the boundary condition is $\partial p/\partial n = 0$. Equation (6.108) shows that this corresponds to zero velocity in the direction normal to the boundary. At the inlet we have either a prescribed pressure $p = p_{\text{in}}$, or a prescribed normal velocity $V_{\text{in}} = -(S/h)\partial p/\partial n$.

If the fluid is Newtonian and isothermal, then its viscosity μ is constant. Equation (6.109) then gives the familiar parabolic velocity profile in the thickness direction, and S takes on the value

$$S = h^3/3\mu \tag{6.110}$$

Then Eq. (6.108) reduces to a familiar-looking result (cf. Section 5.2.1):

$$\bar{v}_x = -\frac{h^2}{3\mu}\frac{\partial p}{\partial x}, \quad \bar{v}_y = -\frac{h^2}{3\mu}\frac{\partial p}{\partial y} \tag{6.111}$$

If we also have a uniform gap height (constant h) then S is constant, and the Hele–Shaw governing equation for pressure, Eq. (6.106), reduces to Laplace's equation in two dimensions:

$$\frac{\partial^2 p}{\partial x^2} + \frac{\partial^2 p}{\partial y^2} = 0 \tag{6.112}$$

This version of the Hele–Shaw equations has exactly the same form as the equations for two-dimensional steady-state heat conduction: pressure replaces temperature, the quantities $h\bar{v}_x$ and $h\bar{v}_y$ replace the components of the heat flux vector, and S replaces the thermal conductivity. For boundary conditions, prescribed pressure is the analog of prescribed temperature, a sealed edge of the mold is like an insulated boundary, and a prescribed inlet velocity is like a prescribed heat flux. This correspondence is more than a curiosity, because any solution to a two dimensional steady-state heat conduction problem is also a solution to some Hele–Shaw problem.

Although one can find analytical solutions for a handful of geometries, the major application of the Hele–Shaw approach lies in numerical solutions for more complex shapes. A typical mold-filling simulation for thin cavities uses an explicit timestepping algorithm that goes as follows.

1. Start with a small part of the cavity near the inlet filled.
2. Apply boundary conditions on the current boundaries and solve Eq. (6.106) to find the pressure distribution $p(x, y)$.
3. Differentiate the pressure solution and use Eq. (6.108) to find the gapwise average velocities $\bar{v}_x$ and $\bar{v}_y$.
4. Choose a time step Δt and advance the flow front to its new position by moving points on the front by $\Delta x = \bar{v}_x \Delta t$ and $\Delta y = \bar{v}_y \Delta t$.
5. Repeat the cycle starting from Step 2, until the mold is full.

Detailed descriptions of numerical schemes that use the Hele–Shaw formulation for injection molding are given by Wang and Lee (1989) and by Kennedy (1995).

Although finding a precise pressure distribution for some complex geometry may require a numerical solution, we can take advantage of the mathematical structure of the Hele–Shaw equations to make some qualitative estimates. In the x–y plane one can sketch the isobars and the streamlines corresponding to the average velocities $\bar{v}_x$ and $\bar{v}_y$ using the following rules. Equation (6.108) tells us that the velocity is parallel to the pressure gradient but opposite in direction. That is, in a Hele–Shaw flow, fluid moves directly from high pressure to low pressure. This means that the streamlines will always be perpendicular to the isobars. The flow front is the $p = 0$ isobar, and if the inlet boundary condition is a constant pressure, the inlet is also an isobar. The isobars must meet the boundaries of the mold at right angles to satisfy the $\partial p/\partial n = 0$ boundary condition, and these mold edges must themselves be streamlines. All of these conditions constrain the solution enough that it is often easy to sketch the isobars and streamlines, and thus to get a good general idea of the flow pattern. Figure 6.12 shows an example.

The Hele–Shaw formulation cannot capture physical phenomena that rely on the terms we dropped when simplifying the equations. This makes the Hele–Shaw solution unrealistic at the edges of the mold and at the flow front. Near the flow front there will be a fountain flow, as discussed in Section 5.3.4. In that region v_z will have

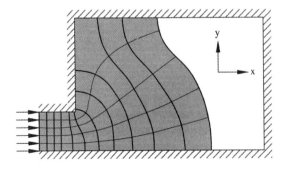

Figure 6.12: Example of an approximate solution for a Hele–Shaw flow. The heavy lines are isobars, and the light lines are streamlines.

the same order of magnitude as v_x and v_y, and Eq. (6.109) will no longer give accurate velocities. At the edges of the mold the Hele–Shaw boundary condition requires zero velocity normal to the boundary, but places no restriction on the velocity parallel to the boundary. In fact, there will always be a pressure gradient parallel to the wetted edges of the mold (see Fig. 6.12 as an example), so the Hele–Shaw solution always gives non-zero tangential velocity on the boundary. In reality the no-slip condition will apply along the vertical surface at the edges of the mold, and the local velocity will differ from the Hele–Shaw solution. This effect, as well as the fountain effect at the flow front, are confined to boundary layers (when viewed in the x–y plane) whose width is of order h. If $h \ll L$, then these boundary layers are a small part of the total flow domain.

6.6.3 DERIVATION OF THE HELE–SHAW FORMULATION

Scaling Analysis

We now proceed with the derivation of Eqs. (6.106)–(6.109). Consider a geometry such as that shown in Fig. 6.11, with $h/L \ll 1$. Assume that we have a generalized Newtonian fluid with constant density. To start, we scale the variables according to

$$x^* = x/L, \quad y^* = y/L, \quad z^* = z/h \qquad (6.113)$$

$$v_x^* = v_x/V, \quad v_y^* = v_y/V, \quad v_z^* = v_z/V_z \qquad (6.114)$$

Here we have assumed that the x and y directions have the same characteristic length L and velocity V, whereas the z-direction length and velocity scales are h and V_z, respectively. At this point V_z is unknown. Applying these scales to the mass balance equation gives

$$\frac{V}{L}\left(\frac{\partial v_x^*}{\partial x^*} + \frac{\partial v_y^*}{\partial y^*}\right) + \frac{V_z}{h}\frac{\partial v_z^*}{\partial z^*} = 0 \qquad (6.115)$$

The scaled derivatives are now $\mathcal{O}(1)$, so we conclude that

$$V_z = \frac{h}{L}V \qquad (6.116)$$

We are interested in cases where $h/L \ll 1$, so we have learned that $V_z \ll V$, or v_z is small compared to v_x and v_y. However, although v_z is unimportant, we cannot drop the $\partial v_z/\partial z$ term. We will return to the mass balance equation in a moment.

To scale the momentum balance equations, we follow the procedure presented in Section 5.1.1 when we introduced the lubrication approximation, and scale each component equation separately. For a generalized Newtonian fluid we substitute $\tau = 2\eta(\dot{\gamma})\mathbf{D}$ into the general form of the momentum balance, and we obtain

$$\rho\left(\frac{\partial \mathbf{v}}{\partial t} + \mathbf{v}\cdot\nabla\mathbf{v}\right) = -\nabla p + \nabla\cdot[2\eta(\dot{\gamma})\mathbf{D}] + \rho\mathbf{b} \qquad (6.117)$$

Before nondimensionalizing this equation, we introduce additional scaled variables for time and pressure:

$$t^* = t/t_c, \quad p^* = p/\Delta p_c \qquad (6.118)$$

The characteristic values t_c and Δp_c will be chosen shortly. Body forces were assumed to be negligible, so that term is dropped. Nondimensionalizing the x-direction momentum equation gives

$$\frac{\rho V}{t_c}\frac{\partial v_x^*}{\partial t^*} + \frac{\rho V^2}{L}\left(v_x^*\frac{\partial v_x^*}{\partial x^*} + v_y^*\frac{\partial v_x^*}{\partial y^*}\right) + \frac{\rho V V_z}{h}v_z^*\frac{\partial v_x^*}{\partial z^*}$$
$$= -\frac{\Delta p_c}{L}\frac{\partial p^*}{\partial x^*} + \frac{V}{L^2}\left\{\frac{\partial}{\partial x^*}\left(2\eta(\dot\gamma)\frac{\partial v_x^*}{\partial x^*}\right) + \frac{\partial}{\partial y^*}\left(\eta(\dot\gamma)\left[\frac{\partial v_x^*}{\partial y^*} + \frac{\partial v_y^*}{\partial x^*}\right]\right)\right\}$$
$$+ \frac{1}{h}\frac{\partial}{\partial z^*}\left(\eta(\dot\gamma)\left[\frac{V}{h}\frac{\partial v_x^*}{\partial z^*} + \frac{V_z}{L}\frac{\partial v_z^*}{\partial x^*}\right]\right) \quad (6.119)$$

We select $t_c = L/V$ and use the fact that $V_z = hV/L$ to simplify this expression. Because the viscosity η varies with position, we introduce a scaled viscosity variable η^* in the usual way, using a characteristic viscosity value η_0:

$$\eta^* = \eta/\eta_0 \quad (6.120)$$

Here η_0 would be the viscosity at the characteristic shear rate V/h. As with all scaled variables, η^* is of the order of one. Now the x-direction momentum equation is

$$\left[\frac{\rho V^2}{L}\right]\left(\frac{\partial v_x^*}{\partial t^*} + v_x^*\frac{\partial v_x^*}{\partial x^*} + v_y^*\frac{\partial v_x^*}{\partial y^*} + v_z^*\frac{\partial v_x^*}{\partial z^*}\right)$$
$$= -\left[\frac{\Delta p_c}{L}\right]\frac{\partial p^*}{\partial x^*} + \left[\frac{V\eta_0}{L^2}\right]\left\{\frac{\partial}{\partial x^*}\left(2\eta^*\frac{\partial v_x^*}{\partial x^*}\right) + \frac{\partial}{\partial y^*}\left(\eta^*\left[\frac{\partial v_x^*}{\partial y^*} + \frac{\partial v_y^*}{\partial x^*}\right]\right)\right\}$$
$$+ \left[\frac{V\eta_0}{h^2}\right]\frac{\partial}{\partial z^*}\left(\eta^*\frac{\partial v_x^*}{\partial z^*}\right) + \left[\frac{V\eta_0}{L^2}\right]\frac{\partial}{\partial z^*}\left(\eta^*\frac{\partial v_z^*}{\partial x^*}\right) \quad (6.121)$$

Multiplying through by $h^2/V\eta_0$ gives

$$\left\{\frac{\rho V h}{\eta_0}\frac{h}{L}\right\}\left(\frac{\partial v_x^*}{\partial t^*} + v_x^*\frac{\partial v_x^*}{\partial x^*} + v_y^*\frac{\partial v_x^*}{\partial y^*} + v_z^*\frac{\partial v_x^*}{\partial z^*}\right)$$
$$= -\left\{\frac{\Delta p_c h^2}{LV\eta_0}\right\}\frac{\partial p^*}{\partial x^*} + \frac{\partial}{\partial z^*}\left(\eta^*\frac{\partial v_x^*}{\partial z^*}\right) \quad (6.122)$$
$$+ \left\{\frac{h^2}{L^2}\right\}\left(\frac{\partial}{\partial x^*}\left(2\eta^*\frac{\partial v_x^*}{\partial x^*}\right) + \frac{\partial}{\partial y^*}\left(\eta^*\left[\frac{\partial v_x^*}{\partial y^*} + \frac{\partial v_y^*}{\partial x^*}\right]\right) + \frac{\partial}{\partial z^*}\left(\eta^*\frac{\partial v_z^*}{\partial x^*}\right)\right)$$

Now we recognize that the dimensionless group on the left-hand side, which contains the generalized Reynolds number $\rho V h/\eta_0$ and the aspect ratio h/L, is very small, so that the inertial terms can be neglected. We can also drop the terms on the right-hand side multiplied by $\{h^2/L^2\}$. The characteristic pressure difference is then

$$\Delta p_c = \frac{LV\eta_0}{h^2} \quad (6.123)$$

and we are left with

$$0 = -\frac{\partial p^*}{\partial x^*} + \frac{\partial}{\partial z^*}\left(\eta^*\frac{\partial v_x^*}{\partial z^*}\right) \quad (6.124)$$

This is certainly a substantial simplification.

6.6 THE GENERALIZED HELE–SHAW APPROXIMATION

The scaling of the y-momentum equation proceeds identically to the scaling of the x-momentum equation, and we find that

$$0 = -\frac{\partial p^*}{\partial y^*} + \frac{\partial}{\partial z^*}\left(\eta^* \frac{\partial v_y^*}{\partial z^*}\right) \tag{6.125}$$

The z-momentum equation comes out a little differently. The derivation follows precisely the same path as the derivation of Eq. (5.21), and therefore it is not repeated here. The result is

$$\frac{\partial^2 v_z^*}{\partial z^{*2}} = \left\{\frac{L^2}{h^2}\right\}\frac{\partial p^*}{\partial z^*} \tag{6.126}$$

The discussion surrounding Eq. (5.21) also applies here, and we therefore conclude that $\partial p^*/\partial z^* \ll \partial p^*/\partial x^*$ or $\partial p^*/\partial y^*$. This is a key result, because it allows us to solve for a two-dimensional pressure distribution even though the problem is physically three dimensional.

Velocity Solution

At this point no dimensionless parameters remain in the simplified balance equations, and it is clearer to return to the use of dimensional variables. The dimensional x- and y-direction momentum equations have been reduced to

$$\frac{\partial}{\partial z}\left(\eta \frac{\partial v_x}{\partial z}\right) = \frac{\partial p}{\partial x} \tag{6.127}$$

$$\frac{\partial}{\partial z}\left(\eta \frac{\partial v_y}{\partial z}\right) = \frac{\partial p}{\partial y} \tag{6.128}$$

Using $\partial p/\partial z = 0$, we integrate these equations in z to obtain

$$\frac{\partial v_x}{\partial z} = \frac{z}{\eta}\frac{\partial p}{\partial x} \tag{6.129}$$

$$\frac{\partial v_y}{\partial z} = \frac{z}{\eta}\frac{\partial p}{\partial y} \tag{6.130}$$

where we have used the fact that $\partial v_x/\partial z$ and $\partial v_y/\partial z$ vanish at the midplane $z=0$ as a result of symmetry. Under the current simplifications $\tau_{zx} = \eta\, \partial v_x/\partial z$ and $\tau_{zy} = \eta\, \partial v_y/\partial z$, so Eqs. (6.129) and (6.130) tell us that the shear stresses vary linearly across the gap and are zero at the midplane:

$$\tau_{zx} = \frac{\partial p}{\partial x} z \tag{6.131}$$

$$\tau_{zy} = \frac{\partial p}{\partial y} z \tag{6.132}$$

Note that the viscosity η is not a constant, but varies with the shear rate $\dot{\gamma}$ or the shear stress τ. Because shear stress varies with position, we will write the viscosity as $\eta(z)$ to emphasize this dependence; η also varies with x and y, but for the moment we will focus on the z dependence. Now, integrating Eqs. (6.129) and (6.130) once again and using a no-slip condition on $z = h$, we have

$$v_x(z) = -\frac{\partial p}{\partial x} \int_z^h \frac{z' dz'}{\eta(z')} \tag{6.133}$$

$$v_y(z) = -\frac{\partial p}{\partial y} \int_z^h \frac{z' dz'}{\eta(z')} \tag{6.134}$$

Here z' is a dummy variable of integration, and the minus sign comes from interchanging the limits on the integral. The pressure gradients have been taken outside of the integral because the pressure is independent of z, from Eq. (5.21). These integrals will always be positive, so the direction of the velocity will be from high pressure to low pressure. We cannot evaluate the integrals yet, because we have not chosen a constitutive equation.

We next introduce the gapwise average velocity, whose x and y components are

$$\bar{v}_x = \frac{1}{h} \int_0^h v_x \, dz \tag{6.135}$$

$$\bar{v}_y = \frac{1}{h} \int_0^h v_y \, dz \tag{6.136}$$

Substituting the result for v_x and v_y gives

$$\bar{v}_x = \frac{1}{h} \int_0^h \left\{ -\frac{\partial p}{\partial x} \int_z^h \frac{z' dz'}{\eta(z')} \right\} dz \tag{6.137}$$

$$\bar{v}_y = \frac{1}{h} \int_0^h \left\{ -\frac{\partial p}{\partial y} \int_z^h \frac{z' dz'}{\eta(z')} \right\} dz \tag{6.138}$$

Pressure is independent of z, so the derivatives of pressure may again be taken outside the outer integral. We may reverse the order of integration in these equations if we are careful to revise the limits of integration appropriately, as shown in Fig. 6.13. This yields

$$\bar{v}_x = -\frac{1}{h} \frac{\partial p}{\partial x} \int_0^h \left\{ \int_0^{z'} \frac{z' dz}{\eta(z')} \right\} dz' \tag{6.139}$$

$$\bar{v}_y = -\frac{1}{h} \frac{\partial p}{\partial y} \int_0^h \left\{ \int_0^{z'} \frac{z' dz}{\eta(z')} \right\} dz' \tag{6.140}$$

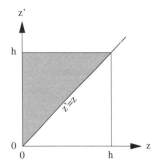

Figure 6.13: Integration limits for Eqs. (6.137)–(6.140). The shaded area represents $\int_0^h (\int_z^h dz') \, dz = \int_0^h (\int_0^{z'} dz) \, dz'$.

6.6 THE GENERALIZED HELE–SHAW APPROXIMATION

Now z' and $\eta(z')$ are fixed in the inner integral, which is taken with respect to z, so the inner integral can be evaluated to give

$$\bar{v}_x = -\frac{1}{h}\frac{\partial p}{\partial x}\int_0^h \frac{z'^2\, dz'}{\eta(z')} \tag{6.141}$$

$$\bar{v}_y = -\frac{1}{h}\frac{\partial p}{\partial y}\int_0^h \frac{z'^2\, dz'}{\eta(z')} \tag{6.142}$$

The latter integral is the same in both expressions, and we assign it a symbol:

$$S(x, y) \equiv \int_0^h \frac{z'^2\, dz'}{\eta(x, y, z')} \tag{6.143}$$

Here we have included the full dependence of viscosity on x, y, and z', to show where S gains its dependence on x and y. Now the gapwise average velocities can be written in a compact form:

$$\bar{v}_x = -\frac{S}{h}\frac{\partial p}{\partial x} \tag{6.144}$$

$$\bar{v}_y = -\frac{S}{h}\frac{\partial p}{\partial y} \tag{6.145}$$

Pressure Solution

Equations (6.144) and (6.145), together with Eqs. (6.133) and (6.134), show how the velocity depends on the pressure gradient, but we still do not have an equation to solve for p. Because we used the momentum balance equation to get this far, the missing information must come from the mass balance equation:

$$\frac{\partial v_x}{\partial x} + \frac{\partial v_y}{\partial y} + \frac{\partial v_z}{\partial z} = 0 \tag{6.146}$$

Unfortunately we have ignored v_z, and we have already shown that the $\partial v_z/\partial z$ term cannot be neglected here, so Eq. (6.146) cannot be used in its present form. To make this equation useful for our purposes, we integrate it from $z = 0$ to $z = h$. The integral of the right-hand side is zero, so we have

$$\int_0^h \left(\frac{\partial v_x}{\partial x} + \frac{\partial v_y}{\partial y} + \frac{\partial v_z}{\partial z} \right) dz = 0 \tag{6.147}$$

We may interchange the order of differentiation and integration in the first two terms,

$$\frac{\partial}{\partial x}\int_0^h v_x\, dz + \frac{\partial}{\partial y}\int_0^h v_y\, dz + \int_0^h \frac{\partial v_z}{\partial z}\, dz = 0 \tag{6.148}$$

and recognize that the first two integrals contain the gapwise average velocities. This gives

$$\frac{\partial}{\partial x}(h\bar{v}_x) + \frac{\partial}{\partial x}(h\bar{v}_x) + v_z|_0^h = 0 \tag{6.149}$$

Now $v_z = 0$ at the midplane ($z = 0$) because of symmetry, and v_z vanishes at $z = h$ because there is no flow through the wall. Thus, the last term is zero, and we have eliminated v_z from the equation. The remaining equation,

$$\frac{\partial}{\partial x}(h\bar{v}_x) + \frac{\partial}{\partial x}(h\bar{v}_x) = 0 \tag{6.150}$$

is sometimes called the *integrated continuity equation*. This is a weaker requirement than the full mass balance equation, Eq. (6.146). Equation (6.150) is equivalent to requiring a mass balance on a control volume that has infinitesimal dimensions Δx and Δy, but it extends from $z = 0$ to $z = h$. If mass enters this control volume near the top of one side and exits at the bottom of the other side, Eq. (6.150) will be satisfied. Ordinarily such a loose continuity requirement would not be acceptable, but it works here because the gap is thin, the velocity in the gapwise direction is small, and we are willing to sacrifice some of the z-direction details to simplify the solution process.

To get an equation for pressure, we simply substitute the expressions for $\bar{v}_x$ and $\bar{v}_y$ from Eqs. (6.144) and (6.145) into the integrated continuity equation to obtain

$$\frac{\partial}{\partial x}\left(S\frac{\partial p}{\partial x}\right) + \frac{\partial}{\partial y}\left(S\frac{\partial p}{\partial y}\right) = 0 \tag{6.151}$$

This is the Hele–Shaw pressure equation.

Because we worked so hard to obtain this result, one would hope that it is useful, and indeed it is! In a general three-dimensional steady fluid flow problem we must solve for four dependent variables, v_x, v_y, v_z, and p, as functions of three independent variables, x, y, and z. But for highly viscous fluids in a narrow gap, the Hele–Shaw formulation allows us to solve one partial differential equation, Eq. (6.151), for one independent variable, p, which only depends on x and y. This is an enormous simplification, and it is no wonder that the Hele–Shaw approximation has long been the preferred way to model injection mold filling.

Once the pressure distribution has been found, the gapwise average velocities can be recovered from Eqs. (6.144) and (6.145), and the detailed velocity profiles from Eqs. (6.133) and (6.134). These are simple integrations, which are much easier than solving a partial differential equation. One can even recover v_z, by substituting the detailed velocities $v_x(x, y, z)$ and $v_y(x, y, z)$ into the full continuity equation, Eq. (6.146), and integrating with respect to z.

Flow Conductance

To finish the derivation, we need to compute the flow conductance S. To understand the issues here, it is sufficient to consider an isothermal power law fluid. Recall that the viscosity of this fluid is given by

$$\eta = m\dot{\gamma}^{n-1} \tag{6.152}$$

In general the scalar strain rate is

$$\dot{\gamma} = \sqrt{2D_{ij}D_{ji}} \tag{6.153}$$

However, for the Hele–Shaw flow, we found that $\partial v_x/\partial z$ and $\partial v_y/\partial z$ were larger than the other velocity gradients by a factor of (L/h), so the expression for strain rate

6.6 THE GENERALIZED HELE–SHAW APPROXIMATION

simplifies to

$$\dot{\gamma} = \sqrt{\left(\frac{\partial v_x}{\partial z}\right)^2 + \left(\frac{\partial v_y}{\partial z}\right)^2} \tag{6.154}$$

Substituting Eqs. (6.129) and (6.130) reduces this to

$$\dot{\gamma} = \frac{z}{\eta}\sqrt{\left(\frac{\partial p}{\partial x}\right)^2 + \left(\frac{\partial p}{\partial y}\right)^2} \tag{6.155}$$

We may define the magnitude of the pressure gradient as

$$\Lambda \equiv \sqrt{\left(\frac{\partial p}{\partial x}\right)^2 + \left(\frac{\partial p}{\partial y}\right)^2} \tag{6.156}$$

Λ is a function of x and y, but not of z. Now the strain rate is

$$\dot{\gamma} = \frac{\Lambda z}{\eta} \tag{6.157}$$

For the power law fluid, the viscosity is then

$$\eta = m\left(\frac{\Lambda z}{\eta}\right)^{n-1} \tag{6.158}$$

which we may solve for η to find

$$\eta = m^s \Lambda^{1-s} z^{1-s} \tag{6.159}$$

where, as usual, $s \equiv 1/n$. Finally, substituting this into Eq. (6.143), we can evaluate S for the power law fluid as

$$S = \frac{\Lambda^{s-1} h^{s+2}}{(s+2)m^s} \tag{6.160}$$

In the special case in which the pressure gradient is constant, S is a constant. However, in general, S depends on the pressure gradient, and this makes the Hele–Shaw equation, Eq. (6.151), a nonlinear partial differential equation for p. This nonlinearity is not severe, and many numerical schemes will converge to the solution. The simplest method, successive substitution, is to start with trial values of $S(x, y)$, solve Eq. (6.151) as if these S values were fixed to obtain a trial pressure distribution $p(x, y)$, and then use these pressures to update S. The iteration continues until the solution converges.

For models other than the power law fluid there is no simple formula like Eq. (6.160), but Eq. (6.157) still gives the strain rate, and successive substitution or some other iterative numerical method can be used to find the solution. If the problem is nonisothermal then one must also solve the energy equation to find the temperature field and use a temperature-dependent model for viscosity. Because we already assumed that η varies with x, y, and z, this presents no new complications in the Hele–Shaw equations, and the solution proceeds as before. This is the calculation performed in commercial injection mold-filling simulations.

Commercial mold-filling software also handles nonplanar geometries. This is easy, as long as the part is thin. The part is modeled by using finite elements with a shell-like geometry. The elements are triangles or quadrilaterals that reside in three-dimensional space, and each element has a thickness as an element property. Each element is planar, so the Hele–Shaw equations are applied to each element in a local coordinate system, where the z axis is directed across the thickness of the element. Elements are joined along the edges, where continuity of pressure is enforced. Complete details are given by Kennedy (1995) and by Wang and Lee (1989).

6.7 SUMMARY

Non-Newtonian fluid behavior is the rule, rather than the exception, for many classes of materials, especially polymer melts, semisolid metals, and foods. Although Newtonian solutions may give rough qualitative insights into the flow of these materials, a non-Newtonian model is essential for quantitative predictions. When non-Newtonian flow problems are solved, the balance equations for mass and momentum are unchanged, and the fluid properties are represented by a non-Newtonian constitutive equation.

This chapter emphasized a simple but useful class of non-Newtonian models, the generalized Newtonian fluid. The power law fluid is one important member of this family, and it can represent shear-thinning or shear-thickening behavior. A variety of analytical solutions exist for the power law fluid in simple geometries, and these can be combined to create approximate models for more complex geometries. Other members of the GNF family include the Bingham fluid and the Cross and Carreau models. The Bingham fluid is used to model materials that will not flow until the applied stress exceeds a critical value, called the *yield stress*. The Cross and Carreau models go smoothly from Newtonian behavior at low shear rate to power law behavior at high shear rate, and they give a good quantitative match to the viscosity of polymer melts. All of these models are readily extended to make viscosity a function of temperature.

Polymeric fluids are also viscoelastic, exhibiting some solidlike properties at short times. GNF models do not capture the many interesting behaviors of viscoelastic fluids, and much more complicated constitutive equations must be used. This makes the solution of viscoelastic flow problems a challenging task.

The principles of scaling, estimation, and model simplification apply to non-Newtonian fluids, and they can be used to develop useful simplifications such as the generalized Hele–Shaw model. This model dramatically simplifies the equations of motion and continuity for a viscous fluid flowing in a narrow gap, and it is widely used in modeling polymer injection molding.

BIBLIOGRAPHY

F. P. T. Baaijens. Mixed finite element methods for viscoelastic flow analysis: A review. *J. Non-Newtonian Fluid Mech.* 79:361–385, 1998.

EXERCISES

F. P. T. Baaijens, S. H. A. Selen, H. P. W. Baaijens, G. W. M. Peters, and H. E. H. Meijer. Viscoelastic flow past a confined cylinder of a low density polyethylene melt. *J. Non-Newtonian Fluid Mech.* 68:173–203, 1997.

H. A. Barnes, J. F. Hutton, and K. Walters. *An Introduction to Rheology*. Elsevier, Amsterdam, 1989.

R. B. Bird, R. C. Armstrong, and O. Hassager. *Dynamics of Polymeric Liquids. Volume 1: Fluid Mechanics*. Wiley, New York, 1987.

J. M. Dealy and K. F. Wissbrun. *Melt Rheology and its Role in Plastics Processing: Theory and Applications*. Van Nostrand Reinhold, New York, 1990.

B. Debbaut, T. Avalosse, J. Dooley, and K. Hughes. On the development of secondary motions in straight channels induced by the second normal stress difference: experiments and simulations. *J. Non-Newtonian Fluid Mech.* 69:255–271, 1997.

H. S. Hele-Shaw. The motion of a perfect liquid. *Proc. Roy. Inst.* 16:49–64, 1899.

C. A. Hieber and H. H. Chiang. Shear-rate-dependence modeling of polymer melt viscosity. *Polymer Eng. Sci.* 32:931–938, 1992.

C. A. Hieber and S. F. Shen. A finite-element/finite-difference simulation of the injection-molding filling process. *J. Non-Newtonian Fluid.* 7:1–32, 1980.

P. Kennedy. *Flow Analysis for Injection Molds*. Carl Hanser Verlag, Munich, 1995.

R. Keunings. Simulation of viscoelastic fluid flow. In C. L. Tucker III, editor, *Fundamentals of Computer Modeling for Polymer Processing*. Carl Hanser Verlag, Munich, 1989.

C. W. Macosko. *Rheology: Principles, Measurements, and Applications*. VCH Publishers, New York, 1994.

T. C. Papanastasiou. Flows of materials with yield. *J. Rheology* 31:385–404, 1987.

H Schlichting. *Boundary-Layer Theory*. McGraw-Hill, New York, 6th edition, 1968.

R. I. Tanner. *Engineering Rheology*. Clarendon Press, Oxford, 1985.

H. P. Wang and H. S. Lee. Numerical techniques for free and moving boundary problems. C. L. Tucker III, editor, *Fundamentals of Computer Modeling for Polymer Processing*. Carl Hanser Verlag, Munich, 1989.

M. L. Williams, R. F. Landel, and J. D. Ferry. The temperature dependence of relaxation mechanisms in amorphous polymers and other glass-forming liquids. *J. Amer. Chem. Soc.* 77:3701–3707, 1955.

EXERCISES

1. *Fitting a power law model to data.* A viscometer is used to measure the following data for polypropylene at 190°C. Fit the data to a power law fluid model. Plot the data and your fitted model together on a log–log plot. Is the power law model a good fit for this data? Would you expect it to give a good value of viscosity at $\dot{\gamma} = 20{,}000$ s^{-1}?

$\dot{\gamma}$ (s^{-1})	η (Pa s)
15	750
36	520
100	320
180	210
400	150
1000	85
2000	50
4000	26

2 **Fitting generalized Newtonian models to data.** Fit the following data for shear viscosity versus strain rate by using

(a) a power law model and

(b) the Carreau model.

Use any convenient curve-fitting procedure or program. If you do not have access to a program that can fit these functions, you may plot the data and fit it "by eye." Regardless of the fitting method you use, plot the data and both fitted curves on a single plot with log–log axes. Comment on the quality of the fit. Which model is more likely to be accurate at a strain rate of 1000 s^{-1}? What viscosity is predicted by the power law model as the strain rate goes to zero?

$\dot{\gamma}$ (s^{-1})	η (Pa s)
0.2	2.00E+04
0.5	1.90E+04
1.0	1.40E+04
2.0	1.00E+04
5.0	5.50E+03

3 **Types of homogeneous deformation.** A *homogeneous deformation* is one in which every point in a body has the same velocity gradient, and thus experiences the same rate of deformation and accumulates the same strain over time. Listed below are velocity fields for four different types of homogeneous deformation. For each case, find

- the velocity gradient tensor **L**,
- the rate-of-deformation tensor **D**,
- the vorticity tensor **W**, and
- the three invariants of **D**.

Also, for each case sketch the deformed shape of a block of material that was initially a cube. The flow fields are as follows.

(a) Simple shear: $v_1 = Gx_2$, $v_2 = v_3 = 0$.

(b) Uniaxial elongation: $v_1 = Ex_1$, $v_2 = -(E/2)x_2$, $v_3 = -(E/2)x_3$.

(c) Biaxial elongation: $v_1 = Bx_1$, $v_2 = Bx_2$, $v_3 = -2Bx_3$.

(d) Planar elongation: $v_1 = Px_1$, $v_2 = -Px_2$, $v_3 = 0$.

4 **Elongational viscosity for generalized Newtonian fluids.** The elongational viscosity $\bar{\eta}$ is defined as

$$\bar{\eta} \equiv \frac{\tau_{11} - \tau_{33}}{\dot{\epsilon}}$$

when the velocity field is steady uniaxial elongation,

$$v_1 = \dot{\epsilon}x_1, \quad v_2 = -\frac{1}{2}\dot{\epsilon}x_2, \quad v_3 = -\frac{1}{2}\dot{\epsilon}x_3$$

This is the fluid version of a tensile test: the sample is a slender, cylindrical thread that is stretched along its axis, and $\bar{\eta}$ equals (F/A) divided by $\dot{\epsilon}$, where

EXERCISES

F is the pulling force, A is the cross-sectional area, and $\dot{\epsilon} = (d\ell/dt)/\ell$, with $\ell(t)$ begin the sample length.

(a) Show that for a Newtonian fluid $\bar{\eta} = 3\mu$.

(b) Derive a similar relationship for a power law fluid, expressing $\bar{\eta}$ in terms of m, n, and $\dot{\epsilon}$.

5 **Power law fluid in planar elongation.** A power law fluid is subjected to a homogeneous planar elongational flow. That is, the velocity field is

$$v_x = \dot{\epsilon}_p x$$
$$v_y = -\dot{\epsilon}_p y$$
$$v_z = 0$$

where $\dot{\epsilon}_p$ is the elongation rate.

(a) Indicate the shape change associated with this velocity field by sketching a small cube of material before and after deformation.

(b) Find the extra stress tensor τ for a power law fluid in this flow. Express your answer in terms of m, n, and $\dot{\epsilon}_p$.

6 **Pressure driven flow of a power law fluid.** This problem is analogous to the problem done in Section 6.2.3, where we analyzed the flow of a power law fluid in a long circular tube. Consider instead the flow of the fluid between parallel plates, as sketched below. You may consider the flow to be two dimensional.

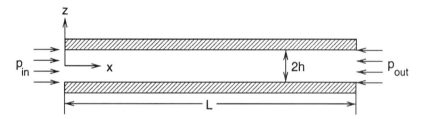

(a) Use V_x as the characteristic velocity in the x direction and V_z as the characteristic velocity in the z direction. Scale the mass balance equation and estimate the relative magnitudes of V_x and V_z. State any other assumptions you make.

(b) Scale the expressions for the various components of **D**, and then write a scaled expression for the scalar strain rate $\dot{\gamma}$ and estimate its characteristic value in terms of V_x and other relevant parameters.

(c) Using scaling, identify a dimensionless parameter that describes the relative importance of inertia and viscosity. How does your parameter compare to the result for a Newtonian fluid in Section 5.1?

(d) Assuming steady, fully developed flow and negligible inertia and gravity, solve for the velocity profile $v_x(z)$. Important: use a symmetry boundary condition at $z = 0$ and solve for only half the domain.

(e) Find the flow rate Q. Let W be the width of the channel.

(f) Plot velocity profiles for $n = 1.0, 0.5$, and 0.1, for a constant value of the average velocity. Show all of the curves on a single plot.

7. **Power law fluid in axial annular drag flow.** A power law fluid is confined in the annular space between two concentric cylinders. The outer cylinder has radius R and the inner cylinder radius is κR, where $0 < \kappa < 1$. The inner cylinder translates in the axial (z) direction with velocity V, and the outer cylinder is stationary. The cylinders are long compared to their radii, so you can assume that the flow is fully developed. Also assume that the flow is steady, and that the density of the fluid is constant. There is no pressure gradient in the z direction, so this is purely a drag flow.

   (a) Find the velocity profile $v_z(r)$ for a power law fluid in this geometry.

   (b) Find the volume flow rate of fluid Q.

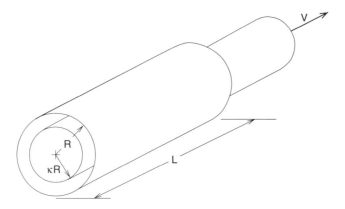

8. **Power law fluid in tangential annular drag flow.** A power law fluid is contained between two long concentric cylinders of radii R and κR, with $0 < \kappa < 1$. The outer cylinder is stationary and the inner cylinder is rotated with an angular velocity Ω. Assume that all the motion is tangential and ignore end effects, so that $v_\theta = v_\theta(r)$ with $v_r = v_z = 0$.

   (a) Find the shear stress distribution $\tau_{r\theta}(r)$ and the velocity distribution $v_\theta(r)$.

   (b) Plot the velocity distribution for $\kappa = 1/3$ and $n = 1.0, 0.5$, and 0.2. Also plot the shear stress distribution $\tau_{r\theta}(r)$. Explain why the velocity distribution changes shape as n decreases.

9. **Power law fluid in radial flow between disks.** A power law fluid is pumped at volume flow rate Q through a central hole, into the space between two parallel disks. The polymer flows radially outward through a gap whose total height is $2h$. The flow enters at a radius R_i and exits at radius R_0. The gap between the disks is small compared to the radius; that is, $h \ll R_0$. The flow is axisymmetric, so $v_\theta = 0$ and $\partial/\partial\theta = 0$. This is problem has no analytic solution for the full equations of motion, but you will be able to find an excellent approximate solution when the gap is small compared to the radius ($h \ll R_0$).

   Assume a steady flow with a known volume flow rate Q, an incompressible generalized Newtonian fluid, negligible inertia and body forces, and no-slip boundary conditions. The flow is axisymmetric and pressure equals p_0 at the outer radius R_0. You may also assume that p is a function of r only, that is, that $\partial p/\partial z \ll \partial p/\partial r$. (This assumption can be justified by a scaling argument, provided the gap is small.)

(a) Use the continuity equation and scaling to show that v_z is much smaller than v_r.

(b) Show that any velocity distribution of the form $v_r = f(z)/r$ and $v_z = 0$ will satisfy the continuity equation exactly.

(c) Using the r-direction equation of motion, show by scaling that the only significant viscous term will be the one containing τ_{zr}.

(d) Solve your simplified r-direction momentum equation for v_r, treating $\partial p/\partial r$ as an (unknown) function of r (i.e., just let it ride along in the equations for now).

(e) Integrate the velocity solution to find an expression for the volume flow rate Q. In fact Q is known, so this equation gives an expression for $\partial p/\partial r$. Substitute that back into the velocity solution to get an answer in terms of only known quantities. Verify that this velocity solution satisfies the continuity equation, provided $v_z = 0$.

(f) Integrate your expression for $\partial p/\partial r$ to find the pressure distribution $p(r)$. Find the pressure at $r = R_i$ and write an equation relating the total pressure drop Δp to the flow rate Q.

10 *A power law fluid in a mixed flow.* A geometry that is used in some extrusion dies is a thin annular gap between concentric cylindrical surfaces. The inner cylinder may also be rotated to promote a more uniform distribution of wall thickness around the circumference. The real geometry looks something like the sketch on the left below.

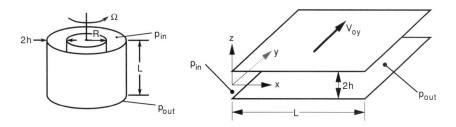

If the gap width $2h$ is much smaller than the radius R (not the way the picture is drawn), then very little error is introduced by ignoring the curvature of the gap and treating the problem as a flow between parallel plates. This can be justified by estimation procedures. The idealized geometry looks like the sketch on the right.

In this problem you are to analyze the flow of a power law fluid in this situation. The flow in the x direction is being driven by a known pressure gradient $\partial p/\partial x = (p_{in} - p_{out})/L$, and the flow in the y direction is caused by the movement of the upper plate (or, if you like, the inner cylinder), which has velocity $V_{oy} = \Omega R$. There is no pressure gradient in the y or z directions. Assume that inertia and body forces are negligible.

(a) Assuming that v_x and v_y are functions of z only, work out the components of the rate-of-deformation tensor. Identify which components are zero and write an expression for $\dot{\gamma}$, the scalar magnitude of the strain rate.

(b) Write the equations of motion for this case, simplified for the assumptions stated above. Also eliminate any terms containing τ_{ij}'s that are zero, based on your results from part (a).

(c) Solve the equations of motion for τ_{zx} and τ_{zy}. Your solutions will have some undetermined constants that will have to be evaluated later.

(d) Write the constitutive equations for τ_{zx} and τ_{zy} and substitute to obtain expressions in terms of velocity only.

(e) Combine your results from parts (c) and (d) to get a set of equations for the velocity distributions $v_x(z)$ and $v_y(z)$. These will be coupled differential equations. Do not attempt to solve them yet.

(f) Specialize your equations for the case in which $v_y \ll v_x$. This could be achieved in practice by rotating the inner cylinder very quickly. You will have to scale the various terms and decide which ones can be dropped.

(g) Solve your specialized equations to obtain the velocity distributions $v_x(z)$ and $v_y(z)$. What is interesting or unusual about the answer? Can you explain it qualitatively?

11 *Flow in an extrusion die.* A power law fluid with $m = 4.0 \times 10^4$ N s^n/m^2 and $n = 0.20$ is extruded through the die described in Section 5.2.9, using an inlet pressure of 35 MPa.

(a) Find the overall flow rate Q through the die. Use the general approach shown in Section 5.2.9 and make use of the power law solutions in Section 6.3.

(b) Find the speed of the polymer sheet as it exits the die. Assume that downstream of the die the thickness of the extruded sheet equals the gap height at the die exit.

(c) Find the shear rate at the wall at the beginning and end of the second segment, and in the die lip region (the third segment).

12 *Runner balancing.* Revisit the runner balancing problem (Chapter 5, Exercise 14), this time for a power law fluid.

(a) A tube of length L and radius R, initially empty, is filled from one end with a constant inlet pressure p_{in}. Derive an expression for $\ell(t)$, the filled length of the tube, and find the time required to fill the tube. Equation (6.37) should prove useful.

(b) Solve part (e) of Exercise 14 from Chapter 5, this time for a power law fluid. Is the proper runner design sensitive to the rheology of the fluid? That is, if the runners are balanced for one polymer, will they be balanced for other polymers, which will have different values of m and n?

13 *The capillary rheometer.* A common instrument for measuring polymer viscosity is the *capillary rheometer*. A piston and heated cylinder contain the polymer

EXERCISES

and push it through a small-diameter tube (the capillary), which has length L and radius R. Some instruments control the volume flow rate Q and measure the pressure drop Δp; others control the pressure and measure the flow rate. By conducting experiments over a range of flow rates, one can determine the viscosity at several different shear rates. Flow in the capillary can be analyzed as fully developed, pressure-driven flow in a tube. For viscosity measurement purposes, the analysis usually focuses on the shear stress and the shear rate at the capillary wall.

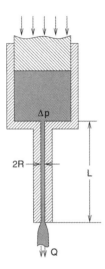

(a) Using the general equation of motion, show that τ_w, the shear stress τ_{rz} at the wall of the capillary, equals $\Delta p R / 2L$. That is, the wall shear stress is determined *only* by the applied pressure and the tube dimensions. This means that τ_w can be measured directly.

(b) The true wall shear rate $\dot{\gamma}_w$ is defined as $\partial v_z / \partial r$ evaluated at $r = R$. This is certainly related to the flow rate (which can be measured), but the exact relationship depends on the velocity profile. The exact velocity profile is *not* known because we do not know if the fluid is Newtonian, power law, or has some other behavior. A common approximation is to use the "apparent" wall shear rate $\dot{\gamma}_a$. This is defined as the wall shear rate for a Newtonian fluid at the given flow rate Q. Derive an expression for $\dot{\gamma}_a$ in terms of Q and the dimensions of the capillary. You may use the solution for Newtonian flow in a tube.

(c) Using the solution for pressure-driven flow of a power law fluid in a tube, find the relationship between the true wall shear rate $\dot{\gamma}_w$ and the apparent wall shear rate $\dot{\gamma}_a$ for a power law fluid. Show that the ratio of these two quantities is independent of flow rate. This means that a power law fluid will show a power law relationship between τ_w and $\dot{\gamma}_a$, even though this is not the true relationship between shear stress and shear rate for the material.

(d) A rheologist measures Q versus Δp over a range of shear rates, computes the values τ_w and $\dot\gamma_a$, and fits the data with

$$\tau_w = (5 \times 10^3 \text{ N s/m}^2)\dot\gamma_a^{0.30}$$

If the relationship between τ_w and $\dot\gamma_w$ can be written as

$$\tau_w = m\dot\gamma_w^n$$

what are m and n for this material? Because τ_w and $\dot\gamma_w$ are the true stress and true strain rate at the same point, m and n are the correct power law coefficients for the material.

14 *Radial flow between disks: the Hele–Shaw formulation.*

  (a) The Generalized Hele–Shaw governing equation for pressure, written in Cartesian coordinates, is

  $$\frac{\partial}{\partial x}\left(S\frac{\partial p}{\partial x}\right) + \frac{\partial}{\partial y}\left(S\frac{\partial p}{\partial y}\right) = 0$$

  Rewrite this equation in cylindrical coordinates, assuming that z is the gapwise direction and $p = p(r, \theta)$. Suggestion: Treat this as an exercise in translating between coordinate systems. First translate this expression into Gibbs notation, which applies in any coordinate system. Then translate the Gibbs notation into cylindrical coordinates. Look at the heat conduction equation in various coordinate systems for guidance.

  (b) Solve for the pressure distribution $p(r)$ and the overall flow rate Q when a Newtonian fluid is injected radially between parallel disks. The total gap height is $2h$, and the fluid is injected at pressure p_i at an inner radius R_i. Pressure equals zero at an outer radius R_f.

  (c) Repeat part (b) for a power law fluid. Here S is given by Eq. (6.160).

  (d) For a fixed Q, examine the behavior of p_i as $R_i \to 0$, for both the Newtonian and the power law solutions.

  (e) For the power law fluid consider the mold-filling problem, where R_f is a function of t and the injection pressure p_i is constant. Find the motion of the flow front $R_f(t)$. Sketch $R_f(t)$ and also sketch $p(r)$ for several different times.

15 *Hele–Shaw equations for compression molding.* In the compression molding process, flow is generated by squeezing the two halves of the mold together. If the cavity is thin ($h \ll L$) then the Hele–Shaw formulation can be adapted to treat this case. In this problem you will set up the Hele–Shaw equations for compression molding. Let the x and y coordinates define the midplane of the part, with the upper and lower mold surfaces at $z = \pm h$. Note that h is a function of time. Let the closing speed of the mold be defined by $s = -dh/dt$. Note that $dh/dt < 0$ when the mold is closing, so our definition makes $s > 0$. Also, placing $z = 0$ at the midplane means that the upper mold half is moving down and the lower mold half is moving up, relative to our coordinate system. This maintains symmetry about the midplane.

  (a) Use a scaling analysis on the continuity equation to show that $v_z \ll v_x$ or

v_y. As a consequence of this, the rest of the scaling results we developed for Hele–Shaw are still valid, so the Hele–Shaw momentum equation analysis can be reused without any rederivation. In particular, we can still use

$$\bar{v}_x = -\frac{S}{h}\frac{\partial p}{\partial x}, \quad \bar{v}_y = -\frac{S}{h}\frac{\partial p}{\partial y}$$

(b) Redo the integration of the continuity equation, accounting for the fact that we no longer have $v_z = 0$ at $z = h$. Your integrated continuity equation should have all the old terms plus one new term that contains s.

(c) Combine the momentum equation results from part (a) with your integrated continuity equation from part (b) to get a governing equation for the pressure distribution.

(d) Convert the Hele–Shaw compression-molding equation to cylindrical coordinates (see Exercise 14; you should be able to do this with minimal derivation).

(e) Solve for the pressure distribution in a disk of radius R with total thickness $2h$, when the disk consists of a Newtonian fluid and temperature is constant. Also find the velocity distribution $\bar{v}_r(r)$.

16 *Paint on a wall.* After some types of paint are applied to a vertical wall, they drip or "sag" before they dry, flowing down the wall in an uneven pattern. To avoid this behavior, paint manufacturers try to introduce a yield stress in their products.

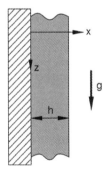

(a) Consider a layer of paint with uniform thickness h and density ρ applied to a vertical wall. Find the shear stress distribution $\tau_{zx}(x)$; g is the acceleration that is due to gravity.

(b) If the paint can be modeled as a Bingham fluid with a yield stress τ_Y, find the combination of parameters (density, coating thickness, etc.) for which the paint will stay on the wall with no motion.

(c) Show that τ, the scalar magnitude of the extra stress tensor $\boldsymbol{\tau}$, equals $|\tau_{zx}|$ for this problem.

(d) Assume that the problem parameters are such that the maximum value of τ in the fluid layer is greater than τ_Y. Find the value of x, call it x_Y, where $\tau = \tau_Y$.

(e) Solve for the velocity field in the region where the fluid is yielding, $0 \leq x < x_Y$.

(f) Solve for the velocity field in the region where the fluid is not yielding, $x_Y \leq x \leq h$. Use the fact that the fluid velocity must be continuous at x_Y to find an appropriate boundary condition.

(g) Make a careful sketch of the complete velocity profile for the case $x_Y = h/2$.

17 *Pressure dependent viscosity.* The viscosity of a polymer melt depends on pressure, a factor that can be important in injection molding and high-speed extrusion where the pressures can easily reach 10,000 psi. A reasonable model is an exponential dependence on pressure. For a Newtonian fluid this would be

$$\mu(p) = \mu_a e^{\beta p}$$

Here p is gauge pressure, μ_a is the viscosity at atmospheric pressure ($p = 0$), and β is a material constant.

Consider a fluid that follows this constitutive equation, flowing in a long tube of length L and radius R, with $L \gg R$. We impose a volume flow rate Q at the entrance ($z = 0$), and at the exit the pressure is atmospheric, $p = 0$.

Using the same type of scaling arguments we made for the Hele–Shaw flow, we can show that p, and thus μ, do not vary with radius. Then the classical solution for Newtonian flow in a tube applies at every location z along the axial direction, using the local viscosity. In particular, we have

$$Q = -\frac{\pi R^4}{8\mu(p)} \frac{dp}{dz}$$

(a) Using this equation, find the pressure at the tube entrance when the viscosity is constant, $\mu = \mu_a$. Call this pressure p_0.

(b) Now find the pressure distribution $p(z)$ when the viscosity depends on pressure. The algebra will be cleaner if you introduce p_0 to replace the group of constants that gives a characteristic pressure. Note that dp/dz is *not* constant along the tube.

(c) Express your solution in dimensionless form, as p/p_0 versus z/L. You should find that βp_0 appears as a dimensionless parameter. Give a physical interpretation of this parameter.

(d) Plot the dimensionless pressure versus position for $\beta p_0 = 0.01, 0.5$, and 0.8. Give a physical interpretation of the results.

CHAPTER SEVEN

Heat Transfer with Fluid Flow

When we considered heat conduction in Chapter 4, we assumed that all velocities were zero. This allowed us to deal with the energy equation by itself, unaffected by the momentum balance equation. Such a representation is essentially limited to stationary solids. In this chapter we relax this restriction and consider problems in which there is simultaneous fluid flow and heat transfer. This combination is important in many materials processing applications, include injection molding, metal casting, welding, and crystal growth. We continue the pattern of the previous chapters, starting with a simple example and then considering increasingly complex phenomena.

We continue to focus on condensed phases, so we once again write the energy balance equation by using temperature and pressure as the primitive thermodynamic variables:

$$\rho \left(c_p \frac{DT}{Dt} + L_p \frac{Dp}{Dt} \right) = \nabla \cdot (\mathbf{k} \cdot \nabla T) + \boldsymbol{\tau} : \mathbf{D} + \rho \dot{R} \qquad (7.1)$$

Here c_p is the specific heat at constant pressure, L_p is the latent heat of pressure change, $\mathbf{k}$ is the thermal conductivity tensor, $\dot{R}$ is the rate of internal heat generation, and $\mathbf{D}$ and $\boldsymbol{\tau}$ are the rate-of-deformation and extra stress tensors, respectively.

In the first section we consider problems in which the velocity field is independent of the temperature field. In this class of problems, called *uncoupled advection*, velocity affects the temperature distribution but temperature does not affect velocity. We then proceed to situations in which the temperature and velocity solutions are strongly coupled. This usually arises when fluid properties change with temperature, and we discuss two such cases: temperature-dependent viscosity, and temperature-dependent density.

7.1 UNCOUPLED ADVECTION

In this first section we will assume that the velocity field is given and is unaffected by the presence of temperature variations. Later sections will show how to determine if this assumption is valid. We also assume that all material properties are constant, that the pressure variation contribution $[L_p(D_p/D_t)]$ is negligible, and that the thermal conductivity is isotropic as well as constant. The reduced form of the energy balance equation is then

$$\rho c_p \left(\frac{\partial T}{\partial t} + \mathbf{v} \cdot \nabla T \right) = k \nabla^2 T + \boldsymbol{\tau} : \mathbf{D} + \rho \dot{R} \tag{7.2}$$

Here the material derivative has been expanded to emphasize the presence of both the transient and the advection terms. Note that the advection term is sometimes referred to as a "convection" term; this can cause confusion with a surface heat transfer coefficient, which is often referred to as a "convective" boundary condition, or with buoyancy-driven flow, which is often called "natural convection." In this book we will use "advection" to refer to the $\rho c_p \mathbf{v} \cdot \nabla T$ term in the energy equation, and similar terms in other transport equations.

7.1.1 CONSTANT VELOCITY

A very simple velocity field is one where the velocity is constant everywhere, but even such a simple flow field leads to interesting solutions of the energy equation. Consider a segment of a continuous solid slab, entering the domain at $x = 0$ with velocity V in the x direction, at constant initial temperature T_0. The slab is cooled (or heated) from both sides, as shown in Fig. 7.1, by an ambient temperature T_1. This configuration is a prototype for continuous cooling of extruded polymer sheets, continuous heating of steel slabs for rolling, and numerous other problems in materials processing. Usually, one is interested in the distance required to reach a constant through-thickness temperature.

We will assume that the temperature distribution is steady, that is, unchanging in time, and that the problem is two dimensional. We also assume that there is no internal heat generation. Because the velocity is constant the rate of deformation $\mathbf{D}$ is zero, and therefore the viscous dissipation is zero as well. The energy equation thus reduces to

$$\rho c_p V \frac{\partial T}{\partial x} = k \left(\frac{\partial^2 T}{\partial x^2} + \frac{\partial^2 T}{\partial y^2} \right) \tag{7.3}$$

Note that the advection term in the y direction is dropped because $v_y = 0$.

To solve this problem, we will also need boundary conditions at $y = \pm H$. Suppose for now that the temperature at these boundaries is $T = T_1$. We can later make this boundary condition more realistic by assigning a convective heat transfer coefficient h at the boundaries.

We now proceed to scale the equation. There is no obvious length scale in the x direction for this continuous process, so for the moment we leave this as simply L.

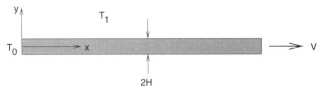

Figure 7.1: A continuously cooled (or heated slab) traveling at constant velocity V in the x direction. The ambient temperature is T_1.

7.1 UNCOUPLED ADVECTION

Choose the following scaled variables:

$$x^* = \frac{x}{L}, \quad y^* = \frac{y}{H}, \quad \theta = \frac{T - T_0}{T_1 - T_0} \tag{7.4}$$

Substituting these dimensionless variables into Eq. (7.3) yields

$$\left[\frac{\rho c_p V}{L}\right] \frac{\partial \theta}{\partial x^*} = k \left(\frac{1}{L^2} \frac{\partial^2 \theta}{\partial x^{*2}} + \frac{1}{H^2} \frac{\partial^2 \theta}{\partial y^{*2}}\right) \tag{7.5}$$

Figure 7.1 suggests that $H \ll L$, so we divide through by k/H^2. (This is equivalent to guessing that conduction across the slab thickness is important.) We also recall that the thermal diffusivity α is defined as $k/\rho c_p$. This leads to

$$\left\{\frac{VH}{\alpha} \frac{H}{L}\right\} \frac{\partial \theta}{\partial x^*} = \left\{\frac{H^2}{L^2}\right\} \frac{\partial^2 \theta}{\partial x^{*2}} + \frac{\partial^2 \theta}{\partial y^{*2}} \tag{7.6}$$

If $H \ll L$, conduction in the x direction will be negligible relative to conduction in the y direction. A new dimensionless group has appeared, which is called the *Péclet number*, Pe:

$$\text{Pe} = \frac{VH}{\alpha} \tag{7.7}$$

The Péclet number represents the ratio of heat transferred by advection to heat transferred by conduction. The product of Pe with H/L occurs in problems in which one characteristic length is much smaller than the others; this is called the *Graetz number*, Gz.

We consider first the case in which the Péclet number is large, Pe $\gg$ 1. We continue to assume that $H/L \ll 1$, so the dimensionless energy equation reduces to

$$\text{Pe} \frac{H}{L} \frac{\partial \theta}{\partial x^*} = \frac{\partial^2 \theta}{\partial y^{*2}} \tag{7.8}$$

Notice that the term containing the highest derivative in x has been eliminated, so we should expect some sort of boundary layer to be present. Also, because the term with the highest-order derivative has been discarded, we lose the ability to satisfy one of the boundary conditions. In this case, using Eq. (7.8) means that we cannot impose any boundary condition on the downstream side of the domain.

Because there is no downstream boundary condition to satisfy, we can view this as an initial value problem, beginning at $x = 0$ and traveling along the x axis at velocity V. To make this more clear, define a pseudotime variable t^*:

$$t^* = \frac{1}{\text{Pe}} \frac{L}{H} x^* \tag{7.9}$$

Substituting t^* for x^* in Eq. (7.8) yields

$$\frac{\partial \theta}{\partial t^*} = \frac{\partial^2 \theta}{\partial y^{*2}} \tag{7.10}$$

This has the same form as the one-dimensional transient heat conduction equation from Chapter 4. We have shown that, for large Péclet numbers, the moving slab problem is equivalent to the transient conduction in a stationary slab. We can imagine following a slice of the slab as it enters the cooling (or heating) zone. This slice enters with uniform temperature $\theta = 0$, and the temperatures on its outer surfaces are immediately changed to $\theta = 1$. The temperature distribution $\theta(y^*, t^*)$ is identical to the transient solution given in Eq. (4.99) (with y^* here replacing x^* in that equation), when we make the correspondence between t^* and our x^* by using Eq. (7.10).

Our original question was, "What is the distance in the axial direction where the temperature becomes uniform?" We may now estimate this distance. Recall that the characteristic time for conduction was given by ℓ^2/α, where ℓ is a characteristic length. We have scaled both ℓ and α to be one in Eq. (7.10), so the characteristic pseudotime t^* over which the temperature becomes uniform is approximately one. From the definition of t^* we can convert this to a scaled distance:

$$x^* = \text{Pe}\frac{H}{L} \qquad (7.11)$$

The temperature will be uniform across the thickness when x^* is $\mathcal{O}(1)$, so we must have

$$L = \text{Pe}\, H \qquad (7.12)$$

Thus, we see that the distance for the temperature to become uniform across the cross section increases linearly with Pe, and therefore linearly with V. This scaling is valid for Pe > 1.

A second limiting case occurs when Pe $\ll$ 1. In this case the advection term is negligible, and we are left with

$$0 = \frac{H^2}{L^2}\frac{\partial^2 \theta}{\partial x^{*2}} + \frac{\partial^2 \theta}{\partial y^{*2}} \qquad (7.13)$$

We now invoke our usual argument that both terms must be $\mathcal{O}(1)$, and therefore $L = H$. No matter how small the velocity V, the length scale for cooling is never less than H.

The general version of this problem must be solved numerically. Numerical solutions for several values of Pe are are shown in Fig. 7.2. Note the small difference between the solutions for Pe = 0 and Pe = 1. As the Péclet number increases, the velocity seems to sweep the isotherms downstream, and indeed this is the general

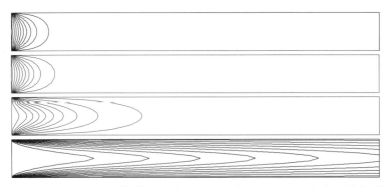

Figure 7.2: Temperature distributions for a slab moving at constant velocity with fixed temperature boundary conditions. From top to bottom: Pe = 0, 1, 10, and 100. The isotherms range from $\theta = 0.05$ to $\theta = 0.95$ in increments of 0.1.

7.1 UNCOUPLED ADVECTION

characteristic of advection problems. At Pe = 100 the isotherms are clustered near the horizontal surfaces. We expect that higher Péclet numbers will continue this trend, leading to thin thermal boundary layers.

We can use scaling to estimate the thickness of the thermal boundary layer at high Péclet numbers. Let δ represent the local boundary layer thickness, and define a new, rescaled y coordinate as

$$\eta = \frac{y}{\delta} \tag{7.14}$$

We want to relate δ to the known parameters in the problem. Keeping the other scales the same, substitute this expression into Eq. (7.3) and rearrange the terms to get

$$\frac{\partial \theta}{\partial x^*} = \left\{ \frac{\alpha L}{V \delta^2} \right\} \left(\left\{ \frac{\delta^2}{L^2} \right\} \frac{\partial^2 \theta}{\partial x^{*2}} + \frac{\partial^2 \theta}{\partial \eta^2} \right) \tag{7.15}$$

We certainly expect that the boundary layer thickness δ will be small compared to L, so the axial conduction term is again negligible. However, within the boundary layer the conduction term on the right-hand side must be comparable with the advection term on the left-hand side. We therefore set the premultiplier to one,

$$\frac{\alpha L}{V \delta^2} = 1 \tag{7.16}$$

and solve to find $\delta = \sqrt{\alpha L / V}$. In dimensionless form this is

$$\frac{\delta}{L} = \frac{1}{\sqrt{\text{Pe}}} \tag{7.17}$$

Note that this result was obtained without ever specifying L. That means that we may use *any* value of x in place of L, and at that point the boundary layer thickness will scale as

$$\delta = \sqrt{\frac{\alpha x}{V}} \tag{7.18}$$

We have found the important result that the boundary layer thickness grows in proportional to $\sqrt{x}$. The bottom plot in Fig. 7.2 confirms this observation.

Note in Fig. 7.2 that all of the isotherms in each plot converge in the two left-hand corners. At those points the temperature gradient becomes infinite. This singularity is an artifact caused by our choice of boundary conditions, which set the upper and lower surface temperatures to T_1 beginning at that corner. A more realistic boundary condition would use a heat transfer coefficient, and we consider that possibility now.

The boundary conditions at the upper and lower surfaces now become

$$\mp k \frac{\partial T}{\partial y} = h(T - T_1) \quad \text{at } x = \pm H \tag{7.19}$$

where the minus case applies on the upper surface and the plus case applies on the lower surface. The signs arise from the direction of the normal vector at each surface.

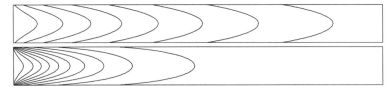

Figure 7.3: Temperature solutions for a slab moving at constant velocity with heat transfer coefficient boundary conditions. Top: Bi = 1, Pe = 10. Bottom: Bi = 10, Pe = 10. Isotherms are as in Fig. 7.2.

Applying the scales introduced earlier for the temperature and y, we find that this becomes

$$\mp \frac{\partial \theta}{\partial y^*} = \text{Bi}\,\theta \tag{7.20}$$

where $\text{Bi} = hH/k$ is the Biot number. The influence of the surface heat transfer coefficient is reflected in the value of the Biot number, and the (dimensionless) solution to this problem depends on two dimensionless parameters, Pe and Bi. The solutions in Fig. 7.2 are for $\text{Bi} \to \infty$.

Figure 7.3 shows solutions for Bi = 1 and 10, at Pe = 10. The solution at Bi = 10 is not very different from the solution for infinite Biot number in Fig. 7.2, except that the isotherms do not all converge at the left-hand corners. For Bi = 1 (smaller heat transfer coefficient), the isotherms are noticeably different, the slab cools more slowly, and we might question our scaling estimate for L. In fact, for sufficiently small values of the heat transfer coefficient we can reuse the small Biot number solution for transient conduction in a stationary slab (Section 4.2.2) and obtain a new scaling for L, provided that the Péclet number remains large.

7.1.2 CONTINUOUS CASTING

An important application in which we may gain insight by assuming that the velocity is constant everywhere is continuous casting. Most of the steel and aluminum produced in the world today is manufactured by using this process. Commercial practices vary with different metals, but the processes have some common features, as well. Molten alloy is introduced into a water-cooled copper mold through a ceramic tube. The mold has an open bottom that is initially closed by a bottom block, onto which the metal begins to freeze. After a short time the bottom block is continuously withdrawn, while the metal level in the mold is maintained constant by using a control rod or sliding gate. The metal continues to freeze through contact with the mold and heat transfer to submold water sprays, and eventually a steady state is established.

Figure 7.4 illustrates the process for both steel and aluminum. The shaded regions in the ingots correspond to solidified material. Steel, which has relatively high strength and ductility at high temperature and relatively low thermal conductivity, is cast at high speed and bent from a vertical to a horizontal orientation while the core is still molten. In contrast, aluminum has high conductivity and is rather weak at high temperatures, and it is cast slowly without bending.

Notice the similarities between the isotherms shown in Fig. 7.2 and the solidification boundary sketched in Fig. 7.4. This similarity is no accident, and it suggests that steel casting processes have a high Péclet number, whereas aluminum casting

7.1 UNCOUPLED ADVECTION

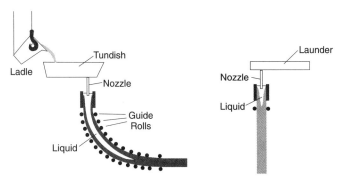

Figure 7.4: Schematics of the continuous casting process. The drawing on the left is typical of steel casting; that on the right is typical for aluminum alloys.

processes have a relatively low Péclet number. The material properties and typical casting parameters for the two alloys are given in Table 7.1. From these data, we compute Péclet numbers for the two processes that confirm our observation.

A more complete analysis of steel casting is possible by using some of the tools we have already developed. When the Péclet number is large, recall from the preceding section that the governing equation could be recast as a pseudotransient problem by replacing the axial coordinate by a pseudotime variable scaled on Pe. We may do the same thing now for this case, but include the latent heat associated with solidification.

The simplest problem is one in which the ingot surface temperature is instantly cooled to T_1. This is not a very realistic approximation, but it does enable us to find an analytical solution. For this case, the pseudotime formulation of the problem reduces to the mold-control case for the Stefan problem in Section 4.3.1. This problem is given as an exercise.

7.1.3 VELOCITY DISTRIBUTIONS AND THE PRANDTL NUMBER

The problems that we have considered so far have all had constant velocity. There are many other important uncoupled advection problems in which the velocity is not uniform.

Table 7.1 Typical Process Parameters and Material Properties for Continuously Cast Steel and Aluminum (at high temperature)

Material	Steel	Aluminum
Section half-width (mm)	100	100
Casting speed (mm/s)	33	1
Thermal conductivity (kW/m K)	0.08	0.211
Density (kg/m^3)	7210	2555
Specific heat (kJ/kg K)	0.67	1.19
Péclet number	199	1.4

As an example, consider a pressure-driven flow between parallel plates where the fluid enters at temperature T_0 and the walls are maintained at T_1. This problem contains some of the important thermal aspects of injection mold filling, where hot polymer is injected into a cold mold. If the fluid viscosity is not affected by temperature (not a very realistic assumption for injection molding), then the velocity profile will be parabolic as in Eq. (5.43). The temperature solution for this problem proceeds along the same lines as any other advection problem, the main difference being that velocity is a function of position, rather than a constant over the entire domain.

The scaling analysis for this problem is identical to the uniform-velocity problem examined in Section 7.1.1, and the general character of the solution is the same. There is a short entrance length at low Péclet number, a longer entrance length at higher Péclet number, and thin thermal boundary layers when the Péclet number is very large.

That scaling analysis also closely parallels to the scaling for velocity entrance length in Section 5.4.2: the estimates for the entrance lengths, Eqs. (7.12) and (5.143), have the same form, as do the formulas for the boundary layer thicknesses, Eqs. (7.18) and (5.141). However, the thermal entrance length is governed by the thermal diffusivity $\alpha = k/\rho c_p$, whereas the momentum entrance length is governed by the kinematic viscosity $\nu = \mu/\rho$. The ratio of these two properties is another classical dimensionless group, the *Prandtl number*, Pr:

$$\Pr = \frac{\nu}{\alpha} = \frac{\mu c_p}{k} \tag{7.21}$$

Taking the ratio of Eqs. (7.12) and (5.143), we find that the ratio between the thermal and the velocity entrance lengths is equal to the Prandtl number:

$$\frac{L_{\text{thermal}}}{L_{\text{velocity}}} = \Pr \tag{7.22}$$

Table 7.2 gives the Prandtl numbers for several materials. Polymers, with their very high viscosities and low thermal conductivities, invariably have $\Pr \gg 1$. Thus, a polymer melt flow will have a fully developed velocity profile in a very short distance,

Table 7.2 Prandtl Number ν/α for Several Fluids

Material	Pr
Lead	0.017
Aluminum	0.030
Air (dry, 100°C)	0.70
Air (dry, 20°C)	0.72
Water (100°C)	1.78
Water (20°C)	7.07
Polycarbonate	$\sim 10^6$

7.1 UNCOUPLED ADVECTION

but it may have a very long thermal development length. Thus, the advection term in the energy equation may be important even when the inertial terms in the momentum balance are negligible. In contrast, liquid metals, with a high conductivity and a low viscosity, have Pr $\ll 1$. In the entrance flow of a liquid metal the temperature profile will be fully developed while the velocity profile is still very flat. At small Prandtl number the advection of heat may be negligible, even when the inertial terms dominate the momentum balance.

7.1.4 TRAVELING HEAT SOURCES: WELDING

Another materials processing application in which the velocity field is everywhere constant is welding. Figure 7.5 depicts a source of heat, such as a welding arc or a laser beam, traveling at constant velocity across the surface of a thick plate. This model geometry approximates the application of energy to the surface of a material as in a welding process. The figure indicates the beam traveling to the right at constant velocity V. The shaded region indicates the heated region that stretches out behind the beam, corresponding to the area where the beam recently passed.

We now make some assumptions to simplify the problem. We consider the energy beam to be a point source that transfers energy to the workpiece at a rate Q. (Q is the power of the beam.) Obviously this point-source model cannot be valid immediately underneath the beam, because a finite amount of power is passing through an area of zero size. However, it should be reasonable at distances greater than a few beam diameters away from the source point. Real beams used in welding have a distribution of energy density over a finite area that is approximately Gaussian. We shall also assume that the plate is infinite in extent in the x and z directions, semi-infinite in the y direction, and has constant material properties. There is one further assumption that we will make after setting up the analysis.

We wish to find the temperature distribution in the plate. To this end, we write the energy balance equation,

$$\frac{DT}{Dt} = \alpha \left(\frac{\partial^2 T}{\partial x^2} + \frac{\partial^2 T}{\partial y^2} + \frac{\partial^2 T}{\partial z^2} \right) \tag{7.23}$$

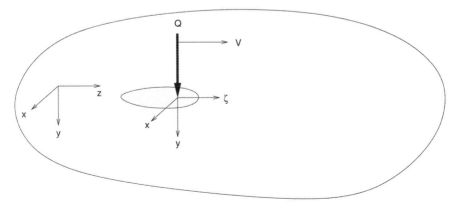

Figure 7.5: A point source of energy moving at constant velocity V over a semi-infinite plate.

and the boundary condition for the energy beam,

$$q(0, 0, z) = Q\delta[z - z_b(t)] \tag{7.24}$$

where $z_b(t)$ is the current location of the beam. Here δ is the Dirac delta function, having infinite amplitude where the argument is zero, zero amplitude otherwise, and an integral equal to one. The beam is moving at constant velocity V, so we have $z_b(t) = z_0 + Vt$ (assuming the beam started at $z = z_0$), and thus

$$q(0, 0, z) = Q\delta(z - z_0 - Vt) \tag{7.25}$$

The far-field boundary conditions at $x, y, z \to \infty$ are simply $T = T_0$.

This formulation of the problem is somewhat inconvenient, because the spatial location of the boundary condition is changing. We therefore transform to a coordinate system fixed on the energy beam itself, that is, translating with velocity V in the z direction. To this end, define $\zeta = z - z_b$, and transform from the (x, y, z) coordinate system to the (x, y, ζ) coordinate system. In this new coordinate system, the axes "see" the material translating with velocity $-V$ in the ζ direction. Recall from Chapter 2 that under this transformation, the material derivative takes the form

$$\frac{DT}{Dt} = \frac{\partial T}{\partial t} + \mathbf{v} \cdot \nabla T = \left.\frac{\partial T}{\partial t}\right|_\zeta - V\frac{\partial T}{\partial \zeta} \tag{7.26}$$

The governing equation and boundary condition at the beam in the transformed coordinate system are then

$$\left.\frac{\partial T}{\partial t}\right|_\zeta - V\frac{\partial T}{\partial \zeta} = \alpha\left(\frac{\partial^2 T}{\partial x^2} + \frac{\partial^2 T}{\partial y^2} + \frac{\partial^2 T}{\partial \zeta^2}\right) \tag{7.27}$$

and

$$q(0, 0, 0) = Q\delta(0) \tag{7.28}$$

The heat source boundary condition is now applied at the origin of the moving coordinate system. The boundary conditions at $x, y, \zeta \to \infty$ are still $T = T_0$.

This form of the equation is particularly helpful if we make one further assumption, namely that the temperature distribution is steady in the *transformed* frame. This is sensible when the beam is far from any edge of the plate, and after it has been traveling for some time.

As we saw with the problems we solved earlier in infinite domains, there is no natural length or time scale in this problem, so we solve it in dimensional form. This solution was first presented by Rosenthal (1941). He used a slightly different form of the boundary condition at the point source, which we now introduce. In the limit as we approach zero, the heat flux from a hemispherical cap of radius r must be equal to the total heat produced by the point source. We therefore write

$$\lim_{r \to 0} 2\pi r^2\left(-k\frac{\partial T}{\partial r}\right) = Q \tag{7.29}$$

7.1 UNCOUPLED ADVECTION

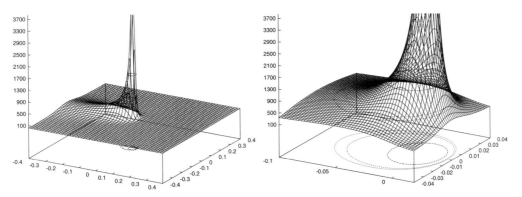

Figure 7.6: Computed temperatures on the plate surface for a moving point source of energy. $T_0 = 25$, $Q/2\pi k = 50$, and $V/2\alpha = 50$. Note the difference of scales in the two figures. Isotherms are shown at 650°C, 750°C, and 1500°C.

with $r^2 = x^2 + y^2 + \zeta^2$. Rosenthal also gave the solution for the temperature, which is

$$T = T_0 + \frac{Q}{2\pi k}\frac{1}{r}\exp\left(-\frac{V(r+\zeta)}{2\alpha}\right) \tag{7.30}$$

Notice that this solution has both a spherical part, embodied in the terms containing r, and an exponential "tail" in the direction of motion of the heat source. Note also that the temperature is infinite at the origin, a nonphysical consequence of assuming the beam to be a point source. The meaning of the phrase "far from any edge of the plate" is now clear: combinations of r and ζ such that the second term in Eq. (7.30) is negligible.

To illustrate the character of the solution, Figs. 7.6 and 7.7 show plots of the temperature distribution on the surface of the plate near the weld beam for two cases. In the first case, the parameters in Eq. (7.30) are chosen such that $T_0 = 25$, $Q/2\pi k = 50$, and $V/2\alpha = 50$. This represents a relatively high speed. The extended, tail of the temperature distribution is clearly evident in Fig. 7.6. Three isotherms are shown: 650°C, 750°C, and 1500°C, corresponding to temperatures near the eutectoid and melting transformation temperatures in iron.

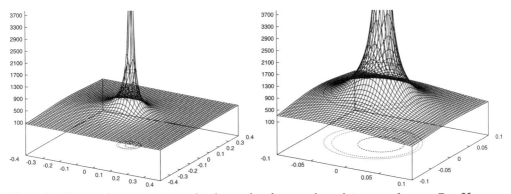

Figure 7.7: Computed temperatures on the plate surface for a moving point source of energy. $T_0 = 25$, $Q/2\pi k = 50$, and $V/2\alpha = 5$. Note the difference of scales in the two figures, and that the length scale on the right is different from Fig. 7.6. Isotherms are shown at 650°C, 750°C, and 1500°C.

Figure 7.7 depicts the temperature distribution for a similar case, except that now $V/2\alpha = 5$, corresponding to a much slower welding speed. One can readily see that the isotherms are closer to circular for this case, and that the heat affected zone around the beam is much larger than in the previous example. As the beam travels more slowly, more heat is carried away by conduction.

A more realistic representation of the energy distribution in a welding beam is a Gaussian distribution. For such a case, the energy density is finite, rather than infinite as for the point source, and the singularity in temperature at the origin no longer occurs. In addition, we should account for the latent heat associated with the melting and freezing of the plate. With these additional complications, these problems must be solved numerically.

A second simplified case that permits an analytical solution is for welding of very thin plates or sheets. In this case the beam passes completely through the sheet and may be thought of as a line source in the y direction, and the temperature distribution is independent of depth in the sheet. The solution for this case, for a sheet of thickness H, was given by Rosenthal (1941) as

$$T = T_0 + \frac{Q}{2\pi k H} \exp\left(-\frac{V\zeta}{2\alpha}\right) K_0\left(\frac{VR}{2\alpha}\right) \tag{7.31}$$

where $R = \sqrt{x^2 + \zeta^2}$ and K_0 is the modified Bessel function of the second kind and order zero. The solutions for this case are qualitatively similar to those of the three-dimensional case discussed above.

7.2 TEMPERATURE-DEPENDENT VISCOSITY AND VISCOUS DISSIPATION

In the preceding section, the velocity affected the temperature distribution through the advection terms of the energy balance equation, but the temperature distribution did not affect the velocity. We now consider one class of problems in which temperature affects the velocity field, here through a temperature-dependent viscosity. We will also discuss viscous dissipation, which is another way that velocity influences temperature. The two phenomena frequently appear together and interact, particularly in polymer melt flows. We first examine each one separately, and then consider their combined effect.

7.2.1 TEMPERATURE-DEPENDENT VISCOSITY

Consider again the drag flow between parallel plates, which we analyzed in Section 5.1. The problem is sketched in Fig. 7.8. A viscous fluid is contained in a

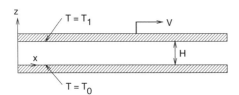

Figure 7.8: Geometry and nomenclature for nonisothermal drag flow between parallel plates.

7.2 TEMPERATURE-DEPENDENT VISCOSITY AND VISCOUS DISSIPATION

channel, whose upper and lower walls are held at temperatures T_1 and T_0, respectively. The upper wall moves with velocity V and the lower wall is stationary. We assume that the flow is two dimensional and fully developed. Also, the fluid has constant density and is Newtonian. The pressure gradient is zero, so all the fluid motion is driven by the moving top plate. We solved the isothermal version of this problem in Section 5.1, but now there are two new factors: the temperature difference between the plates, and a fluid viscosity that depends on temperature.

Several popular models for temperature-dependent viscosity were discussed in Section 6.5.3. Here we choose an exponential dependence of viscosity on temperature,

$$\mu = \mu_0 e^{-a(T-T_0)} \tag{7.32}$$

This form is only accurate over modest ranges of temperature, but it is one of the few realistic forms that is amenable to analytical solution. Here μ_0 and a are material constants and T_0 is a reference temperature at which the viscosity equals μ_0; μ_0 has the units of viscosity (Pa s) and a has units of K^{-1}. The constant a is a measure of the sensitivity of viscosity to temperature, and we can regard $1/a$ as the temperature change that will cause a significant change in viscosity.

We begin the analysis with the continuity equation. The assumptions of constant density and two-dimensional flow imply that

$$\frac{\partial v_x}{\partial x} + \frac{\partial v_z}{\partial z} = 0 \tag{7.33}$$

The first term is zero for a fully developed flow, and the homogeneous boundary conditions on v_z imply that $v_z \equiv 0$.

Next consider the energy balance equation,

$$\rho c_p \left(\frac{\partial T}{\partial t} + v_x \frac{\partial T}{\partial x} + v_y \frac{\partial T}{\partial y} + v_z \frac{\partial T}{\partial z} \right) = k \left(\frac{\partial^2 T}{\partial x^2} + \frac{\partial^2 T}{\partial y^2} + \frac{\partial^2 T}{\partial z^2} \right) + \boldsymbol{\tau} : \mathbf{D} + \rho \dot{R} \tag{7.34}$$

Our assumptions make many terms negligible:

- steady flow: $\partial T/\partial t \to 0$;
- fully developed flow: $\partial/\partial x \to 0$;
- two dimensional: $\partial/\partial y \to 0$;
- $v_z = 0$: $v_z \partial T/\partial z \to 0$; and
- no internal heat generation: $\rho \dot{R} \to 0$.

We also assume that viscous dissipation $\boldsymbol{\tau} : \mathbf{D}$ is negligible. This assumption is made primarily to keep the problem simple and to emphasize the effect of temperature-dependent viscosity. A later section will examine the effects of retaining this term.

This reduces the energy equation to a simple form:

$$\frac{\partial^2 T}{\partial z^2} = 0 \tag{7.35}$$

Integrating twice gives a linear temperature profile,

$$T = c_1 z + c_2 \tag{7.36}$$

and the constants of integration are evaluated from the boundary conditions in the usual way to yield the solution:

$$T - T_0 = (T_1 - T_0)\frac{z}{H} \tag{7.37}$$

This is identical to the solution for steady heat conduction across a solid slab. For this particular problem the velocity is perpendicular to the temperature gradient, so $\mathbf{v} \cdot \nabla T$ equals zero and advection does not influence the temperature solution.

We now use this result to determine the velocity profile. We have already learned that the only nonzero velocity component is v_x, so the flow is one dimensional. The x-direction momentum balance equation is

$$\rho\left(\frac{\partial v_x}{\partial t} + v_x\frac{\partial v_x}{\partial x} + v_y\frac{\partial v_x}{\partial y} + v_z\frac{\partial v_x}{\partial z}\right) = -\frac{\partial p}{\partial x} + \frac{\partial \tau_{xx}}{\partial x} + \frac{\partial \tau_{yx}}{\partial y} + \frac{\partial \tau_{zx}}{\partial z} + \rho b_x \tag{7.38}$$

Note that, just as for non-Newtonian fluids, we must start with the momentum equation in terms of τ from Appendix B.5; the usual Navier–Stokes equations, Appendix B.6, apply for a Newtonian fluid of *constant* viscosity, which is not the case here.

Invoking the following assumptions and results,

- steady flow: $\partial v_x/\partial t \to 0$,
- fully developed flow: $\partial/\partial x \to 0$,
- two-dimensional flow: $\partial/\partial y \to 0$,
- $v_z = 0$: $v_z(\partial v_x/\partial z) \to 0$,
- no axial pressure gradient: $\partial p/\partial x \to 0$, and
- no body force: $b_x \to 0$,

leaves

$$\frac{d\tau_{zx}}{dz} = 0 \tag{7.39}$$

We have switched from a partial to full derivative because we know that τ_{zx} is independent of x and y. Integrating Eq. (7.39) gives the solution for shear stress:

$$\tau_{zx} = c_5 \tag{7.40}$$

In this simple problem the shear stress is constant across the gap.

A constitutive equation is needed to go further. For a Newtonian fluid $\tau = 2\mu\mathbf{D}$, and the stress component we care about is

$$\tau_{zx} = \mu\left(\frac{\partial v_x}{\partial z} + \frac{\partial v_z}{\partial x}\right)$$

$$= \mu_0 e^{-a(T-T_0)}\frac{dv_x}{dz} \tag{7.41}$$

Again the ordinary derivative reflects the one-dimensional nature of the flow. We substitute this expression into Eq. (7.40) along with the temperature solution,

7.2 TEMPERATURE-DEPENDENT VISCOSITY AND VISCOUS DISSIPATION

Eq. (7.37) to obtain

$$\frac{dv_x}{dz} = \frac{c_5}{\mu_0} e^{a(T_1-T_0)z/H} \tag{7.42}$$

The solution is obtained by integrating this equation, and we find

$$v_x = \frac{c_5 H}{\mu_0 a(T_1 - T_0)} e^{a(T_1-T_0)z/H} + c_6 \tag{7.43}$$

The constants c_5 and c_6 are determined by using the boundary conditions $v_x(0) = 0$ and $v_x(H) = V$. This yields the solution for the velocity profile,

$$\frac{v_x}{V} = \frac{e^{a(T_1-T_0)z/H} - 1}{e^{a(T_1-T_0)} - 1} \tag{7.44}$$

In a departure from our usual approach, we scale the equations after the fact. This works for the present case because our assumptions eliminated enough terms in the governing equations to make an analytical solution possible. The choice of dimensionless variables is obvious,

$$v_x^* = \frac{v_x}{V}, \quad z^* = \frac{z}{H} \tag{7.45}$$

and we define a new dimensionless parameter, the *Pearson number*, Pn:

$$\text{Pn} = a(T_1 - T_0) \tag{7.46}$$

Rewriting Eq. (7.44) in terms of these quantities, we get the dimensionless solution:

$$v_x^* = \frac{e^{\text{Pn}\, z^*} - 1}{e^{\text{Pn}} - 1} \tag{7.47}$$

This solution is shown graphically for various values of Pn in Fig. 7.9. When the Pearson number is zero (i.e., the viscosity is constant) then we recover the linear

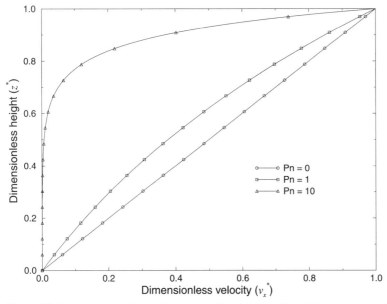

Figure 7.9: Computed velocity profiles for parallel plate drag flow with temperature-dependent viscosity. Pn is the Pearson number, given in Eq. (7.46).

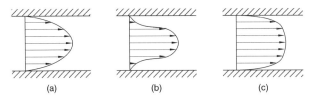

Figure 7.10: Qualitative velocity profiles for nonisothermal pressure-driven flow between parallel plates: (a) isothermal flow; (b) hot fluid, cold walls; (c) cold fluid, hot walls.

velocity profile of the isothermal problem. As the Pearson number increases, the flow is increasingly confined to a layer near the hot plate where the viscosity is low. For very large values of Pn the upper plate appears to slip over the surface, with all of the shearing confined to a very thin layer near the hot plate. Many readers will have seen this type of flow, when a lump of cold butter is placed in a hot frying pan. A thin layer of butter next to the pan melts, and the remaining cold material slips over this low-viscosity layer.

We can regard the Pearson number as the ratio between a temperature difference imposed by operating conditions and a temperature difference (equal to $1/a$) required to change the viscosity:

$$\text{Pn} = \frac{\Delta T_{\text{operating}}}{\Delta T_{\text{viscosity}}} \tag{7.48}$$

When $\text{Pn} \ll 1$ the characteristic operating temperature difference is too small to cause any significant change in viscosity, and the effects of temperature dependent viscosity can be ignored. If $\text{Pn} \geq \mathcal{O}(1)$ then the temperature dependence of viscosity is important.

For pressure-driven flow between parallel plates the shear stress is no longer constant across the gap thickness, but the solution procedure is unchanged. Figure 7.10 shows the types of velocity profiles that occur when fluid at T_0 flows into a channel where both walls are at T_1. This is a thermal entry problem, and we consider the case when $\text{Pe} > 1$ and $\text{Pr} \gg 1$. Then the temperature near the walls is close to T_1 whereas the temperature in the center is close to T_0, like the high Péclet solution in Fig. 7.2. The details of each velocity profile will depend on the rheology of the fluid, as well as on the Pearson and Péclet numbers. When the walls are cold, the viscosity near the walls is high as we see a bell-shaped velocity profile like Fig. 7.10(b). When the walls are hot, the viscosity near the walls is low, the fluid slips on these layers, and the velocity profile is flattened as in Fig. 7.10(c).

7.2.2 VISCOUS DISSIPATION

Consider the drag flow between parallel plates again, but this time assume that the viscosity is unaffected by temperature changes, and that viscous dissipation is significant. This is a somewhat idealized problem, but the solution we will develop is frequently used when modeling melting in single screw extruders.

Figure 7.8 still gives the geometry and nomenclature, and we retain all of the other assumptions of the previous section. Because the viscosity is constant, the temperature distribution does not affect the velocity solution and we can use the

7.2 TEMPERATURE-DEPENDENT VISCOSITY AND VISCOUS DISSIPATION

isothermal velocity profile developed in Section 5.1:

$$v_x = \frac{Vz}{H}, \quad v_y = v_z = 0 \tag{7.49}$$

For future use, the velocity gradient is

$$\frac{\partial v_x}{\partial z} = \frac{V}{H} \tag{7.50}$$

Now consider the energy balance equation. Written out in full, this is

$$\rho c_p \left(\frac{\partial T}{\partial t} + v_x \frac{\partial T}{\partial x} + v_y \frac{\partial T}{\partial y} + v_z \frac{\partial T}{\partial z} \right) = k \left(\frac{\partial^2 T}{\partial x^2} + \frac{\partial^2 T}{\partial y^2} + \frac{\partial^2 T}{\partial z^2} \right) + \boldsymbol{\tau} : \mathbf{D} + \rho \dot{R} \tag{7.51}$$

A number of terms can be eliminated by invoking the assumptions of steady, fully developed, two-dimensional flow without internal generation, and the equation reduces to

$$0 = k \frac{\partial^2 T}{\partial z^2} + \boldsymbol{\tau} : \mathbf{D} \tag{7.52}$$

For the viscous dissipation term, the only nonzero component of the velocity gradient in this flow is given in Eq. (7.50), so that

$$\mathbf{D} = \frac{V}{2H} \begin{bmatrix} 0 & 0 & 1 \\ 0 & 0 & 0 \\ 1 & 0 & 0 \end{bmatrix} \tag{7.53}$$

For a Newtonian fluid $\boldsymbol{\tau} = 2\mu \mathbf{D}$, so it is easy to show that

$$\boldsymbol{\tau} : \mathbf{D} = \frac{\mu V^2}{H^2} \tag{7.54}$$

Note that the rate of viscous dissipation is uniform over the cross section in this case, because the shear rate is uniform across the gap. The energy equation now reduces to

$$\frac{\partial^2 T}{\partial z^2} = -\frac{\mu V^2}{k H^2} \tag{7.55}$$

This may be integrated twice to find

$$T = -\frac{\mu V^2}{2k H^2} z^2 + c_3 z + c_4 \tag{7.56}$$

We find c_3 and c_4 by using the boundary conditions $T = T_0$ at $z = 0$ and $T = T_1$ at $z = H$. The solution for the temperature distribution is then

$$T - T_0 = (T_1 - T_0) \frac{z}{H} + \frac{\mu V^2}{2k} \left(\frac{z}{H} - \frac{z^2}{H^2} \right) \tag{7.57}$$

This is the sum of two terms, a linear profile like Eq. (7.37) caused by the temperature difference between the plates, and a parabolic profile associated with the viscosity μ. This latter term is the effect of viscous dissipation.

As in the previous example, we transform the solution to dimensionless form after the fact. Scaling of Eq. (7.57) is straightforward, and the dimensionless variables are

$$\theta = \frac{T - T_0}{T_1 - T_0}, \quad z^* = \frac{z}{H} \tag{7.58}$$

Substituting these definitions into Eq. (7.57) gives the dimensionless solution as

$$\theta = z^* + \left\{\frac{\mu V^2}{k(T_1 - T_0)}\right\} \left(\frac{z^* - z^{*2}}{2}\right) \tag{7.59}$$

The new dimensionless group is called the *Brinkman number*, Br:

$$\mathrm{Br} = \frac{\mu V^2}{k(T_1 - T_0)} \tag{7.60}$$

The Brinkman number represents the ratio of the rate of heat generation by viscous dissipation to the rate of heat conduction.

Graphs of the dimensionless temperature solution are shown in Fig. 7.11 for several values of Br. When Br = 0 (no viscous dissipation) the temperature profile is linear. As the Brinkman number increases, the parabolic component becomes important. For Br ≫ 1 the maximum temperature occurs close to the midplane between the plates, and it can be much higher than the plate temperatures. Note in Fig. 7.11 that at Br = 10 the sign of the temperature gradient at the upper plate has reversed, and heat actually flows from the fluid into the hot plate.

To get an idea of how important viscous dissipation can be, note that if $T_1 = T_0$ in Eq. (7.57) then the maximum temperature rise $T - T_0$ equals $\mu V^2/8k$. If we choose a reasonable velocity, say $V = 1$ m/s, then the remaining quantities are all

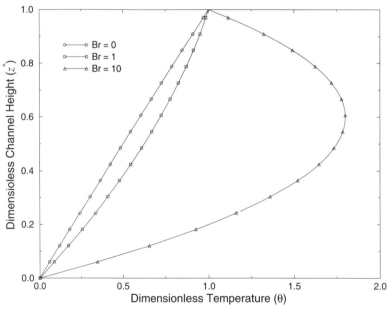

Figure 7.11: Temperature distributions for drag flow between parallel plates. Br is the Brinkman number, given in Eq. (7.60).

material properties. Using typical material properties for molten polycarbonate gives a maximum temperature rise of 100 K. The same calculation for molten aluminum gives a maximum temperature rise less than 10^{-5} K. Clearly viscous dissipation is primarily an issue with high-viscosity, low-conductivity materials such as polymer melts, unless the velocity gradient is very large.

This discussion highlights a convenient physical interpretation of the Brinkman number: it is the ratio of the temperature rise from viscous dissipation to the operating temperature difference.

$$\mathrm{Br} = \frac{\Delta T_{\text{dissipation}}}{\Delta T_{\text{operating}}} \tag{7.61}$$

7.2.3 COMBINED EFFECTS

We have examined temperature dependent viscosity and viscous dissipation alone, but given the large ΔT's that are possible in polymer melt flows, it would not be surprising to encounter these two effects together. To see what how this combination affects the model and its solution, let us continue with the parallel plate drag flow example. We retain the previous geometry and assumptions, except that now we include both temperature-dependent viscosity and viscous dissipation.

Equation (7.39) is still the reduced momentum balance equation, because we did not make any assumptions about the temperature distribution $T(z)$ in deriving it, and Eq. (7.41) is still the appropriate constitutive equation. Combining these two gives an equation for the velocity distribution:

$$\frac{dv_x}{dz} = \frac{c_5}{\mu_0} e^{a(T-T_0)} \tag{7.62}$$

The constant c_5 arises from integrating Eq. (7.39) and is equal to the shear stress within the gap. We cannot integrate Eq. (7.62) yet, because the temperature distribution $T(z)$ is not yet known.

For the energy equation we can again use Eq. (7.52), because we made no assumptions about the velocity profile when deriving it. For the viscous dissipation term we must be a bit more general than the previous section, allowing for the possibility that the shear rate dv_x/dz is not uniform across the gap. The correct expression is found by consulting Appendix B.8:

$$\boldsymbol{\tau} : \mathbf{D} = \tau_{zx} \frac{\partial v_x}{\partial z} \tag{7.63}$$

Substituting the constitutive equation for τ_{zx}, we find that this becomes

$$\boldsymbol{\tau} : \mathbf{D} = \mu_0 e^{-a(T-T_0)} \left(\frac{dv_x}{dz}\right)^2 \tag{7.64}$$

and using this expression in Eq. (7.52) gives the energy equation for this problem:

$$0 = k \frac{d^2 T}{dz^2} + \mu_0 e^{-a(T-T_0)} \left(\frac{dv_x}{dz}\right)^2 \tag{7.65}$$

Looking at Eqs. (7.62) and (7.65), we see that we need the temperature to find the velocity, and the velocity to find the temperature. The two equations are *strongly coupled*. In contrast, the problems in the two previous sections had only weak coupling: we could either find the temperature without knowing the velocity, or find the velocity without knowing the temperature. When the equations are strongly coupled the solution process becomes more difficult, and numerical solutions are often required. Even without viscous dissipation the temperature-dependent viscosity problem will be coupled unless the advection terms are negligible.

The present problem has no easy analytical solution, so we will scale the equations to find the important parameters, and we will look qualitatively at the solutions. Let us consider the case when $T_1 = T_0$, so that the only source of temperature change is viscous dissipation. The scaled variables are

$$v_x^* = \frac{v_x}{V}, \quad z^* = \frac{z}{H}, \quad \theta = \frac{T - T_0}{\Delta T_c} \tag{7.66}$$

Here ΔT_c is a characteristic temperature difference whose value is as yet unknown. Introducing these variables into Eqs. (7.62) and (7.65) gives

$$\left[\frac{V}{H}\right]\frac{dv_x^*}{dz^*} = \left[\frac{c_5}{\mu_0}\right]e^{\{a\Delta T_c\}\theta} \tag{7.67}$$

$$0 = \left[\frac{k\Delta T_c}{H^2}\right]\frac{d^2\theta}{dz^{*2}} + \left[\frac{\mu_0 V^2}{H^2}\right]e^{-\{a\Delta T_c\}\theta}\left(\frac{dv_x^*}{dz^*}\right)^2 \tag{7.68}$$

From the latter equation we equate the two expressions in brackets and find the characteristic temperature difference to be

$$\Delta T_c = \frac{\mu_0 V^2}{k} \tag{7.69}$$

This is the scale for temperature change that is due to viscous dissipation. The equations also contain a new dimensionless group, the *Nahme–Griffith number*, Na:

$$\text{Na} = a\Delta T_c = \frac{a\mu_0 V^2}{k} \tag{7.70}$$

We will see its significance in a moment. The constant c_5 in Eq. (7.67) is still unknown, so to finish scaling the equations we define a scaled constant,

$$c_5^* = \frac{c_5 H}{\mu_0 V} \tag{7.71}$$

The scaled equations are now

$$\frac{dv_x^*}{dz^*} = c_5^* e^{\text{Na}\theta} \tag{7.72}$$

$$0 = \frac{d^2\theta}{dz^{*2}} + e^{-\text{Na}\theta}\left(\frac{dv_x^*}{dz^*}\right)^2 \tag{7.73}$$

The Nahme–Griffith number is the only parameter that appears in these governing equations. (The constant c_5^* is determined by a boundary condition.) The

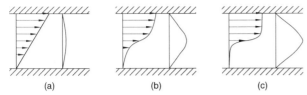

Figure 7.12: Velocity profiles (left) and temperature profiles (right) for drag flow between parallel plates, when both temperature-dependent viscosity and viscous dissipation are important: (a) Na = 0, (b) Na = $\mathcal{O}(1)$, (c) Na $\gg$ 1.

Nahme–Griffith number can be interpreted as the ratio of the temperature rise that is due to viscous dissipation, and the temperature rise needed to change the viscosity:

$$\text{Na} = \frac{\Delta T_{\text{dissipation}}}{\Delta T_{\text{viscosity}}} \tag{7.74}$$

Comparing this to Eqs. (7.48) and (7.61), we see that

$$\text{Na} = \text{Br} \times \text{Pn} \tag{7.75}$$

so among the Nahme–Griffith, Brinkman, and Pearson numbers, only two are independent.

The general character of the solutions is shown in Fig. 7.12. If Na $\ll$ 1 then the viscosity is constant, the velocity profile is linear, and the temperature profile is parabolic. When Na = $\mathcal{O}(1)$ the high temperature in the center reduces the viscosity there, and the shear rate dv_x/dz must be higher in the center, because the shear stress is constant. This in turn increases the rate of dissipation in the center [see Eq. (7.63)], so the temperature profile becomes more sharply peaked. For very large Nahme–Griffith number all the shearing, and thus all the dissipation, is concentrated in a thin layer near the midplane.

For a pressure-driven flow with Na = $\mathcal{O}(1)$, the velocity profile is similar to Fig. 7.10(b) and the temperature profile is similar to Fig. 7.12(b). At large Nahme–Griffith number a pressure-driven flow may have two solutions: a slow, cold flow in which the viscous dissipation rate is small, and a hot, fast flow in which the viscous dissipation rate is large. This phenomenon, which is explored in one of the exercises, can lead to unstable filling of injection molds.

7.3 BUOYANCY-DRIVEN FLOW

We now turn our attention to problems in which the transfer of heat, and the attendant temperature differences, drive the fluid flow. This type of problem is frequently called *natural convection*, or buoyancy-driven flow. We will use the latter term to avoid confusion with the use of the word *convection* in other contexts. Three ingredients must be present for buoyancy-driven flow to occur: a fluid whose density depends on temperature, a temperature difference; and a body force, usually gravity. Local changes in temperature change the local density of the fluid, which alters the body force provided by gravity. These spatial variations in body force drive the fluid motion. Note that the fluid motion itself may alter the temperature distribution through the advection terms, so normally one must solve both the momentum and the energy balance equations simultaneously to deal with a buoyancy-driven flow.

Our goals will be to show how to model buoyancy-driven flow, to identify the important dimensionless groups, to discuss some typical behaviors, and to identify where this phenomenon is important in materials processing.

7.3.1 CONSTITUTIVE EQUATION FOR DENSITY

When a fluid of uniform density is supported in a container and acted upon by gravity, it develops a hydrostatic pressure gradient as discussed in Section 2.3.5, but no velocity. For buoyancy-driven flow to occur we must have a nonuniform density. Thus, an essential part of the formulation is a constitutive equation for density.

Recall from Chapter 2 that for any material a thermodynamic equation of state will give one of the three state variables (density, temperature, and pressure) in terms of the other two. We choose density as the dependent variable, and we expand the equation of state in a Taylor series about some reference pressure p_0 and reference temperature T_0:

$$\rho(p, T) = \rho_0 + \left.\frac{\partial \rho}{\partial T}\right|_{p_0, T_0} (T - T_0) + \left.\frac{\partial \rho}{\partial p}\right|_{p_0, T_0} (p - p_0) + \text{higher-order terms} \tag{7.76}$$

where $\rho_0 = \rho(p_0, T_0)$ is called the reference density. For small changes in temperature and pressure the higher-order terms are negligible. The variation of density with pressure can also be neglected in many cases. This last statement is sometimes made as an assertion about the material, saying that the fluid is "incompressible." If we then use ρ_0 to denote the density at the reference conditions, and define the volumetric thermal expansion coefficient β as

$$\beta \equiv -\frac{1}{\rho_0} \left.\frac{\partial \rho}{\partial T}\right|_{p_0, T_0} \tag{7.77}$$

then Eq. (7.76) simplifies to

$$\rho(T) = \rho_0 - \rho_0 \beta (T - T_0) \tag{7.78}$$

We will use this as our constitutive equation for density in modeling buoyancy-driven flows.

Table 7.3 gives values of β for a few different materials. Note that for all of these materials, a small temperature excursion, say $(T - T_0) < 10°\text{C}$, produces a change in

Table 7.3 Values of the Volumetric Thermal Expansion Coefficient β for Several Fluids

Material	$\beta (\text{K}^{-1})$
Sodium (94°C)	2.7×10^{-4}
Water (20°C)	2.1×10^{-4}
Water (100°C)	7.1×10^{-4}
Air (dry, 100°C)	2.7×10^{-3}
Air (dry, 20°C)	3.4×10^{-3}

7.3 BUOYANCY-DRIVEN FLOW

density that is small compared with the reference density ρ_0. More formally, if ΔT_c is a characteristic temperature difference and $\Delta \rho_c$ is a characteristic change in density, then

$$\frac{\Delta \rho_c}{\rho_0} = \{\beta \Delta T_c\} \tag{7.79}$$

If $\{\beta \Delta T_c\} \ll 1$ then the temperature-induced change in density will be small compared to the average value. We will use this in a moment to simplify the governing equations.

7.3.2 BALANCE EQUATIONS WITH VARIABLE DENSITY

We now proceed to derive the balance equations for buoyancy-driven flows. Our approach will be to scale the equations according to the normal procedure and to simplify them for the case of $\{\beta \Delta T_c\} \ll 1$. This will lead to a classical formulation of the problem called the *Boussinesq approximation*. In what follows we consider only steady flows, but the analysis for transient flows follow exactly the same pattern.

Momentum Balance

Start with the general form of the momentum balance equation, assuming only that the flow is steady. Writing the body force that is due to gravity, Eq. (2.91), and substituting Eq. (7.78) for the density gives

$$[\rho_0 - \rho_0 \beta (T - T_0)] (\mathbf{v} \cdot \nabla \mathbf{v}) = -\nabla p + \nabla \cdot \boldsymbol{\tau} - \rho_0 g \mathbf{e}_z + \rho_0 \beta (T - T_0) g \mathbf{e}_z \tag{7.80}$$

The next-to-last term on the right-hand side represents the effect of gravity on the reference density, whereas the last term represents gravity acting on the density variations. We treat the reference density part exactly as we treated the constant-density case, using Eq. (2.92) to replace $\mathbf{e}_z$ by ∇h and then combining this term with the pressure gradient. This requires that we define the modified pressure to be

$$\hat{p} \equiv p + \rho_0 g h \tag{7.81}$$

With this substitution the momentum balance becomes

$$[\rho_0 - \rho_0 \beta (T - T_0)] (\mathbf{v} \cdot \nabla \mathbf{v}) = -\nabla \hat{p} + \nabla \cdot \boldsymbol{\tau} + \rho_0 \beta (T - T_0) g \mathbf{e}_z \tag{7.82}$$

We now proceed with the scaling. In a typical case the length scale L is known, and we will use the same length scale for all three axes. Also, one typically knows a reference temperature T_0, a characteristic temperature difference ΔT_c, and a reference pressure p_0. However, the characteristic velocity V is unknown. We also introduce an unknown pressure difference Δp_c. We then define the usual dimensionless variables,

$$x^* = \frac{x}{L}, \quad y^* = \frac{y}{L}, \quad z^* = \frac{z}{L}, \quad \mathbf{v}^* = \frac{\mathbf{v}}{V}$$

$$\hat{p}^* = \frac{\hat{p}}{\Delta p_c}, \quad \theta = \frac{T - T_0}{\Delta T_c}, \quad \boldsymbol{\tau}^* = \frac{L}{\mu V} \boldsymbol{\tau} \tag{7.83}$$

where the scaling for $\boldsymbol{\tau}$ is consistent with a Newtonian fluid of viscosity μ. Also, as a notational convenience, we define a dimensionless gradient operator ∇^* as the gradient in the dimensionless coordinates,

$$\nabla^* = \frac{\partial}{\partial x^*}\mathbf{e}_x + \frac{\partial}{\partial y^*}\mathbf{e}_y + \frac{\partial}{\partial z^*}\mathbf{e}_z \qquad (7.84)$$

Substituting these variables into Eq. (7.82), we find

$$\left[\frac{\rho_0 V^2}{L}\right](1 - \{\beta\Delta T_c\}\theta)(\mathbf{v}^* \cdot \nabla^*\mathbf{v}^*) = -\left[\frac{\Delta p_c}{L}\right]\nabla^*\hat{p}^* + \left[\frac{\mu V}{L^2}\right]\nabla^* \cdot \tau^*$$
$$+ [\rho_0 g\beta\Delta T_c]\theta\mathbf{e}_z \qquad (7.85)$$

Reading from left to right, the terms in this equation represent inertia, pressure gradient, viscous forces, and buoyant forces, respectively. In most buoyancy-driven flows the pressure gradient is not a dominant effect. Two special cases then arise: one in which the buoyant forces are balanced by inertial forces, and another in which the buoyant forces are balanced by viscous forces.

Consider the case in which the velocity is determined by a balance between body forces from density variations and inertia. Because we expect inertial forces to be $\mathcal{O}(1)$, we divide the equation by $[\rho_0 V^2/L]$ to obtain

$$(1 - \{\beta\Delta T_c\}\theta)(\mathbf{v}^* \cdot \nabla^*\mathbf{v}^*) = -\left\{\frac{\Delta p_c}{\rho_0 V^2}\right\}\nabla^*\hat{p}^* + \left\{\frac{\mu}{\rho_0 V L}\right\}\nabla^* \cdot \tau^*$$
$$+ \left\{\frac{g\beta\Delta T_c L}{V^2}\right\}\theta\mathbf{e}_z \qquad (7.86)$$

Looking first at the left-hand side, we now use our assumption that $\{\beta\Delta T_c\} \ll 1$, so the effect of density variations can be omitted from the inertia terms. The unknown pressure scale Δp_c is found by setting the term multiplying $\nabla^*\hat{p}^*$ to one, yielding $\Delta p_c = \rho_0 V^2$, which is a typical result for inertia-dominated flows. We identify the term in braces, multiplying $\nabla^* \cdot \tau^*$, as the inverse of the Reynolds number:

$$\text{Re} = \frac{\rho_0 V L}{\mu} \qquad (7.87)$$

The last term in Eq. (7.86) is used to determine the unknown velocity scale, because we have assumed that this is the main driving force for the buoyancy-driven flow. Setting that term's dimensionless group equal to one, we find the characteristic velocity scale to be

$$V = (g\beta\Delta T_c L)^{1/2} \qquad (7.88)$$

This is an important result of the scaling analysis. Note that this scaling for velocity is valid only for $\text{Re} \geq 1$; for small Reynolds number one must rescale, balancing the viscous terms against the body force. This is left as an exercise.

Using this velocity scale, we can write the Reynolds number in terms of the known parameters of the problem:

$$\text{Re} = \frac{(g\beta\Delta T_c)^{1/2} L^{3/2}}{\nu_0} \qquad (7.89)$$

Here $\nu_0 = \mu/\rho_0$ is a kinematic viscosity based on the average density. The square of this Reynolds number is widely used in characterizing natural convection problems;

it is called the *Grashof number*, Gr:

$$\text{Gr} \equiv \frac{g\beta \Delta T_c L^3}{\nu_0^2} = \text{Re}^2 \tag{7.90}$$

There is no reason to prefer the Grashof number over the Reynolds number, except possibly to avoid taking square roots, and the Grashof number has the property of "appearing" to be larger than it is. For example, a Grashof number of 10^4 seems huge, but this corresponds to a Reynolds number of 10^2, which is large but still in the laminar flow regime. Still, the Grashof number is the standard dimensionless group found in the literature on this subject.

With the simplifications we have made, the scaled momentum balance equation becomes

$$\mathbf{v}^* \cdot \nabla^* \mathbf{v}^* = -\nabla^* \hat{p}^* + \left\{\frac{1}{\text{Re}}\right\} \nabla^* \cdot \boldsymbol{\tau}^* + \theta \mathbf{e}_z \tag{7.91}$$

or, in dimensional form,

$$\rho_0 \mathbf{v} \cdot \nabla \mathbf{v} = -\nabla \hat{p} + \nabla \cdot \boldsymbol{\tau} + \rho_0 g \beta (T - T_0) \mathbf{e}_z \tag{7.92}$$

In this dimensional equation, the main simplification over Eq. (7.82) is that the average density ρ_0 is used in the inertia term.

Mass Balance

We next derive the appropriate form for the continuity equation. Because density varies with temperature, we must start with the general form of the continuity equation. This is conveniently written as

$$\frac{D\rho}{Dt} + \rho (\nabla \cdot \mathbf{v}) = 0 \tag{7.93}$$

or, in a steady flow,

$$\mathbf{v} \cdot \nabla \rho + \rho (\nabla \cdot \mathbf{v}) = 0 \tag{7.94}$$

Take the material derivative of the equation of state, Eq. (7.78), to find that

$$\frac{D\rho}{Dt} = -\rho_0 \beta \frac{DT}{Dt} \tag{7.95}$$

Using this result, the continuity equation, Eq. (7.94), for steady flow becomes

$$-\rho_0 \beta (\mathbf{v} \cdot \nabla T) + \rho (\nabla \cdot \mathbf{v}) = 0 \tag{7.96}$$

Now use Eq. (7.78) for the density, and scale the equation by substituting in the dimensionless variables from Eqs. (7.83), to obtain

$$-\left[\frac{\rho_0 \beta V \Delta T_c}{L}\right] \mathbf{v}^* \cdot \nabla^* \theta + \left[\frac{\rho_0 V}{L}\right] (1 - \{\beta \Delta T_c\} \theta) \nabla^* \cdot \mathbf{v}^* = 0 \tag{7.97}$$

Then divide this equation by $[\rho_0 V/L]$ and find

$$-\{\beta \Delta T_c\} \mathbf{v}^* \cdot \nabla^* \theta + (1 - \{\beta \Delta T_c\} \theta) \nabla^* \cdot \mathbf{v}^* = 0 \tag{7.98}$$

As before we assume that $\{\beta\Delta T_c\} \ll 1$, so the continuity equation simplifies to

$$\nabla^* \cdot \mathbf{v}^* = 0 \tag{7.99}$$

In dimensional terms this is

$$\nabla \cdot \mathbf{v} = 0 \tag{7.100}$$

which we recognize as the continuity equation for a fluid with constant density. As long as $\{\beta\Delta T_c\} \ll 1$, the changes in density introduced by temperature changes are small enough compared with the average density that we can neglect them in the continuity equation.

Energy Balance

To complete the formulation we must analyze the energy balance equation. As before, we consider steady-state problems, and we assume constant conductivity and specific heat. Note also that the constitutive equation was made density independent of pressure, so the work associated with density changes is identically zero. We also assume that there are no internal heat sources and that viscous dissipation is negligible. This latter point could be justified by a scaling argument, which we omit here in the interest of brevity. The energy balance equation with these assumptions retains only the advection and diffusion terms:

$$\rho c_p \mathbf{v} \cdot \nabla T = k \nabla^2 T \tag{7.101}$$

We scale this equation and find, as in the other balance equations, that for $\{\beta\Delta T_c\} \ll 1$ we can use the reference density ρ_0 on the left-hand side. Skipping some steps, we find that the scaled energy equation is

$$\mathbf{v}^* \cdot \nabla^* \theta = \left\{\frac{1}{\mathrm{Pe}}\right\} \nabla^{*2} \theta \tag{7.102}$$

where the Péclet number has the usual definition,

$$\mathrm{Pe} = \frac{VL}{\alpha_0} \tag{7.103}$$

and α_0 is the thermal diffusivity based on the reference density ρ_0. Using the derived scale for velocity, Eq. (7.88), we see that the Péclet number becomes

$$\mathrm{Pe} = \frac{(g\beta\Delta T_c)^{1/2} L^{3/2}}{\alpha_0} \tag{7.104}$$

The square of this quantity is sometimes used in natural convection problems and is called the *Boussinesq number*, Bo.

Summary: The Boussinesq Approximation

This analysis has shown that we can model buoyancy-driven flows by treating the density as constant in almost all parts of the equations – using the reference density ρ_0 in the inertial and advection terms and adopting the continuity equation for constant density – but allowing density to depend on temperature in the body force term in the z-direction momentum balance. This simplification is valid as long as $\{\beta\Delta T_c\} \ll 1$.

The resulting set of balance equations represents a classical and widely used approach for modeling buoyancy-driven flows, called the *Boussinesq approximation*. One can conveniently obtain the Boussinesq approximation balance equations from the usual balance equations by taking these steps.

1. Use ρ_0 for the density wherever it appears.
2. Replace p by $\hat{p}$, Eq. (7.81), to account for the effect of gravity on average density.
3. Use a body force of $\rho_0 g \beta (T - T_0)$ in the z-direction momentum balance.

7.3.3 MORE ON DIMENSIONLESS GROUPS

Our scaling analysis yielded two familiar dimensionless groups, the Reynolds and Péclet numbers, and a characteristic velocity scale, Eq. (7.88). These are the meaningful quantities for this problem. However, the literature on buoyancy-driven flow uses two other dimensionless groups, the Prandtl number, Pr,

$$\Pr \equiv \frac{\nu_0}{\alpha_0} \tag{7.105}$$

which was introduced in Section 7.1.3, and the *Rayleigh number*, Ra,

$$\mathrm{Ra} \equiv \frac{g \beta \Delta T_c L^3}{\nu_0 \alpha_0} \tag{7.106}$$

Comparing these groups to Eqs. (7.89) and (7.104), we see that the Prandtl and Rayleigh numbers are simply different ways to express the information contained in the Reynolds and Péclet numbers:

$$\Pr = \frac{\mathrm{Pe}}{\mathrm{Re}} \tag{7.107}$$

$$\mathrm{Ra} = \mathrm{Re}\,\mathrm{Pe} \tag{7.108}$$

That is, specifying the Rayleigh and Prandtl numbers for a problem is equivalent to specifying the Reynolds and Péclet numbers.

One advantage of these traditional dimensionless groups is that the Prandtl number is a material property. Thus, once the fluid has been chosen, all of the effects of problem size and operating conditions are contained in a single parameter, the Rayleigh number.

Table 7.2 (p. 246) gave the Prandtl numbers for several fluids. Note that the Prandtl number is the ratio of the diffusivity of inertia to the diffusivity of heat. We can also think of the Prandtl number as the ratio of the advection terms in the energy balance to the inertial terms in the momentum balance. Liquid metals have low viscosity but high density and thermal conductivity, so their Prandtl numbers are small. In a liquid metal it is possible for the Reynolds number to be large and the Péclet number to be small. Air and water have Prandtl numbers of the order of one, so in these fluids the Reynolds and Péclet numbers will always be comparable. Polymer melts, with their very large viscosities and low conductivities, have extremely large Prandtl numbers. In these materials the Péclet number can be very large and the Reynolds number can be tiny. (Remember that for Re < 1 we need a different scaling for the buoyancy-driven velocity than the one developed here.)

7.3.4 BUOYANT CONVECTION IN A THIN LAYER

The formulation outlined above is best understood by studying an example. Although there are only a few problems of this type that can be solved analytically, some of them are quite relevant to materials processing. Consider the idealized system below, which shows a layer of fluid held between two parallel, insulated plates. The right wall is maintained at high temperature T_h, while the left wall is maintained at lower temperature T_c. Gravity acts vertically downward. This problem is a prototype for crystal growth in a horizontal ampoule, so we will refer to the domain as an ampoule.

We anticipate a buoyancy-driven flow that rises along the hot wall, traverses the ampoule, falls along the cold wall, and returns along the bottom. This flow is illustrated schematically in Fig. 7.13. Our objective is to analyze the flow to determine the velocity and temperature profile within the fluid layer.

We make a number of assumptions:

1. The velocity and temperature distributions are steady and two dimensional.
2. The flow is laminar, and away from the ends the velocity field is fully developed.
3. The upper and lower walls are insulated, and the fluid does not slip on these boundaries.
4. The Boussinesq approximation is valid, and $\beta \Delta T_c \ll 1$.
5. The fluid is Newtonian and incompressible.
6. Viscous dissipation is negligible.

We begin the analysis with the continuity equation:

$$\frac{\partial v_x}{\partial x} + \frac{\partial v_y}{\partial y} + \frac{\partial v_z}{\partial z} = 0 \qquad (7.109)$$

The assumptions that the flow is two dimensional and fully developed imply that the first two terms on the left-hand side are zero, leaving

$$\frac{\partial v_z}{\partial z} = 0 \qquad (7.110)$$

Because v_z is also independent of y and z by the same assumptions, and v_z vanishes on the upper and lower boundaries, we must have $v_z \equiv 0$. We could have reached the same conclusion for a slender ampoule, $H/L \ll 1$, without assuming a fully

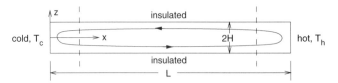

Figure 7.13: A horizontal ampoule with differentially heated ends and insulated top and bottom. Between the dashed lines, the flow should be fully developed.

7.3 BUOYANCY-DRIVEN FLOW

developed flow, by following the line of reasoning that we used for the lubrication approximation. Remember that this solution applies away from the ends of the ampoule, because v_z must be significant at the ends as the flow turns around.

Next, we consider the momentum balance equations. The z-momentum balance equation, from Eq. (7.92), is

$$\rho_0 \left(v_x \frac{\partial v_z}{\partial x} + v_y \frac{\partial v_z}{\partial y} + v_z \frac{\partial v_z}{\partial z} \right) = -\frac{\partial \hat{p}}{\partial z} + \mu \left(\frac{\partial^2 v_z}{\partial x^2} + \frac{\partial^2 v_z}{\partial y^2} + \frac{\partial^2 v_z}{\partial z^2} \right)$$
$$+ \rho_0 g \beta (T - T_c) \quad (7.111)$$

where we have taken the reference temperature to be T_c. Because $v_z \equiv 0$, this equation reduces to

$$\frac{\partial \hat{p}}{\partial z} = \rho_0 g \beta (T - T_c) \quad (7.112)$$

Unlike the narrow-gap flows we have examined before, this problem has a significant pressure gradient in the z direction, which is associated with temperature differences. However, there is still no velocity in the z direction.

Consider next the x-momentum balance equation,

$$\rho_0 \left(v_x \frac{\partial v_x}{\partial x} + v_y \frac{\partial v_x}{\partial y} + v_z \frac{\partial v_x}{\partial z} \right) = -\frac{\partial \hat{p}}{\partial x} + \mu \left(\frac{\partial^2 v_x}{\partial x^2} + \frac{\partial^2 v_x}{\partial y^2} + \frac{\partial^2 v_x}{\partial z^2} \right) \quad (7.113)$$

All of the terms on the left-hand side are zero, from our assumptions and previous analysis. The right-hand side can also be reduced, because two-dimensional flow implies that $\partial^2 v_x / \partial y^2 = 0$, and fully developed flow implies that $\partial^2 v_x / \partial x^2 = 0$. This leaves

$$\frac{\partial \hat{p}}{\partial x} = \mu \frac{\partial^2 v_x}{\partial z^2} \quad (7.114)$$

Before going further, we find it convenient to scale the remaining equations. We define the usual dimensionless variables:

$$x^* = \frac{x}{L}, \quad z^* = \frac{z}{H}, \quad \theta = \frac{T - T_c}{T_h - T_c} = \frac{T - T_c}{\Delta T}, \quad v_x^* = \frac{v_x}{V}, \quad p^* = \frac{\hat{p}}{\Delta p_c}$$
$$(7.115)$$

The characteristic scales V and Δp_c are as yet unknown. Inserting the scaled variables into Eqs. (7.112) and (7.114) gives

$$\frac{\partial p^*}{\partial x^*} = \left[\frac{\mu V L}{H^2 \Delta p_c} \right] \frac{\partial^2 v_x^*}{\partial z^{*2}} \quad (7.116)$$

$$\frac{\partial p^*}{\partial z^*} = \left[\frac{\rho_0 g \beta \Delta T H}{\Delta p_c} \right] \theta \quad (7.117)$$

The scales for V and Δp_c are now clear,

$$\Delta p_c = \rho_0 g \beta \Delta T H, \quad V = \frac{\rho_0 g \beta \Delta T H^3}{\mu L} \quad (7.118)$$

and the final scaled momentum balance equations are

$$\frac{\partial p^*}{\partial x^*} = \frac{\partial^2 v_x^*}{\partial z^{*2}} \qquad (7.119)$$

$$\frac{\partial p^*}{\partial z^*} = \theta \qquad (7.120)$$

We can make some progress in solving for the velocity field by using these equations. We eliminate p^* between the two equations by differentiating the first with respect to z^* and the second with respect to x^*, and by equating the right-hand sides. This gives

$$\frac{\partial^3 v_x^*}{\partial z^{*3}} = \frac{\partial \theta}{\partial x^*} \qquad (7.121)$$

At this point we need to think about the nature of the temperature distribution. We have assumed that the velocity is fully developed, so that v_x^* depends only on z^*. We might expect a similar sort of result for the temperature distribution, except that temperature must certainly change along the x^* axis, to connect the cold end on the left to the hot end on the right. A reasonable guess is that the temperature solution is the sum of two functions, one depending on x^* and another depending on z^*:

$$\theta(x^*, z^*) = f_1(x^*) + f_2(z^*) \qquad (7.122)$$

Taking $\partial/\partial x^*$ of this equation eliminates the z^* function, so the right-hand side of Eq. (7.121) must depend only on x^*. The left-hand side of that equation depends only on z^*, so both terms must equal a constant, which we will call c_1. We now have two useful results:

$$\frac{\partial^3 v_x^*}{\partial z^{*3}} = c_1 \qquad (7.123)$$

$$\frac{\partial \theta}{\partial x^*} = c_1 \qquad (7.124)$$

Because v_x^* is independent of x^* and y^*, the partial derivative in the first equation can be replaced by an ordinary derivative, and we integrate directly to find v_x^*:

$$v_x^* = \frac{c_1}{6} z^{*3} + \frac{c_2}{2} z^{*2} + c_3 z^* + c_4 \qquad (7.125)$$

Here c_2–c_4 are constants of integration. The velocity must vanish on the upper and lower boundaries ($z^* = \pm 1$), which requires that c_2 and c_4 be zero. There must also be zero net flow rate across any cross section, so

$$\int_{-1}^{1} v_x^* dz^* = 0 \qquad (7.126)$$

This boundary condition requires that $c_3 = -c_1/6$, so we have

$$v_x^* = \frac{c_1}{6}(z^{*3} - z^*) \qquad (7.127)$$

Except for the constant c_1, we now know the velocity profile.

7.3 BUOYANCY-DRIVEN FLOW

We now turn to the energy balance equation. In dimensional form this is

$$\frac{\partial T}{\partial t} + v_x \frac{\partial T}{\partial x} + v_y \frac{\partial T}{\partial y} + v_z \frac{\partial T}{\partial z} = \alpha_0 \left(\frac{\partial^2 T}{\partial x^2} + \frac{\partial^2 T}{\partial y^2} + \frac{\partial^2 T}{\partial z^2} \right) \qquad (7.128)$$

The first, third, and fourth terms on the left-hand side are zero based on our assumption of steady, two-dimensional flow and the fact that $v_z = 0$. On the right-hand side the first term is zero because a constant $\partial \theta / \partial x^*$ means that $\partial^2 T / \partial x^2$ must equal zero, and the second term is zero because the flow is two dimensional. This leaves

$$v_x \frac{\partial T}{\partial x} = \alpha_0 \frac{\partial^2 T}{\partial z^2} \qquad (7.129)$$

Substituting the scaled variables and rearranging terms gives

$$\left\{ \frac{\rho_0 g \beta \Delta T H^5}{\mu \alpha_0 L^3} \right\} v_x^* \frac{\partial \theta}{\partial x^*} = \frac{\partial^2 \theta}{\partial z^{*2}} \qquad (7.130)$$

Now we introduce the Rayleigh number,

$$\text{Ra} = \frac{\rho_0 g \beta \Delta T H^3}{\mu \alpha_0} \qquad (7.131)$$

and recall that $\partial \theta / \partial x^* = c_1$ to rewrite Eq. (7.130) as

$$c_1 \text{Ra} \frac{H^2}{L^2} v_x^* = \frac{\partial^2 \theta}{\partial z^{*2}} \qquad (7.132)$$

Finally, we substitute the result for v_x^* from Eq. (7.127) to get

$$\frac{c_1^2 \text{Ra}}{6} \frac{H^2}{L^2} (z^{*3} - z^*) = \frac{\partial^2 \theta}{\partial z^{*2}} \qquad (7.133)$$

For the moment, it is convenient to lump all of constants on the left-hand side together into a single constant, Ω:

$$\Omega \equiv \frac{c_1^2 \text{Ra}}{6} \frac{H^2}{L^2} \qquad (7.134)$$

Integrating Eq. (7.133) twice with respect to z^* gives the temperature distribution:

$$\theta = \Omega \left(\frac{z^{*5}}{20} - \frac{z^{*3}}{6} \right) + c_5(x^*) z^* + c_6(x^*) \qquad (7.135)$$

Because we have integrated a partial derivative, $c_5(x^*)$ and $c_6(x^*)$ may be functions x^*. However, we know that c_5 must be constant; otherwise, $\partial \theta / \partial x^*$ would not be constant. By the same reasoning, we conclude that $c_6(x^*) = c_1 x^* + c_7$, where c_7 is constant. Next we apply the boundary condition of no flux at the top and bottom surfaces,

$$\frac{\partial \theta}{\partial z^*}(y = \pm 1) = 0 \qquad (7.136)$$

which yields $c_5 = \Omega/4$. The temperature solution now has the form

$$\theta = \Omega \left(\frac{z^{*5}}{20} - \frac{z^{*3}}{6} + \frac{z^*}{4} \right) + c_1 x^* + c_7 \qquad (7.137)$$

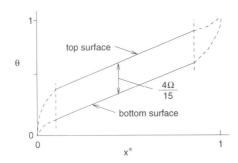

Figure 7.14: Schematic temperature distribution in the horizontal ampoule, showing a core where the temperature difference between the top and bottom plates is constant, and boundary layers on the ends. The analytical solution does not apply outside the vertical dashed lines.

This solution has two remaining constants, c_1 and c_7, so we need two more boundary conditions. Ordinarily, we would use boundary conditions at $x^* = 0$ and $x^* = 1$ to evaluate these last two constants. However, our solution assumed a fully developed velocity and temperature distribution and is not valid near the ends of the domain. However, we can still use these end conditions if we think about what the temperature solution looks like. Equation (7.137) tells us that the difference in temperature between the top and bottom surfaces is the same for any value of x^*:

$$\theta|_{z^*=1} - \theta|_{z^*=-1} = \frac{4\Omega}{15} \tag{7.138}$$

Figure 7.14 shows what the temperature solution along the top and bottom surfaces should look like. Away from the ends the top and bottom temperature curves are straight and parallel, and they are separated by $4\Omega/15$. At the ends the boundary conditions must be satisfied, and the dashed lines indicate the general form of the solution near the ends. At the hot end the temperature must rise steeply at the bottom to reach $\theta = 1$, while the temperature of the top surface reaches the same temperature smoothly. Similarly, at the cold end the bottom surface reaches $\theta = 0$ smoothly, while the top temperature drops rapidly to that value. A reasonable approximation is that the top curve extrapolates linearly to $\theta = 1$ at the hot end, and the bottom curve extrapolates linearly to $\theta = 0$ on the cold end. This gives us the boundary conditions we need:

$$\theta(x^* = 0, z^* = -1) = 0, \quad \theta(x^* = 1, z^* = 1) = 1 \tag{7.139}$$

The boundary condition on the cold end allows us to solve for c_7:

$$\theta(x^* = 0, z^* = -1) = 0 = -\frac{2\Omega}{15} + c_7 \tag{7.140}$$

$$c_7 = \frac{2\Omega}{15} \tag{7.141}$$

The boundary condition on the hot end is related to c_1:

$$\theta(x^* = 1, z^* = 1) = 1 = \frac{4\Omega}{15} + c_1 \tag{7.142}$$

However, this is not a complete answer yet, because Ω contains c_1. Substituting the definition of Ω, Eq. (7.134), into the boundary condition results in a quadratic

7.3 BUOYANCY-DRIVEN FLOW

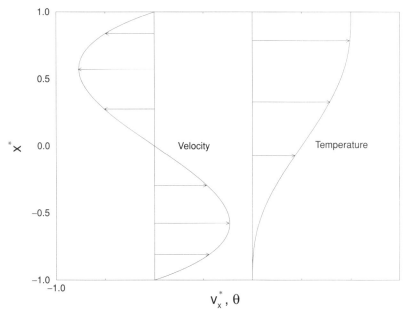

Figure 7.15: Temperature and velocity at an arbitrary cross section in the horizontal ampoule, in the fully developed region.

equation for c_1, with the solution

$$c_1 = \frac{-1 + \sqrt{1 + \frac{8\mathrm{Ra}}{45}\frac{H^2}{L^2}}}{\frac{4\mathrm{Ra}}{45}\frac{H^2}{L^2}} \tag{7.143}$$

The positive root is physically meaningful because c_1 equals the temperature gradient in the x^* direction, and this must be positive. The final form for the temperature distribution is then

$$\theta = c_1 x^* + \frac{c_1^2 \mathrm{Ra}}{360}\frac{H^2}{L^2}(3z^{*5} - 10z^{*3} + 15z^* + 8) \tag{7.144}$$

with c_1 given by Eq. (7.143).

Figure 7.15 shows the shapes of the velocity and temperature solutions at any cross section in the fully developed region. The velocity profile has this shape everywhere and is reminiscent of a pressure-driven flow to the left on the upper half and a pressure-driven flow to the right on the lower half. The shape of the temperature variation with z is given, but the values will be shifted by a constant that depends linearly on x^*. Heat conduction in the z direction keeps the temperature distribution smooth, but it is temperature differences that drive the flow.

7.3.5 BUOYANT CONVECTION IN A SQUARE CAVITY

There are many situations in materials processing in which buoyant convection is important but the geometry is quite different from the slender ampoule analyzed in the preceding section. These include several crystal growth processes, ingot solidification, and melting furnaces. As a representative geometry, we analyze

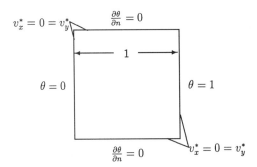

Figure 7.16: A square cavity, insulated on the top and bottom, with the two sides maintained at different temperatures.

buoyancy-driven flow in a square cavity. The results from this idealized problem can be used to help understand and analyze real buoyant convection problems.

The problem domain and boundary conditions are sketched in Fig. 7.16. The fluid is contained in a square cavity with an insulated floor and ceiling, and the two side walls are maintained at different temperatures.

The Boussinesq approximation is assumed to apply, and we have already chosen a scaled geometry. We will consider parameters values where buoyant forces are balanced by inertia, so our previous scaling analysis is still valid. The scaled governing equations are

$$\nabla^* \cdot \mathbf{v}^* = 0$$

$$\mathbf{v}^* \cdot \nabla^* \mathbf{v}^* = -\nabla^* \hat{p}^* + \left\{\frac{1}{\text{Re}}\right\} \nabla^* \cdot \boldsymbol{\tau}^* + \theta \mathbf{e}_z$$

$$\mathbf{v}^* \cdot \nabla^* \theta = \left\{\frac{1}{\text{Pe}}\right\} \nabla^{*2} \theta$$

There are no further simplifications available, and no analytical solution to this problem, so we will examine numerical solutions for several different cases. This problem is taken from the examples manual for the commercial program FIDAP (Engelman, 1993), and the details of the numerical solution method are given in that reference.

Like most commercial programs, FIDAP is written to solve the governing equations in dimensional form. We leave it as an exercise to show that the code will solve the scaled equations if we provide the following input: a density equal to Re, a specific heat equal to Pe/Re (= Pr), and all other material properties equal to one. In the results that follow the Prandtl number is fixed at 0.71, which is appropriate for air. The solutions are shown for various values of the Reynolds number, which we can regard as a measure of the strength of the convective effect. Equation (7.89) shows that one may increase the Reynolds number by increasing the temperature difference between the hot and cold walls.

For a small Reynolds number, Re = 1, the buoyant forces are weak. Figure 7.17 shows that this produces a slow circulation of fluid, which rises near the hot wall and falls near the cold wall. The temperature solution is a slightly disturbed version of the pure conduction solution, which would have parallel vertical isotherms. As the Reynolds number increases to 10 and then to 100, in Figs. 7.18 and 7.19 respectively, the velocity increases, and the flow eventually becomes confined to a boundary layer

7.3 BUOYANCY-DRIVEN FLOW

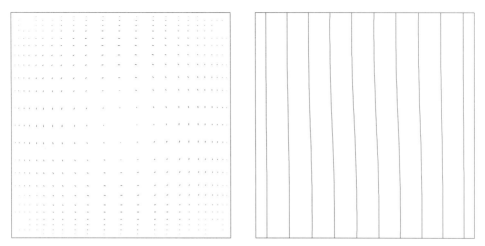

Figure 7.17: Velocity vectors and temperature contours for buoyant convection in a square cavity at Re = 1. The maximum velocity vector plotted has a magnitude of 0.394. The isotherms are evenly spaced from zero to one.

near the walls. The isotherms also become increasingly distorted until, at Re = 100, they are nearly perpendicular to the side walls in the central region of the cavity. Note the clustering of isotherms (large temperature gradient) near the walls at Re = 100. The Pèclet numbers for these three cases are 0.71, 7.1, and 71, and the distortion of the isotherms at high Pe is consistent with the advection problems considered earlier in this chapter.

The temperatures on the upper and lower boundaries are plotted in Fig. 7.20 for Re = 100. Notice the boundary layers near the ends, and the central core where the temperature difference between the top and bottom surfaces is essentially constant. Compare this to Fig. 7.14, where we guessed that the temperature solution in a slender cavity would look very much like this.

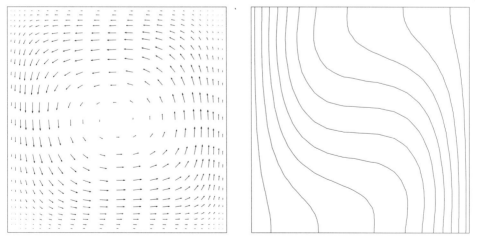

Figure 7.18: Velocity vectors and temperature contours for buoyant convection in a square cavity at Re = 10. The maximum velocity vector plotted has a magnitude of 0.223. The isotherms are evenly spaced from zero to one.

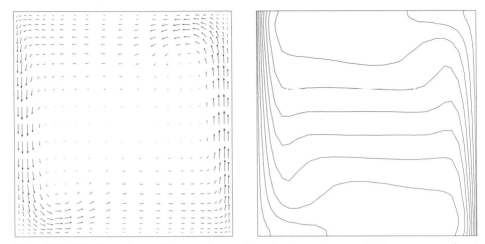

Figure 7.19: Velocity vectors and temperature contours for buoyant convection in a square cavity at Re = 100. The maximum velocity vector plotted has a magnitude of 0.251. The isotherms are evenly spaced from zero to one.

If the Reynolds number is increased still more, the buoyancy-driven flow eventually becomes turbulent. This typically occurs in the vicinity of Ra = 10^8–10^{10}. Detailed modeling of turbulent natural convection is subject to the same difficulties as other turbulent flow models, as discussed in Section 5.4.3. In addition, the velocity perturbations that transport momentum, and are responsible for the Reynolds stress, also transport heat. The usual modeling approach time averages the energy equation, in which case a turbulent heat flux term appears, in analogy to the Reynolds stresses.

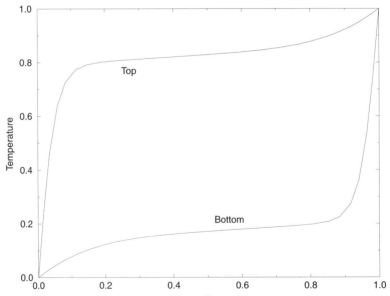

Figure 7.20: Computed temperature distributions on the upper and lower surfaces of the square cavity for Re = 100 (cf. Fig. 7.14).

7.4 SUMMARY

This chapter considered combined fluid flow and heat transfer. In all such problems, advection provides an important mechanism for transporting energy. The importance of advection relative to conduction is described by the Péclet number. If Pe $\ll 1$ then advection is unimportant, and the temperature solution will be dominated by heat conduction. If Pe $\gg 1$ then advection dominates, and each fluid particle tends to carry its temperature with it. However, even at very high Péclet numbers, conduction will always be important in thin boundary layers, wherever the boundary conditions impose a temperature different from the upstream fluid. Scaling can estimate the thickness of these boundary layers.

Problems in which a heat source moves over a solid body are frequently encountered in welding, heat sealing, and similar processes. These problems are conveniently analyzed in a coordinate system in which the heat source is stationary and the solid body moves. Some classical solutions to this problem were presented, and these can be used to predict the width and depth of welds and heat-affected zones, and their dependence on processing parameters.

When the fluid viscosity changes significantly with temperature, the flow and heat transfer problems become strongly coupled and must be solved simultaneously. The velocity profiles differ from the isothermal profiles in Chapter 5, sometimes dramatically. When the temperature difference is imposed by boundary conditions, the importance of temperature-dependent viscosity is described by the Pearson number. Another source of temperature rise in high-viscosity fluids is viscous dissipation, and the Brinkman number gives the magnitude of dissipation relative to heat conduction. When both effects occur simultaneously, the Nahme–Griffith number describes the influence of viscous heating on the viscosity.

A second type of strongly coupled problem is buoyancy-driven flow (natural convection). This occurs when the fluid density depends on temperature, a temperature difference is imposed, and gravity is present. A standard formulation, the Boussinesq approximation, is the proper limit of the full governing equations when the change in density is small compared with the average density. Scaling analysis provides an estimate for the characteristic velocity in terms of the imposed ΔT, fluid properties, and length scales. This estimate leads to particular forms of the Reynolds and Péclet numbers for buoyant convection problems. The Prandtl and Rayleigh numbers are an alternate, but equivalent, set of parameters. Few buoyant convection problems have analytical solutions, but some example solutions provided insights into the phenomena.

BIBLIOGRAPHY

M. S. Engelman. *FIDAP Examples Manual*. Fluent, Inc., Evanston, IL, 1993.
D. Rosenthal. *Welding J.* 20:220, 1941.

EXERCISES

1. **Melting on a heated grid.** A semicrystalline polymer is melted by pushing a bar of solid material against a wire screen that is heated by electrical resistance heating. The polymer melts some distance ahead of the screen, so that by the time it reaches the screen it is a liquid that can pass through the screen. We wish to know the temperature distribution in the polymer at steady state.

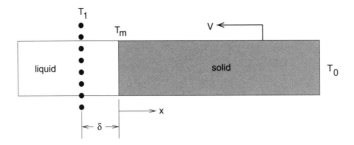

   We will choose a coordinate system with $x = 0$ at the solid–liquid interface. The grid maintains the polymer at T_1 at the location $x = -\delta$, but the value of δ is initially unknown. Far upstream of the grid the temperature is T_0. The polymer moves into the grid at a steady speed V in the negative x direction. The polymer has a heat of fusion L_f.

   (a) Write the energy equation and boundary conditions that govern this process. Assume that the process is steady, the properties are constant, k and c_p are the same in the liquid and the solid, and temperature is a function of x only. Use scaling to find the characteristic length scale (characteristic value of x) in this problem.

   (b) Write the boundary conditions for this problem. The solid–liquid interface is a free boundary (because its position is initially unknown), so you should have two boundary conditions there.

   (c) Solve for the temperature distributions $T_s(x)$ in the solid and $T_\ell(x)$ in the liquid. Use the boundary conditions at $x = 0$ to match the two solutions, and determine an explicit expression for δ.

2. **Moving heat source.** A laser beam is sometimes used to surface harden steel parts for abrasion resistance, for example in turbine blades and gear teeth. The hardening depth is usually small compared to the size of the part, so we can model this as a semi-infinite moving point source problem. Here we wish to find how deep the hardened layer is for a given set of operating conditions.

   The temperature distribution in the frame moving with the laser beam is given by the Rosenthal solution, Eq. (7.30). The beam is traveling with velocity V in the z direction, and ζ is a coordinate that points in the z direction but is fixed on the beam. Consider the case with data given in the table below.

EXERCISES

Quantity	Symbol	Value	Units
Initial temperature	T_0	25	°C
Thermal conductivity	k	30	W/mK
Thermal diffusivity	α	1×10^{-5}	m²/s
Beam velocity	V	1×10^{-3}	m/s
Beam power	Q	10,000	W

Graph the temperature distribution in the part on a vertical plane directly under the laser beam, that is, on the y–z or y–ζ plane (see Fig. 7.5). Note that r has a simpler form on this plane. In particular, plot the isotherms at 1500°C (melting) and at 750°C and 650°C (spanning the eutectoid temperature).

3 *Hardness after welding.* Shown below are isotherms on the top surface of a very thick steel plate welded by a 10-kW laser. The weld beam speed was 0.002 m/s and 0.04 m/s for the two cases shown. Also shown below is a Jominy curve for the steel being welded. The hardness of a steel that has been cooled from above the eutectoid temperature correlates quite well with its cooling rate at 700°C. The Jominy curve can be measured experimentally for any steel alloy, and the data can be interpreted as final hardness versus cooling rate at 700°C.

Make plots for the two cases, showing the hardness distribution transverse to the welding direction after the process is complete. For each distance from the weld, estimate the cooling rate at 700°C by using the time to cool from 750°C to 650°C from the graphs below.

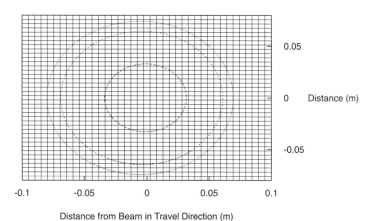

Isotherms at 650°C, 750°C, and 1500°C for a welding speed of 0.002 m/s.

4 *Velocity distribution caused by temperature differences.* A Newtonian fluid with temperature-dependent viscosity flows between parallel plates whose total separation is $2h$, driven by a known pressure gradient $-\partial p/\partial x$. The lower plate is maintained at T_0 and the upper plate is held at a higher temperature T_1. The plates are very long in the flow direction, so the temperature and velocity attain fully developed profiles (i.e., Gz $\ll$ 1).

(a) Find the temperature and velocity distributions $T(z)$ and $v_x(z)$. Assume that viscous dissipation has a negligible effect on the temperature distribution (Br $\ll$ 1). Use an exponential temperature dependence for the viscosity:

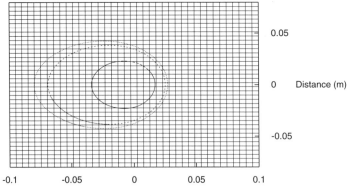

Isotherms at 650°C, 750°C, and 1500°C for a welding speed of 0.002 m/s.

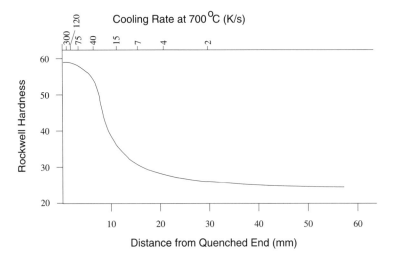

$$\mu = \mu_0 e^{-a(T-T_0)}$$

(b) Find a characteristic value for the velocity in terms of the problem parameters. Express your answer for the velocity profile in terms of the dimensionless Pearson number, $\text{Pn} = a(T_1 - T_0)$.

(c) Plot velocity profiles for $\text{Pn} = 1$ and 5.

5   *Velocity profiles in nonisothermal Hele–Shaw flow.* The Generalized Hele–Shaw model provides a convenient equation for the velocity profile $v_x(z)$ in flow where the viscosity varies with z, as a result of temperature variations. Consider the case in which an initially cold polymer is flowing between two hot walls. As a rough approximation, suppose that the polymer is Newtonian, and that heat transfer from the walls has warmed a layer near the walls to viscosity μ_1, whereas in the cold core the viscosity is μ_0. The core extends from $z = -h^*$ to $z = h^*$, as shown in the sketch.

EXERCISES

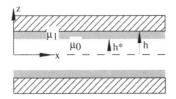

(a) Solve for the velocity distribution $v_x(z)$, assuming that the pressure gradient $-\partial p/\partial x$ is known. Plot the velocity distribution for $h^* = 0.9h$ and $\mu_1 = 0.1\mu_0$.

(b) The situation of a cold, high-viscosity core and hot, low-viscosity material near the walls occurs in compression molding of thermoset polymers. Some experiments with that process show "preferential flow," in which the hot material near the walls flows faster than the core (see the sketch). Can a velocity profile with preferential flow ever be predicted by using a Generalized Hele–Shaw model? Explain your answer.

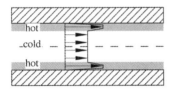

6 *Temperature distribution caused by viscous dissipation.* A Newtonian fluid is pumped at a constant flow rate Q between parallel plates with total separation $2h$ and width W. The width is large compared to the gap height, $W \gg h$, so the flow is two dimensional. Both plates are maintained at a constant temperature T_0. The plates are very long in the flow direction, so the temperature attains a fully developed profile (Gz $\ll$ 1). Find the temperature distribution across the gap $T(z)$ caused by viscous dissipation. Assume that the fluid viscosity is constant (Na $\ll$ 1).

7 *Viscous dissipation for a power law fluid.* For a power law fluid in a pressure-driven flow in a narrow gap, the velocity distribution is given by Eq. (6.41).

(a) Retrace the derivation done in Section 7.2.2 to find the temperature distribution in the power law fluid when the advection terms are negligible (Pe $\ll$ 1). Also assume that viscosity is independent of temperature, Pn $\ll$ 1.

(b) Identify the Brinkman number for this problem. (This will look a little different than it did for the Newtonian case.) Compute the value of the Brinkman number for these data: $m = 8 \times 10^4$ N s^n/m, $n = 0.2$, $\Delta T = 30$ K, $\rho = 10^3$ kg/m^3, $k = 0.25$ W/m K, $H = 3 \times 10^{-3}$ m, $L = 0.1$ m, and $V = 0.075$ m/s.

(c) Plot the temperature distribution across the gap.

(d) Often, the viscosity of a polymer varies strongly with temperature. How might this affect the temperature field, and how might one solve for the temperature in this case?

8 *Injection molding flow at a large Nahme–Griffith number.* Consider a polymer injection-molding process. In such operations, the cavity typically has a high aspect ratio, the melt viscosity is high, and the thermal conductivity is low. It is therefore reasonable to model such a process assuming that Gz ≪ 1, Na ≫ 1, and Br ≫ 1. In physical terms, the flow is slow enough to be thermally fully developed, there is significant viscous heating, and the polymer viscosity is strongly affected by the temperature rise that is due to viscous dissipation. In this case, viscous dissipation keeps the polymer fluid near the center of the cavity, and a frozen layer at the wall provides insulation.

Assume that the only flow is in the x direction and solve for the local temperature and velocity distributions, $T(z)$ and $v_x(z)$. Assume that the thermal properties are constant and are identical in the liquid and solid. Model the polymer as a Newtonian fluid, with a constant value μ above some melting temperature T_m and infinite below that temperature. (This is too sharp a change to represent real polymer behavior, but it makes the problem tractable.) Assume thermal symmetry about $z = 0$ in a cavity of total thickness $2h$. The wall temperature is fixed at T_w, which is lower than T_m. Let h_m be the value of z at the liquid–solid interface. That is, the situation looks something like the sketch below.

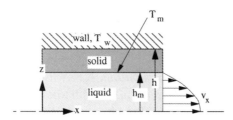

Solve the problem by using the following steps.

(a) Solve for the velocity profile $v_x(z)$ for $0 \leq z \leq h_m$. Note that the velocity is zero for $z > h_m$, and that the value of h_m is not yet known.

(b) Simplify the energy equation, and then solve for the temperature distribution in the liquid. At this point you should not have any unknown constants other than h_m.

(c) Solve for the temperature distribution in the solid layer, $h_m \leq z \leq h$.

(d) Find one more condition to match the temperature solutions for parts (b) and (c). Use this to evaluate h_m.

(e) Rephrase your answer for h_m in dimensionless form (i.e., h_m/h). What dimensionless parameter(s) of the problem control this answer? How does the solution behave for different values of the control parameter(s)?

9 *Natural convection in a vertical narrow enclosure.* In a long, slender cavity such as one finds in thermopane windows, natural convection occurs and can reach a steady state, fully developed flow. This phenomenon may also occur in crystal growth and other melt-processing problems. The figure below shows the geometry. (The figure has been turned sideways to fit the page; note that z points up). The flow is two dimensional and steady and both the temperature and the

velocity are fully developed. $\beta \Delta T \ll 1$, so the Boussinesq approximation is valid.

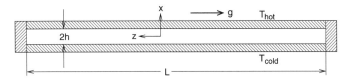

(a) Use the continuity equation to show that $v_x = 0$.

(b) Write out the the x and z components of the momentum balance equation, assuming a Newtonian fluid with constant viscosity, and using the form that contains $\hat{p}$ (as in the Boussinesq approximation). Show that, given the assumptions made, the inertial terms in both equations vanish, and that $\hat{p}$ depends only on z.

(c) Scale the z-direction equation and find a characteristic velocity V. To do this, assume that the pressure gradient $\partial p/\partial z$ is small compared to other terms.

(d) Write the energy balance equation, simplify it, and determine the temperature distribution, assuming that the material properties are constant and that viscous dissipation is negligible. (The convective terms should vanish.)

(e) Substitute the computed temperature distribution into your simplified z-direction momentum equation and compute the velocity distribution. Note that $d\hat{p}/dz$ is a constant (because the flow is fully developed); the value of this constant can be determined by requiring that the flow rate over any cross-section be zero. Plot your final velocity profile.

CHAPTER EIGHT

Mass Transfer and Solidification Microstructures

In previous chapters we considered homogeneous materials, but there are many significant problems in materials processing in which more than one chemical species is involved. One technologically important case is the solidification of metal alloys. When more than one species is present, we must address two additional considerations: chemical equilibrium between the various phases, and transport of the species by diffusion. This chapter begins by setting up the governing equations for diffusion and shows that species transport is mathematically similar to energy transport. The use of the diffusion equations is demonstrated in several example applications. We then look at solute partitioning in alloy solidification, consider the stability of the solid–liquid interface in solidification, and show how these phenomena control microstructure development.

8.1 GOVERNING EQUATIONS FOR DIFFUSION

When two soluble materials are put into contact, they mix through the action of molecular diffusion. If one waits long enough, a final state is achieved in which the local concentration of each species is uniform. As an example, think about pouring milk into hot coffee. Eventually, the coffee reaches a uniform light color, which we associate with complete mixing. We can make "eventually" arrive sooner by stirring the coffee, which transports material by advection, but diffusion operates whether we stir or not.

Our discussion of diffusion will be confined to systems in which the species are completely miscible for any composition that is present in the problem. Some important metallurgical systems have limited solubility, and we discuss these later in the context of microstructure development. However, we leave the discussion of diffusion in such systems to advanced texts.

Suppose that we have two species present, and call them A and B. If A is the minor component and B is the major component, then we call A the *solute* and B the *solvent*. We wish to keep track of the relative amounts of the species that are present at any point. We will use the mass concentration $\hat{C}_A$ to denote the mass of species A per unit volume of the mixture (e.g., grams per cubic centimeter). There are other possible measures for concentration, including the mass fraction C_A (grams A per gram of mixture) and the atomic fraction (moles of A per mole of mixture). Interconversion between these measures can be readily performed by using the molecular weight and density of each species. For example, the mass fraction C_A is related to the concentration $\hat{C}_A$ through

$$C_A = \frac{\hat{C}_A}{\rho} \tag{8.1}$$

where ρ is the mixture density. C_A is sometimes expressed as a percentage instead of a fraction.

In this development we will treat mixture density as a constant, which simplifies the derivations considerably. This treatment is appropriate for dilute systems. A more complete treatment would require allowing ρ to be a function of the mixture concentration.

Balance Equations

A balance equation for species transport can be constructed in the same manner as the balance equations for mass, momentum, and energy. We will start by doing this in terms of $\hat{C}_A$, and then we convert the equation to use C_A. Consider a material volume V with a surface S and a unit normal vector $\mathbf{n}$ at the surface (Fig. 2.1). Species A may cross the surface S by diffusing relative to the surrounding material. We denote the diffusive flux by $\hat{\mathbf{J}}_A$, which has units of mass per unit area per unit time. Species A may also be created or destroyed in the volume by chemical reactions, and we denote the production rate by $\hat{R}_A$, with units of mass of A per unit volume and time. The balance statement of species for A is then

$$\frac{d}{dt} \int_V \hat{C}_A \, dt = -\int_S \hat{\mathbf{J}}_A \cdot \mathbf{n} \, dS + \int_V \hat{R}_A \, dV \tag{8.2}$$

We use the arguments surrounding the Reynolds transport theorem (Section 2.2.5) to introduce the material derivative on the left-hand side, and we use the divergence theorem to transform the surface integral into a volume integral:

$$\int_V \frac{D\hat{C}_A}{Dt} \, dt = -\int_V \nabla \cdot \hat{\mathbf{J}}_A \, dV + \int_V \hat{R}_A \, dV \tag{8.3}$$

We then invoke the argument that the control volume is arbitrary to write the balance equation in differential form:

$$\frac{D\hat{C}_A}{Dt} = -\nabla \cdot \hat{\mathbf{J}}_A + \hat{R}_A \tag{8.4}$$

This is a basic balance equation for species A. To complete the formulation we need a constitutive relation that relates the flux $\hat{\mathbf{J}}_A$ to the concentration. Fick proposed a linear relation between the rate of species transport and the local concentration gradient,

$$\hat{\mathbf{J}}_A = -D_{AB} \nabla \hat{C}_A \tag{8.5}$$

where D_{AB} is the diffusion coefficient of solute A in solvent B, also known as the intrinsic chemical diffusivity. This equation is known as *Fick's first law*. Note the similarity to Fourier's law for heat conduction, Eq. (2.137). Substitution of Eq. (8.8) into Eq. (8.4) yields the *diffusion equation* for species A,

$$\frac{D\hat{C}_A}{Dt} = \nabla \cdot (D_{AB}\nabla \hat{C}_A) + \hat{R}_A \tag{8.6}$$

This form can be used to solve diffusion problems. Analogous equations can be written for each species present. In order to conserve mass, the sum of the production terms for all species (weighted by their molecular weight) must equal zero, which represents a constraint on the system of equations.

If the mixture density ρ is constant then we can conveniently transform Eq. (8.6) into an equation for C_A, the mass fraction, by dividing by ρ:

$$\frac{DC_A}{Dt} = \nabla \cdot (D_{AB}\nabla C_A) + R_A \tag{8.7}$$

Here $R_A = \hat{R}_A/\rho$ is the generation rate per unit mixture mass. We can also define a mass fraction flux $\mathbf{J} = \hat{\mathbf{J}}/\rho$, which has units of velocity, and write Fick's law as

$$\mathbf{J}_A = -D_{AB}\nabla C_A \tag{8.8}$$

For many materials processing problems it is more convenient to work in terms of mass fractions, so Eqs. (8.7) and (8.8) will be our primary tools for diffusion problems.

Diffusivity and Temperature

In condensed phases, diffusion occurs by mechanisms that involve jumping and replacement of atoms, so the diffusion rate is highly temperature dependent. Diffusion coefficients often follow an Arrhenius relation like the one followed by viscosity, Eq. (6.96), and in fact the activation energies for the two processes are often nearly the same. Thus we may write

$$D_{AB} = D_0 \exp\left(-\frac{E_D}{RT}\right) \tag{8.9}$$

where E_D is the activation energy for diffusion, R is the gas constant (8.31 J/mol K), and T is the absolute temperature.

Analogy with Heat Conduction

The diffusion equation, Eq. (8.7) is identical in form to the energy balance equation, Eq. (2.149), when viscous dissipation is negligible. Therefore, all of the solutions we developed for the energy equation are also solutions to the diffusion equation. The diffusion solutions can be generated by simply changing symbols as follows:

$$T \to C_A$$
$$k \to D_{AB}$$
$$\rho c_p \to 1 \tag{8.10}$$
$$\alpha \to D_{AB} \tag{8.11}$$
$$\rho \dot{R} \to R_A$$
$$\mathbf{q} \to \mathbf{J}_A$$

8.2 SOLID-STATE DIFFUSION

Most importantly, the concentration takes the place of the temperature. Also, as ρc_p is replaced by one, the chemical diffusivity D_{AB} also replaces the thermal diffusivity α.

The insights that scaling gave us about the interrelation of length and time scales in energy transport apply equally to diffusion. Thus, the characteristic time for diffusion t_D is related to the characteristic length scale L by

$$t_D = \frac{L^2}{D_{AB}} \qquad (8.12)$$

Important differences between energy and species transport arise, however, because of differences in the magnitudes of the thermal and chemical diffusivities. For metal solidification, where both the solid and liquid are present, we have

$$\begin{aligned} \alpha_s &\approx \alpha_\ell \\ D_\ell &\sim 10^{-4} \alpha_\ell \\ D_s &\sim 10^{-4} D_\ell \end{aligned} \qquad (8.13)$$

In view of Eq. (8.12), this difference in properties produces a hierarchy of time scales where, over a given length scale, heat is transported 10,000 times faster than species in the liquid, and species is transported 10,000 times faster in the liquid than in the solid. An alternate way to view this same information is to consider a fixed characteristic time t_c determined by some external event, and to examine the length scales associated with each process. Then we have

$$L_T = \sqrt{\alpha_\ell t_c} \sim 10^2 L_{D_\ell} \sim 10^4 L_{D_s} \qquad (8.14)$$

Thus, the characteristic length scale for heat conduction is 100 times greater than the characteristic length scale for diffusion in the liquid, which is 100 times greater than the characteristic length scale for diffusion in the solid. We will see that this hierarchy of length and time scales has a profound effect on solidification microstructures.

8.2 SOLID-STATE DIFFUSION

We now consider some basic problems of diffusion in materials processing. These include the diffusion of a surface layer into the bulk (a prototype problem for carburization), diffusion of a thin interior layer into the bulk (a prototype problem for welding), and surface oxidation. The analogy between heat conduction and diffusion shortens the derivations by allowing us to reuse solutions from earlier chapters. Most of the problems are for dilute systems, for which it is sufficient to track just one species, the solute. For these problems we will drop the subscripts on solute concentration C and diffusivity D. We also consider a nondilute system, in which we must account for the possibility that $D_{AB} \neq D_{BA}$.

8.2.1 DIFFUSION OF A SURFACE LAYER

It is often advantageous to put a hard, carbon-rich surface coating on a steel of relatively low carbon content, using a process called *carburization* or *case hardening*. This produces good surface wear characteristics, while at the same time maintaining the toughness of the underlying material. The process is performed by placing the

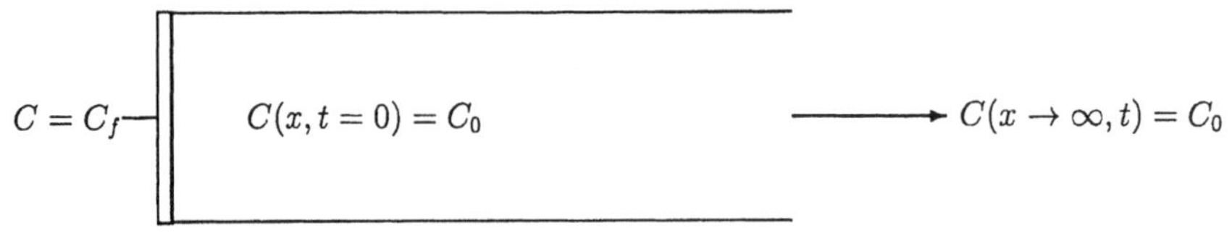

$$C = C_f \quad\quad C(x, t = 0) = C_0 \quad\quad\quad\quad\quad\quad \longrightarrow C(x \to \infty, t) = C_0$$

Figure 8.1: Model of carburization as a one dimensional, semi-infinite diffusion problem with a fixed surface concentration.

steel part in a furnace with a controlled atmosphere, such that a high-carbon surface layer forms on the part. The furnace elevates the temperature so that diffusion of carbon into the steel can take place in a reasonable time.

Consider such a process, idealized as shown in Fig. 8.1. For simplicity of notation, the carbon concentration is denoted by C. We treat this as a one-dimensional diffusion problem into a semi-infinite medium, with a fixed concentration C_f at the surface. The steel is initially at uniform composition C_0, and we also fix the concentration at C_0 as $x \to \infty$. We will see shortly that the diffusion process is so slow that carburization is limited to a very thin layer near the surface, and the assumption of a semi-infinite medium is therefore appropriate.

In addition to the assumption of one-dimensionality, we also assume that the steel is solid (i.e., $\mathbf{v} = 0$), that there is no internal production of solute ($R_C = 0$), and that the diffusion coefficient is constant. Equation (8.7) then reduces to

$$\frac{\partial C}{\partial t} = D \frac{\partial^2 C}{\partial x^2} \tag{8.15}$$

where D is the diffusion coefficient of carbon in iron. The boundary and initial conditions are

$$C(x = 0, t) = C_f \tag{8.16}$$
$$C(x \to \infty, t) = C_0 \tag{8.17}$$
$$C(x, t = 0) = C_0 \tag{8.18}$$

We encountered the same mathematical problem in Section 4.2.1, with the conduction of heat in a semi-infinite body. Because the boundary condition at $x \to \infty$ and the initial condition had the same value, it was possible to use the composite variable $x/2\sqrt{Dt}$ to find a similarity solution, which is

$$C(x, t) = c_1 + c_2 \, \mathrm{erf}\left(\frac{x}{2\sqrt{Dt}}\right) \tag{8.19}$$

The same solution applies here, and we simply use the boundary and initial conditions to evaluate the constants c_1 and c_2, yielding the solution

$$C - C_f = (C_0 - C_f) \, \mathrm{erf}\left(\frac{x}{2\sqrt{Dt}}\right) \tag{8.20}$$

The application of this solution is illustrated in the following example.

EXAMPLE 8.2.1 Surface Carburization

A plate of 1020 steel (0.2 wt.%C) is placed in a carburizing furnace at 700°C. The atmosphere forms a surface layer that is approximately 1.2 wt. %C. How long does it take for the carbon content to reach 0.8 wt. %C at a depth of 1 mm? The diffusion coefficient for C in 1020 steel follows the Arrhenius relation of Eq. (8.9) with $D_0 = 6.2 \times 10^{-7}$ m²/s and $E_D = 80$ kJ/mol K.

Solution:

First compute the diffusion coefficient at the operating temperature:

$$D = (6.2 \times 10^{-7} \text{ m}^2/\text{s}) \exp\left(-\frac{80,000 \text{ J/mol}}{(8.31 \text{ J/mol K})(973 \text{ K})}\right) = 3.13 \times 10^{-11} \text{ m}^2/\text{s}$$

Then substitute numerical values into Eq. (8.20):

$$\frac{0.8 - 1.2}{0.2 - 1.2} = \text{erf}\left(\frac{1 \times 10^{-3} \text{ m}}{2\sqrt{(3.13 \times 10^{-11} \text{ m}^2/\text{s})t}}\right)$$

$$0.40 = \text{erf}\left(\frac{89.4 \text{ s}^{1/2}}{\sqrt{t}}\right)$$

By consulting Table A.1, we find the argument of the error function,

$$\frac{89.4 \text{ s}^{1/2}}{\sqrt{t}} = 0.38$$

and solve for t to find $t = 5.5 \times 10^4$ s, or ~15 h. This may be an impractically long time. Diffusion in solids is a slow phenomenon, and for practical times even relatively small bodies are effectively semi-infinite. This particular process will be more practical at a higher temperature, where the diffusion coefficient is larger.

8.2.2 DIFFUSION OF AN ENCAPSULATED LAYER

We now examine diffusion when a layer of high initial concentration of solute is sandwiched between two much thicker regions with a lower concentration. The thicker regions are large enough that we will consider them to be semi-infinite. This problem represents the concentration profiles one might see after welding a seam in a plate, where the weld material has a different composition than the base material.

Figure 8.2 illustrates the geometry and nomenclature. The governing equation is the one-dimensional transient diffusion equation,

$$\frac{\partial C}{\partial t} = D\frac{\partial^2 C}{\partial x^2}, \quad -\infty < x < \infty \qquad (8.21)$$

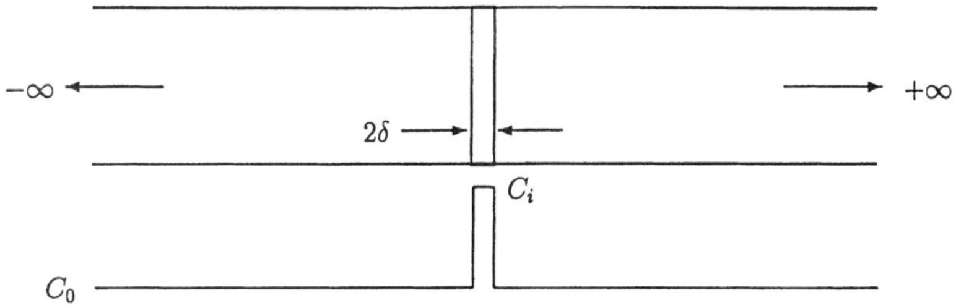

Figure 8.2: A layer of thickness 2δ and initial concentration C_i, sandwiched between two semi-infinite layers with initial concentration C_0.

whereas the boundary and initial conditions are

$$C(x \to \pm\infty, t) = C_0 \tag{8.22}$$

$$C(x, t = 0) = \begin{cases} C_0, & x < -\delta \\ C_i, & -\delta \leq x \leq \delta \\ C_0, & \delta < x \end{cases} \tag{8.23}$$

Concentration is scaled just as we scale temperature, using C_i and C_0, whereas time is scaled by using the diffusivity D and the length scale δ:

$$C^* = \frac{C - C_0}{C_i - C_0}, \quad x^* = \frac{x}{\delta}, \quad t^* = \frac{Dt}{\delta^2} \tag{8.24}$$

The scaled governing equation and boundary conditions are then

$$\frac{\partial C^*}{\partial t^*} = \frac{\partial^2 C^*}{\partial x^{*2}}, \quad -\infty < x^* < \infty \tag{8.25}$$

$$C^*(x^* \to \pm\infty, t^*) = 0 \tag{8.26}$$

$$C^*(x^*, t^* = 0) = \begin{cases} 0, & x^* < -1 \\ 1, & -1 \leq x^* \leq 1 \\ 0, & 1 < x^* \end{cases} \tag{8.27}$$

The solution to this problem is worked out by Weinberger (1965) by using Fourier transforms, for a slightly more general problem in which the initial concentration distribution is given by $C^*(x^*, t^* = 0) = f(x^*)$. The solution is

$$C^*(x^*, t^*) = \frac{1}{2\sqrt{\pi t^*}} \int_{-\infty}^{+\infty} \exp\left(-\frac{(x^* - s)^2}{4t^*}\right) f(s) \, ds \tag{8.28}$$

For our case $f(x^*) = 0$, except in the interval $-1 \leq x^* \leq 1$, where $f(x^*) = 1$, so the limits on the integral in Eq. (8.28) are bounded, and we have

$$C^*(x^*, t^*) = \frac{1}{2\sqrt{\pi t^*}} \int_{-1}^{+1} \exp\left(-\frac{(x^* - s)^2}{4t^*}\right) ds \tag{8.29}$$

To simplify this integral we make the substitution

$$u = \frac{s - x^*}{2\sqrt{t^*}}, \quad du = \frac{ds}{2\sqrt{t^*}} \tag{8.30}$$

and use the fact that $\int_a^b = \int_0^b - \int_0^a$. This gives

$$C^*(x^*, t^*) = \frac{1}{2}\left(\frac{2}{\sqrt{\pi}} \int_0^{(1-x^*)/2\sqrt{t^*}} \exp(-u^2) \, du \right.$$
$$\left. - \frac{2}{\sqrt{\pi}} \int_0^{(-1-x^*)/2\sqrt{t^*}} \exp(-u^2) \, du\right) \tag{8.31}$$

The two terms inside the parentheses are error functions, so we finally obtain

$$C^*(x^*, t^*) = \frac{1}{2}\left(\text{erf}\left(\frac{1 - x^*}{2\sqrt{t^*}}\right) - \text{erf}\left(\frac{-1 - x^*}{2\sqrt{t^*}}\right)\right) \tag{8.32}$$

8.2 SOLID-STATE DIFFUSION

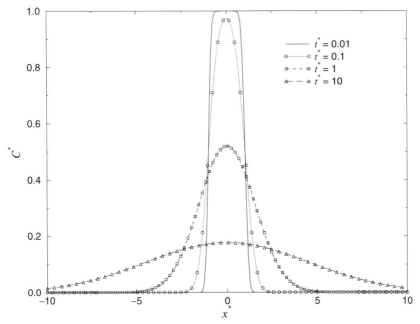

Figure 8.3: Spreading of concentration from an initial layer of material of different composition in an infinite medium.

This solution is shown in Fig. 8.3 for several values of the dimensionless time. At short times, $t^* \leq 0.1$, diffusion smoothes the concentration profile near the boundaries of the inner layer, but the center of the layer is still near its initial concentration. By $t^* = 1$ there has been a substantial drop in the peak concentration and a spread of solution away from the center. This spreading continues over time. The profiles at $t^* = 1$ and 10 look like Gaussian distributions, and indeed the solution approaches a Gaussian shape in the limit of long times.

8.2.3 DIFFUSION WITH A SURFACE REACTION

Some chemical reactions, such as surface oxidation, are controlled by diffusion. We consider this problem in the context of a specific system, the oxidation of nickel. Oxidation occurs by direct reaction between oxygen and nickel ions. The reaction product, NiO, forms a solid surface layer that separates the oxygen and the unreacted nickel, thus inhibiting further reaction. The oxidation rate is thus controlled by diffusion of reactants through the surface layer. We will examine this process with the goal of determining the oxidation rate.

In principle there are two diffusion problems going on here: O^{2-} ions diffuse through the oxide to react with the Ni^{2+} ions, and Ni^{2+} ions diffuse through the oxide to the surface and react with the O^{2-} ions. In practice, however, only the latter reaction proceeds at a significant rate. Table 8.1 explains why this is so. The ionic radius of Ni^{2+} is less than half of that for O^{2-}, so the diffusivity of Ni^{2+} is many orders of magnitude larger than the diffusivity of O^{2-}. We can therefore neglect the diffusion of oxygen into the nickel and assume that all the reaction takes place at the surface.

This is now a one-dimensional diffusion problem for Ni^{2+} ions in NiO, as illustrated in Fig. 8.4. Nickel is the solute, and the concentration of Ni^{2+} is denoted by C.

Table 8.1 Selected Properties of Ni^{2+} and O^{2-} Ions

Property	Units	Ni^{2+}	O^{2-}
Ionic radius	nm	0.069	0.14
D_0 (in NiO)	m²/s	1.83×10^{-7}	1.0×10^{-9}
E_D	kJ/mol	192	226
$D(700°C)$	m²/s	8.9×10^{-18}	7.3×10^{-22}

Just on the NiO side of the Ni–NiO interface, C is assumed to have a constant value C_N; C_N is related to the equilibrium solubility of Ni^{2+} in NiO, but it is difficult to measure. We will return to its determination later. The oxidation reaction between Ni^{2+} and O^{2-} is assumed to be very rapid, so that any Ni^{2+} ions reaching the external surface react immediately to form NiO; the concentration of Ni^{2+} is therefore zero at the surface.

The external surface moves relative to the Ni–NiO interface, because the oxidized layer grows with time as Ni^{2+} ions react at the surface. Let $\delta(t)$ denote the thickness of the oxide layer, and let ρ be its density (assumed to be constant); $d\delta/dt$ equals the volume of oxide being added to the surface per unit area and unit time. If we multiply this by ρ we get the mass of oxide being added. Further multiplying by f_N, the mass fraction of Ni in NiO, gives the mass of Ni begin added to the surface per unit area and time. This must equal the diffusive mass flux $\hat{J}_x$ at $x = \delta$. Recalling that $\hat{J}_x = \rho J_x$, we have

$$f_N \rho \frac{d\delta}{dt} = \rho J_x|_{x=\delta} \tag{8.33}$$

or, writing the flux in terms of the concentration gradient,

$$f_N \frac{d\delta}{dt} = -D \frac{\partial C}{\partial x}\bigg|_{x=\delta} \tag{8.34}$$

This gives an equation to find $\delta(t)$. Grouping the governing equations together, we have

$$\frac{\partial C}{\partial t} = D \frac{\partial^2 C}{\partial x^2}, \quad 0 \leq x \leq \delta \tag{8.35}$$

$$C = C_N, \quad x = 0 \tag{8.36}$$

$$C = 0, \quad x = \delta \tag{8.37}$$

$$\frac{d\delta}{dt} = -D \frac{\partial C}{\partial x}, \quad x = \delta \tag{8.38}$$

$$\delta = 0, \quad t = 0 \tag{8.39}$$

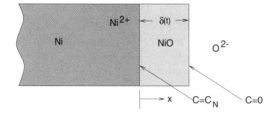

Figure 8.4: Nomenclature for the growth of a nickel oxide layer on nickel.

8.2 SOLID-STATE DIFFUSION

This oxidation problem is mathematically the same as the Stefan problem for solidification that we studied in Section 4.3. Because this problem was discussed at length in Chapter 4, we can work through the solution here in an abbreviated fashion. First, we scale the variables by using

$$C^* = C/C_N, \quad x^* = x/L, \quad t^* = (D/L^2 t), \quad \delta^* = \delta/L \tag{8.40}$$

where L is an arbitrary length, as there is no natural length scale in this problem. The scaled governing equations and boundary conditions then become

$$\frac{\partial C^*}{\partial t^*} = \frac{\partial^2 C^*}{\partial x^{*2}}, \quad 0 \leq x^* \leq \delta^* \tag{8.41}$$

$$C^* = 1, \quad x^* = 0 \tag{8.42}$$

$$C^* = 0, \quad x^* = \delta^* \tag{8.43}$$

$$\frac{d\delta^*}{dt^*} = -g_N \frac{\partial C^*}{\partial x^*}, \quad x^* = \delta^* \tag{8.44}$$

$$\delta^* = 0, \quad t^* = 0 \tag{8.45}$$

Here $g_N \equiv C_N/f_N$ is a dimensionless group that determines the rate of oxide layer growth.

We previously found the solution to this problem to be

$$C^* = 1 - \frac{1}{\operatorname{erf}(\gamma)} \operatorname{erf}\left(\frac{x^*}{2\sqrt{t^*}}\right) \tag{8.46}$$

where γ solves the transcendental equation

$$\gamma e^{\gamma^2} \operatorname{erf}(\gamma) = g_N/\sqrt{\pi} \tag{8.47}$$

The oxide layer grows according to

$$\delta^* = 2\gamma\sqrt{t^*} \tag{8.48}$$

or, in dimensional form,

$$\delta = 2\gamma\sqrt{Dt} \tag{8.49}$$

Clearly, this solution is parameterized by g_N. To examine the behavior of the solution, we give g_N the values of 1, 0.1, and 0.01. For each value of g_N there is a corresponding value for γ, and these are 0.62, 0.22, and 0.0706, respectively. The concentration profiles in the oxide layer can be plotted at any given time t^* for each case, and this is done in Fig. 8.5 for times $t^* = 1$ and $t^* = 10$. The oxide layer grows faster for larger values of g_N. Perhaps more striking, though, is that the concentration profile for each case is very nearly a straight line, from the fixed ion concentration at the Ni–NiO interface down to zero at the surface.

Even though we assumed that C_N is known, it is actually quite difficult to measure C_N directly. For this reason, the usual engineering approach to this problem is rather different from the one we have taken. Usually one *assumes* that the concentration profile is a straight line between C_N and zero. The concentration gradient is then easily computed as the slope of the line, $-C_N/\delta$. Equating the growth rate with the

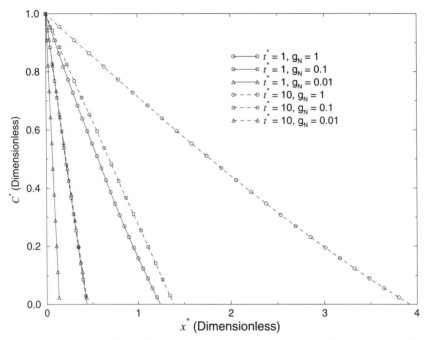

Figure 8.5: Concentration profile in the oxide layer for various values of g_N at $t^* = 1$ and 10.

solute flux, as in Eq. (8.34), we have

$$f_N \frac{d\delta}{dt} = D \frac{C_N}{\delta} \tag{8.50}$$

which is readily integrated to find

$$\delta = \sqrt{\frac{2DC_N t}{f_N}} = \sqrt{2Dg_N t} \tag{8.51}$$

Note that δ is proportional to $\sqrt{t}$ for both solutions. In this form, g_N is called the *Tammann scaling constant*, where "scaling" means "related to the formation of scale (oxide) on the surface."

In practice, scaling constants are determined by measuring the increase in mass of the oxidizing sample over time. It is easier to measure this quantity precisely than to measure the change in thickness. If the surface area of the sample is denoted by A, then the volume of oxide any time t is $A\delta(t)$. Because the nickel mass does not change, the mass gain all comes from the oxygen in the oxide. If the mass fraction of oxygen in the oxide is denoted by f_O $(= 1 - f_N)$, then the mass change ΔM is given by

$$\Delta M = \rho A \delta f_O \tag{8.52}$$

where again ρ is the density of the oxide. Substituting the expression for δ from Eq. (8.51) gives

$$\Delta M = \rho A \underbrace{\sqrt{\frac{2DC_N f_O^2}{f_N}}}_{k_p} \sqrt{t} \tag{8.53}$$

8.2 SOLID-STATE DIFFUSION

The quantity k_p is called the *parabolic scaling constant*. Values for k_p for various metal systems are tabulated in Poirier and Geiger (1994), where the reader will also find a more detailed discussion of this subject.

8.2.4 INTERDIFFUSION OF TWO SPECIES

In all of the preceding sections we limited our attention to problems in which we could consider the diffusion of one species (solute) into another (solvent). This is appropriate for dilute systems, or for problems such surface oxidation, in which the medium is relatively impermeable to one of the species. There is also an important class of problems in which this assumption does not hold. These are known as *diffusion couples*.

If the diffusion coefficient for B in A is equal to the diffusion coefficient for A in B, then the diffusion of the two species should be symmetric about the original interface. That diffusion problem is fairly easy to solve. However, there is no reason that the diffusivities should be equal, and in general they are not. At this point the problem becomes more interesting.

In a classic experiment, Kirkendall examined the interdiffusion of Cu and Zn by using the sample shown in Fig. 8.6. A thick layer of Zn was plated onto an α-brass block (a solid solution of Cu and Zn), with a number of thin, inert tungsten wires at the original interface. The sample was then heated to allow interdiffusion to take place, and the concentrations of Cu and Zn were measured at various times. The surprising result was that the inert markers moved closer together as the experiment progressed! This phenomenon is known as the *Kirkendall effect*, and we will see that it is a consequence of the difference between the intrinsic diffusivities D_{AB} and D_{BA}.

In examining this phenomenon, we must be very careful about the reference frame in which the diffusion is observed. We will use two different frames, one attached to the inert markers (the marker frame) and one in which the overall mass is stationary (the fixed frame). D_{AB} and D_{BA} are the intrinsic diffusion coefficients. Kirkendall's experiment tells us that the marker frame is translating relative to the fixed frame. Our task is to see what the diffusion equation and solutions looks like in the fixed frame.

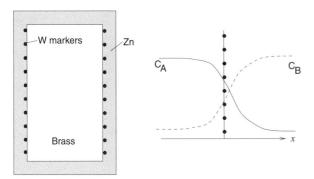

Figure 8.6: The diffusion couple used in the experiment by Kirkendall, and the concentration profiles near the markers.

Referring to the right-hand side of Fig. 8.6, an observer in the marker frame sees A atoms moving to the right and B atoms moving to the left. The flux of each species is described by Fick's law. In the marker frame the fluxes are

$$\mathbf{J}'_A = -D_{AB}\nabla C_A, \quad \mathbf{J}'_B = -D_{BA}\nabla C_B \tag{8.54}$$

Where the primes remind us that these are the fluxes seen in the marker frame. We expect that the two fluxes will have opposite sign.

Consider now the case where $D_{AB} \neq D_{BA}$, so that the magnitudes of the two fluxes are different. An observer fixed on the markers sees a net "drift" of material, as the number of A atoms going to the right is different from the number of B atoms going to the left. The mass of A atoms passing by, per unit time and area, is $\hat{\mathbf{J}}'_A$, whereas the corresponding mass of B atoms passing by is $\hat{\mathbf{J}}'_B$. The vector sum of these two quantities is the net mass flux, and we can divide by the mixture density ρ to get the net volume passing by per unit area and time. This is the drift velocity seen in the marker frame:

$$\mathbf{V}'_{\text{drift}} = \frac{\hat{\mathbf{J}}'_A + \hat{\mathbf{J}}'_B}{\rho} = \mathbf{J}'_A + \mathbf{J}'_B \tag{8.55}$$

We will assume that the density ρ is constant; this is not required, but it makes the derivation easier to follow. Substituting Fick's law in the marker frame, we find

$$\mathbf{V}'_{\text{drift}} = -D_{AB}\nabla C_A - D_{BA}\nabla C_B \tag{8.56}$$

The mass fractions must sum to one, $C_A + C_B = 1$, so we can write $C_B = 1 - C_A$ and $\nabla C_B = -\nabla C_A$. Then the drift velocity is

$$\mathbf{V}'_{\text{drift}} = -(D_{AB} - D_{BA})\nabla C_A \tag{8.57}$$

Because the two mass fractions sum to one everywhere, we only need to write one diffusion equation. The diffusion equation for A in the marker frame is

$$\frac{dC_A}{dt} = \nabla \cdot (D_{AB}\nabla C_A) \tag{8.58}$$

Here we have used d/dt for the time derivative to emphasize that this derivative is taken in a moving frame.

Now let us return to the fixed frame. In the fixed frame the mass is stationary, so the fixed observer sees the markers moving at a speed of

$$\mathbf{V}_{\text{markers}} = -\mathbf{V}'_{\text{drift}} \tag{8.59}$$

To translate the diffusion equation into the fixed frame, we can use Eq. (2.17) to write the time derivative as seen by a moving observer:

$$\frac{dC_A}{dt} = \frac{\partial C_A}{\partial t} + \mathbf{V}_{\text{markers}} \cdot \nabla C_A \tag{8.60}$$

Here $\partial/\partial t$ means the partial time derivative at a point in the fixed frame. Substituting this into Eq. (8.58), we have

$$\frac{\partial C_A}{\partial t} = \nabla \cdot (D_{AB}\nabla C_A) - \mathbf{V}_{\text{markers}} \cdot \nabla C_A \tag{8.61}$$

Now, because the density is constant, we have $\nabla \cdot \mathbf{V}_{\text{markers}} = 0$, and we may combine the terms on the right-hand side to form

$$\frac{\partial C_A}{\partial t} = \nabla \cdot (D_{AB} \nabla C_A - C_A \mathbf{V}_{\text{markers}}) \tag{8.62}$$

Combining Eq. (8.62) with Eqs. (8.57) and (8.59) yields a single diffusion equation for C_A in the fixed frame:

$$\frac{\partial C_A}{\partial t} = \nabla \cdot [(C_B D_{AB} + C_A D_{BA}) \nabla C_A] \tag{8.63}$$

We see then that the apparent diffusivity of species A in the fixed frame is no longer D_{AB}, but rather

$$\tilde{D} \equiv C_B D_{AB} + C_A D_{BA} \tag{8.64}$$

Here $\tilde{D}$ is called the *interdiffusion coefficient*. The diffusion coefficients D_{AB} and D_{BA} are more properly called the *intrinsic diffusion coefficients*. The flux in the fixed frame is computed by using $\tilde{D}$ in Fick's law,

$$\mathbf{J}_A = -\tilde{D} \nabla C_A \tag{8.65}$$

These results tells us that the measurement of concentrations in a diffusion couple does not give the intrinsic diffusivities D_{AB} and D_{BA}, but rather the composite interdiffusion coefficient $\tilde{D}$. Often one wishes to know the intrinsic diffusivities. These can be obtained from an experiment only if one also measures the drift velocity of inert marker particles.

There are many other important and interesting problems relating to diffusion in materials processing, and the reader is referred to the text by Glicksman (2000) for an excellent treatment of this topic.

8.3 SOLIDIFICATION MICROSTRUCTURE DEVELOPMENT

Microstructure development is a very important aspect of metal alloy solidification. The combined effects of heat and mass transfer produce local segregation patterns that ultimately control the properties of the metal. It is important to understand these phenomena, because the microstructure that develops during initial solidification is difficult to modify in subsequent processing.

The classical experiment in this field is *directional solidification*, a process that is also used to grow certain types of crystals. A material sample is placed in a furnace that maintains a temperature difference across the sample. At the hot end of the furnace the sample is liquid, and at the cold end it is solid. The sample is moved slowly toward the cold end, so that it solidifies while the liquid–solid interface remains fixed in space. In this process one can independently control the growth speed of the solid (by choosing the sample velocity) and the temperature gradient (by adjusting the hot and cold temperatures of the furnace). At low growth speed, the solid–liquid interface propagates as a smooth, planar front. As the speed is increased, the interface breaks down into periodic cells. The periodic cells break down further at higher velocity into snowflake-like patterns called *dendrites*. Finally, at extremely high speeds, the interface again becomes smooth and planar again.

We wish to understand the physics that drives these morphological transitions and to examine some of the practical implications. To this end we first discuss equilibrium phase diagrams, and their role in the partitioning of solute during solidification. We then examine the interaction of heat and solute transport in directional solidification to understand something about how the interface morphology develops.

8.3.1 EQUILIBRIUM PHASE DIAGRAMS

The reader should already be familiar with equilibrium phase diagrams for binary systems, and the discussion here is limited to aspects that are necessary for the topics that follow. A thorough presentation of the subject is available in many introductory materials science texts (e.g., Callister, 1997), as well as in texts on solidification (Flemings, 1974; Kurz and Fisher, 1989).

Consider the portion of a binary eutectic phase diagram shown in Fig. 8.7. Now element A is the solvent and element B is the solute. Any particular alloy composition is represented by a vertical line on the diagram, and we choose an overall composition C_0, as shown in the figure. Combinations of temperatures and compositions above the *liquidus* curve correspond to systems that are entirely liquid in equilibrium, whereas temperature-composition combinations below the *solidus* temperature correspond to systems that are entirely solid in equilibrium. At intermediate temperatures, such as the temperature T indicated in the figure, solid and liquid coexist. The solid has a composition C_s given by the intersection of the isotherm with the solidus curve and the liquid has composition C_ℓ given by the intersection of the isotherm with the liquidus curve.

The figure shows the liquidus and solidus curves as straight lines, an approximation that is not essential but that will make our calculations more convenient. Then we can write the equations for the solidus and liquidus curves as

$$T = T_m + m_\ell C_\ell$$
$$T = T_m + m_S C_s \tag{8.66}$$

where T_m is the melting point of the pure solvent, and m_ℓ and m_S are the slopes of

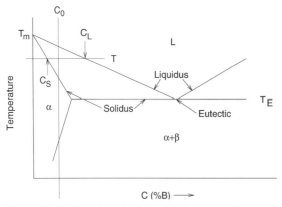

Figure 8.7: Portion of a eutectic equilibrium phase diagram for a binary alloy.

8.3 SOLIDIFICATION MICROSTRUCTURE DEVELOPMENT

the liquidus and solidus curves, respectively. It is also convenient for calculations to define the *partition coefficient* or *segregation coefficient*, k_0, as the ratio of the solid and liquid equilibrium compositions at any given temperature:

$$k_0 \equiv \frac{C_s}{C_\ell} \qquad (8.67)$$

When the liquidus and solidus curves are straight lines as given by Eqs. (8.66), then $k_0 = m_\ell/m_s$, independent of temperature.

It is often important to know the relative amounts of solid and liquid, as well as their compositions. If we assume that solidification proceeds slowly enough that thermodynamic equilibrium is reached in both phases, then these data can be obtained from a species balance. The mass fractions of solid and liquid in the entire sample are denoted by f_s and f_ℓ. Because all of the material present is either solid or liquid, these two quantities must satisfy the equation

$$f_s + f_\ell = 1 \qquad (8.68)$$

At equilibrium the compositions are uniform in each phase, and at any temperature T the total amount of solute must equal the amount in the liquid plus the amount in the solid, so we can write

$$f_s C_s + f_\ell C_\ell = C_0 \qquad (8.69)$$

Substituting Eq. (8.68) and rearranging, we find that the fraction solid is

$$f_s = \frac{C_\ell - C_0}{C_\ell - C_s} \qquad (8.70)$$

This form is referred to as the *inverse lever rule*, or simply the lever rule.

It is often convenient to know f_s as a function of either C_ℓ or T alone. If we substitute $C_s = k_0 C_\ell$ into Eq. (8.70) and rearrange, we find that the liquid composition and the fraction solid are related by

$$C_\ell = \frac{C_0}{1 - f_s(1 - k_0)} \qquad (8.71)$$

To obtain an expression in terms of temperature, substitute Eqs. (8.66) into Eq. (8.70):

$$f_s = \frac{C_\ell - C_0}{C_\ell - C_s} = \frac{\left(\frac{T - T_m}{m_\ell} - \frac{T_L - T_m}{m_\ell}\right)}{\left(\frac{T - T_m}{m_\ell} - \frac{T - T_m}{m_s}\right)} \qquad (8.72)$$

where T_L is the liquidus temperature for the alloy composition C_0, that is, $T_L = T_m + m_\ell C_0$. Rearranging and substituting the definition of k_0 yields the desired result:

$$f_s = \frac{1}{(1 - k_0)} \frac{T_L - T}{T_m - T} \qquad (8.73)$$

As an example, consider an alloy of 4 wt. % Cu in Al. The appropriate portion of the phase diagram is depicted in Fig. 8.8. From the figure, we obtain the

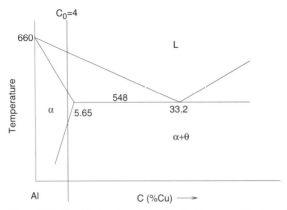

Figure 8.8: The Al–Cu binary equilibrium phase diagram.

following data:

$$m_S = \frac{548 - 660}{5.65} = -19.8$$

$$m_\ell = \frac{548 - 660}{33.2} = -3.37$$

$$k_0 = \frac{5.65}{33.2} = 0.17$$

$$T_L = 660 - 3.37(4) = 646.5$$

$$T_S = 660 - 19.8(4) = 580.8$$

Thus, the expression for the fraction solid in equilibrium solidification is

$$f_s = 1.205 \frac{646.5 - T}{660 - T} \tag{8.74}$$

This curve is plotted in Fig. 8.9. Note that the fraction solid is not a linear interpolation between the liquidus and solidus temperatures. Instead, for Al–Cu alloys, the solid develops more slowly when cooling through the higher temperature ranges, and more rapidly as the solidus temperature is approached.

For numerical simulations of heat transfer in alloy solidification, one often uses an enthalpy-temperature curve for the alloy to account for the gradual release of the heat of fusion between T_L and T_S. Such a curve can easily be derived by writing the definition of enthalpy,

$$H = \int_0^T c_p dT + L_f(1 - f_s(T)) \tag{8.75}$$

and substituting the fraction solid-temperature result, Eq. (8.73), for $f_s(T)$.

8.3.2 NONEQUILIBRIUM SOLIDIFICATION: THE GULLIVER–SCHEIL EQUATION

To derive the lever rule, we assumed that solidification took place slowly enough that equilibrium could be established in both the solid and liquid. However, this assumption is rarely valid. In a typical casting process, the local solidification time (i.e., the time required for the local temperature to traverse the freezing range) varies

8.3 SOLIDIFICATION MICROSTRUCTURE DEVELOPMENT

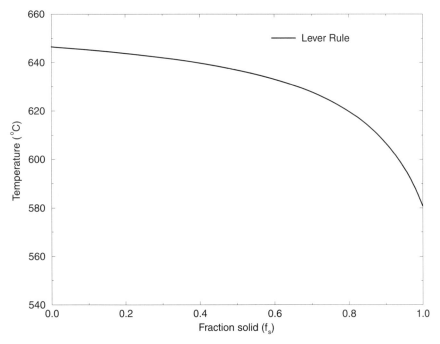

Figure 8.9: Computed equilibrium fraction solid-temperature curve for Al-4 wt. % Cu.

from a few seconds to a few minutes. Is this time sufficient for chemical equilibrium to be established? This is a good question to answer by estimation.

The typical microstructure for a solidified alloy is an array of dendrites, a segment of which is depicted in Fig. 8.10. A dendrite is a treelike structure whose branches follow a crystallographic direction typical of the parent alloy. Reproducing the temperature history during solidification also reproduces the microstructure, and for Al–Cu alloys the following empirical relationship between the secondary dendrite arm spacing (λ_2 in the figure) and the local solidification time t_f has been determined by Bower et al. (1966):

$$\lambda_2 = 9.82 t_f^{1/3} \tag{8.76}$$

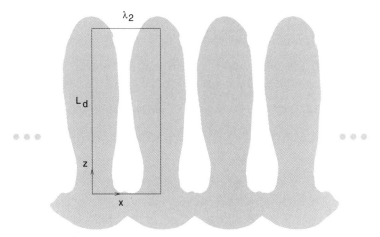

Figure 8.10: Schematic of a dendritic structure during solidification.

Here λ_2 is in micrometers and t_f is in seconds. Also, the length of a typical dendrite arm L_d is large compared to the secondary spacing λ_2.

With this information in hand, we can learn something about the solute redistribution process by scaling the appropriate governing equations. Consider the segment of the dendrite shown in the box in Fig. 8.10, assuming that the structure is two dimensional and periodic with wavelength λ_2. There are no sources of heat or solute within the control volume, so the governing equations for energy and solute transport are

$$\frac{\partial T}{\partial t} = \alpha \left(\frac{\partial^2 T}{\partial x^2} + \frac{\partial^2 T}{\partial z^2} \right) \tag{8.77}$$

$$\frac{\partial C_\ell}{\partial t} = D_\ell \left(\frac{\partial^2 C_\ell}{\partial x^2} + \frac{\partial^2 C_\ell}{\partial z^2} \right) \tag{8.78}$$

$$\frac{\partial C_s}{\partial t} = D_s \left(\frac{\partial^2 C_s}{\partial x^2} + \frac{\partial^2 C_s}{\partial z^2} \right) \tag{8.79}$$

The diffusion equations for the solid and liquid each apply within their respective domains.

We now scale the governing equations. As noted earlier, each equation will have a different characteristic time, but here we choose to scale on the externally imposed solidification time scale t_f. Thus we have

$$x^* = \frac{x}{\lambda_2/2}, \quad z^* = \frac{z}{L_d}, \quad t^* = \frac{t}{t_f}, \quad \theta = \frac{T - T_S}{\Delta T_f} \tag{8.80}$$

where ΔT_f is the freezing range, that is, the difference between the liquidus and solidus temperatures. (This scale is actually unimportant here.) Substituting these variables into the governing equations and rearranging yields

$$\frac{\partial \theta}{\partial t^*} = \left\{ \frac{4\alpha t_f}{\lambda_2^2} \right\} \left(\frac{\partial^2 \theta}{\partial x^{*2}} + \left\{ \frac{\lambda_2^2}{4L_d^2} \right\} \frac{\partial^2 \theta}{\partial z^{*2}} \right) \tag{8.81}$$

$$\frac{\partial C_\ell}{\partial t^*} = \left\{ \frac{4D_\ell t_f}{\lambda_2^2} \right\} \left(\frac{\partial^2 C_\ell}{\partial x^{*2}} + \left\{ \frac{\lambda_2^2}{4L_d^2} \right\} \frac{\partial^2 C_\ell}{\partial z^{*2}} \right) \tag{8.82}$$

$$\frac{\partial C_s}{\partial t^*} = \left\{ \frac{4D_s t_f}{\lambda_2^2} \right\} \left(\frac{\partial^2 C_s}{\partial x^{*2}} + \left\{ \frac{\lambda_2^2}{4L_d^2} \right\} \frac{\partial^2 C_s}{\partial z^{*2}} \right) \tag{8.83}$$

We have assumed that (λ_2/L_d) is small, which tells us that the z^* derivatives are negligible and we have have one-dimensional diffusion of heat and species between the arms of the dendrite:

$$\frac{\partial \theta}{\partial t^*} = \left\{ \frac{4\alpha t_f}{\lambda_2^2} \right\} \frac{\partial^2 \theta}{\partial x^{*2}} \tag{8.84}$$

$$\frac{\partial C_\ell}{\partial t^*} = \left\{ \frac{4D_\ell t_f}{\lambda_2^2} \right\} \frac{\partial^2 C_\ell}{\partial x^{*2}} \tag{8.85}$$

$$\frac{\partial C_s}{\partial t^*} = \left\{ \frac{4D_s t_f}{\lambda_2^2} \right\} \frac{\partial^2 C_s}{\partial x^{*2}} \tag{8.86}$$

8.3 SOLIDIFICATION MICROSTRUCTURE DEVELOPMENT

The dimensionless groups on the right-hand sides of these equations represent ratios between characteristic times for diffusion of heat or solute and the solidification time.

Now consider typical values for Al–Cu alloys, where α is approximately 10^{-6} m^2/s, D_ℓ is approximately 10^{-10} m^2/s and D_s is approximately 10^{-14} m^2/s. Substituting these values into Eq. (8.76), we find that, for t_f in seconds,

$$\frac{4\alpha t_f}{\lambda_2^2} \sim 4 \times 10^4 \, t_f^{1/3}$$

$$\frac{4 D_\ell t_f}{\lambda_2^2} \sim 4 t_f^{1/3}$$

$$\frac{4 D_s t_f}{\lambda_2^2} \sim 4 \times 10^{-4} \, t_f^{1/3}$$

Thus, for freezing times anywhere between 10 s and 1 day,

$$\frac{4\alpha t_f}{\lambda_2^2} \gg 1$$

$$\frac{4 D_\ell t_f}{\lambda_2^2} \gg 1$$

$$\frac{4 D_s t_f}{\lambda_2^2} \ll 1$$

Substituting these results into the scaled governing equations gives

$$0 \approx \frac{\partial^2 \theta}{\partial x^{*2}} \tag{8.87}$$

$$0 \approx \frac{\partial^2 C_\ell}{\partial x^{*2}} \tag{8.88}$$

$$\frac{\partial C_s}{\partial t^*} \approx 0 \tag{8.89}$$

Because the solution is periodic, the derivatives $\partial \theta/\partial x^*$ and $\partial C_\ell/\partial x^*$ must vanish at $x = \lambda/2$, so we conclude that θ and C_ℓ are constant and that C_s does not change with time over the period of solidification.

This scaling analysis has led us to the set of assumptions that Gulliver, and later Scheil, used to derive a simplified expression for the solute redistribution during solidification of binary alloys. They assumed that there was one-dimensional solidification with no diffusion in the solid, complete mixing in the liquid, and no temperature gradient. These assumptions are consistent with the the analysis given above, for a large range of important solidification times. For the analysis to be completed, a relationship between the concentrations in the solid and liquid phases is required. It is usually assumed that the processes at the liquid–solid interface occur fast enough that local thermodynamic equilibrium is established, so at the interface

$$C_s = k_0 C_\ell \tag{8.90}$$

This appears to be an excellent assumption.

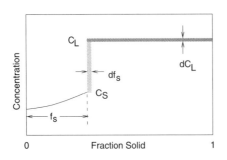

Figure 8.11: Sketch of the solute distribution near the liquid–solid interface in the Scheil model.

We can now derive an expression for solute redistribution, with reference to Fig. 8.11. The assumption of equilibrium at the interface tells us that any newly formed solid will contain less solute than the liquid. That is, solute is *rejected* at the interface during solidification. (This statement applies for alloy systems in which the liquidus and solidus curves slope downward, like that of Fig. 8.7; if the curves slope upward then the solid contains more solute than the liquid, but the analysis still holds.) The solute rejected at the interface by solidification is added to the liquid, changing its concentration. By enforcing a solute balance at the liquid–solid interface, we find that

$$(C_\ell - C_s)(df_s) = f_\ell(dC_\ell) \tag{8.91}$$

Substituting the expression for local equilibrium at the interface, Eq. (8.90), and recognizing that $df_s = -df_\ell$, we have

$$\frac{df_\ell}{f_\ell} = -\frac{dC_\ell}{(1-k_0)C_\ell} \tag{8.92}$$

which we then integrate,

$$\int_1^{f_\ell} \frac{df_\ell}{f_\ell} = \int_{C_0}^{C_\ell} -\frac{dC_\ell}{(1-k_0)C_\ell} \tag{8.93}$$

to obtain

$$C_\ell = C_0 f_\ell^{k_0-1} \tag{8.94}$$

If we now use Eq. (8.66) for the liquidus line to replace C_ℓ by the temperature, we obtain an expression relating f_s and T for nonequilibrium solidification:

$$f_s = 1 - \left(\frac{T-T_m}{T_L-T_m}\right)^{1/(k_0-1)} \tag{8.95}$$

This is called the *Gulliver–Scheil equation*, or more commonly the Scheil equation.

Equation (8.95) has the unphysical property that the fraction solid does not reach unity until the temperature drops to $-\infty$. In reality, when the temperature reaches the eutectic temperature, the liquid concentration is exactly equal to the eutectic composition, and the remaining liquid solidifies as a eutectic. Thus, alloys that should contain no eutectic according to the phase diagram actually form eutectic during nonequilibrium solidification. As an example we return to the Al–4 wt.% Cu alloy considered earlier, and we compare the $f_s(T)$ computed by the Scheil equation and the lever rule in Fig. 8.12. We see that the Scheil equation predicts ∼9% eutectic for this alloy. For this reason precipitation heat treaters must ensure that the temperature

8.3 SOLIDIFICATION MICROSTRUCTURE DEVELOPMENT

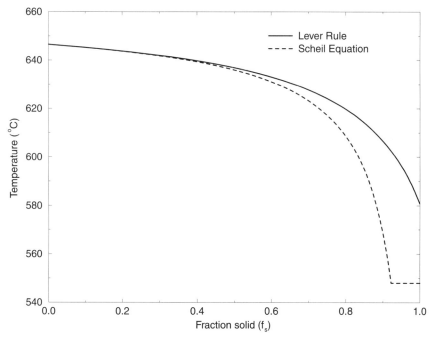

Figure 8.12: Fraction solid-temperature curves for Al–4 wt. % Cu according to the lever rule and the Scheil equation. Note the finite amount of eutectic predicted for the Scheil equation.

for solution heat treatment is always below the eutectic temperature. Otherwise the eutectic portions of the part will melt and lead to porosity in the heat treated part.

Finally, note that the two cases considered thus far, equilibrium solidification and solidification with complete diffusion in the liquid and none in the solid, represent limiting cases. Real processes always lie in between, with a limited amount of diffusion in each phase. Also, there are some alloys in which diffusion coefficient in the solid is much larger than the typical values given above, such as C in Fe. Typically these are alloys in which the solute diffuses interstitially. In these alloys, solute redistribution is actually much closer to the lever rule than to the Scheil equation.

8.3.3 LIMITED DIFFUSION IN THE LIQUID: PLANAR FRONT GROWTH

There are many solidification processes in which the interface rejects solute faster than diffusion can equalize its concentration in the liquid. An important example is Czochralski crystal growth, a process in which silicon crystals are grown from the melt by extracting a seed at constant speed. A more easily analyzed process, which serves as a prototype for Czochralski growth, is directional solidification. When crystals are grown by directional solidification, an encapsulated sample is drawn at constant velocity V through a temperature gradient, which is established by fixing a hot point and a cold point somewhere in space. After an initial transient period, there is a balance between the rate at which solute is rejected at the interface and the rate at which is diffuses away in the liquid. This is suggested by Fig. 8.13.

We assume that the material properties are constant in the solid and the liquid and that we have thermodynamic equilibrium at the liquid–solid interface. The interface is

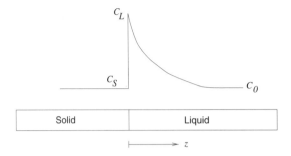

Figure 8.13: The steady-state concentration profile ahead of a moving solidification interface in a binary alloy.

assumed to move at constant velocity V, and the diffusion field is assumed to be steady with respect to the moving interface. We also assume that temperature is a linear function of position. This latter assumption will be justified later when we examine a more complete model of directional solidification. If we choose a coordinate system in which the interface remains stationary, then the material translates with a velocity $-V$ and the governing equation for concentration in the liquid is

$$-V \frac{\partial C_\ell}{\partial z} = D_\ell \frac{\partial^2 C_\ell}{\partial z^2} \qquad (8.96)$$

Far from the interface the liquid has the initial composition, so the far-field boundary is

$$C_\ell = C_0 \quad \text{at } z \to \infty \qquad (8.97)$$

At the liquid–solid interface the assumption of local equilibrium tells us that $C_s = k_0 C_\ell$, but this is not a boundary condition because neither C_s nor C_ℓ is known. The solute rejected ahead of the interface caused by solidification must move away from the interface by diffusion, so we have

$$-D_\ell \frac{\partial C_\ell}{\partial z} = V(C_\ell - C_s) = V(1 - k_0)C_\ell \quad \text{at } z = 0 \qquad (8.98)$$

The solution to Eq. (8.96) with these boundary conditions is

$$C_\ell = C_0 \left(1 + \frac{1 - k_0}{k_0} \exp\left(-\frac{Vz}{D_\ell}\right) \right) \qquad (8.99)$$

At $z = 0$ we note that $C_\ell = C_0/k_0$, and therefore the concentration in the solid is C_0.

The solution shows that there is a concentration boundary layer ahead of the interface, whose characteristic length is $\delta_c \sim D_\ell/V$. For a typical directional solidification experiment, V is 10 μm/s and D_ℓ is approximately 5×10^{-9} m^2/s, so that δ_c is approximately 0.5 mm. The corresponding thermal boundary layer is approximately 10,000 times larger, approximately 5 m in length. Thus it was quite reasonable to neglect deviations from a linear temperature field when considering phenomena on the length scale of the concentration boundary layer.

This steady-state solution can also be used to understand what happens when there is a sudden change of solidification speed. If the speed is suddenly increased in Eq. (8.98), then the rate of solute rejection will exceed the rate of diffusion in the liquid, and the concentration at the interface must increase. Local equilibrium still applies, however, and the concentration in the solid will increase at the same time.

8.3 SOLIDIFICATION MICROSTRUCTURE DEVELOPMENT

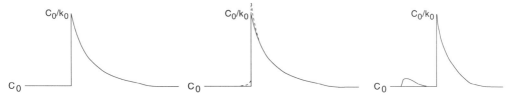

Figure 8.14: A time sequence illustrating the effect of a sudden speed increase on the resulting concentration distribution in the solid and liquid.

The concentration field ahead of the interface eventually adjusts to the new speed, the steady state is reestablished, and the solid concentration at the interface settles back to C_0. The sudden speed change thus causes a bump in the concentration profile within the solid. This process is illustrated in Fig. 8.14. By a similar argument, a sudden decrease in speed leaves behind a solute-depleted zone in the solid. This phenomenon is often observed in continuous casting, in which the interface speed changes suddenly when an air gap forms after initial solidification in the mold, as the casting exits the mold and enters submold water sprays. When growing large single crystals for semiconductors, in which the goal is to have a very uniform concentration of solute (dopants) in the crystal, a speed change causes a serious defect. Crystal growth processes are very carefully regulated to maintain a constant solidification speed.

8.3.4 STABILITY OF THE PLANAR FRONT: PHENOMENOLOGICAL ANALYSIS

We can use the solution for plane front growth to explore the stability of the liquid–solid interface. Note that the concentration in the boundary layer ahead of the liquid–solid interface exceeds C_0. Substituting the equation for the liquidus line, Eq. (8.66), into Eq. (8.99), we find that, at any distance z ahead of the interface, the *local* liquidus temperature T_{liq} is given by

$$T_{\text{liq}}(z) = T_m + m_\ell C_0 \left(1 + \frac{1-k_0}{k_0} \exp\left(-\frac{Vz}{D_\ell}\right)\right) \qquad (8.100)$$

This liquidus temperature distribution is depicted schematically in Fig. 8.15. Also shown in the figure, superimposed on this local liquidus curve, are two possible actual temperature distributions ahead of the interface. For the dashed curve, with a high

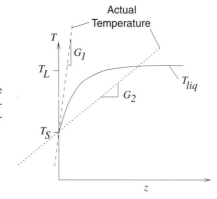

Figure 8.15: The local liquidus temperature ahead of the liquid–solid interface, Eq. (8.100), with two possible temperature distributions; T_S and T_L are the solidus and liquid temperatures for the average concentration C_0.

temperature gradient G_1, the actual temperature is greater than the local liquidus temperature everywhere. This corresponds to a stable interface: if the solid appears ahead of the plane front it will remelt, because the temperature is above the liquidus temperature associated with the local composition.

For the second case, in which the temperature gradient G_2 is low, there is a region ahead of the interface where the actual temperature is *below* the local equilibrium liquidus temperature. The liquid ahead of the interface is sometimes described as being "undercooled because of its constitution," or "constitutionally supercooled." Now the interface is unstable, because any solid that forms ahead of the interface will grow, as it is forming below the liquidus temperature that corresponds to the local composition.

The changeover from a stable to an unstable interface is critically important to crystal growing. The marginal stability condition is obtained when the slope of the actual temperature matches the slope of the local liquidus temperature at the interface, so the stability condition is

$$G \geq \left. \frac{dT_{\text{liq}}}{dz} \right|_{z=0} \tag{8.101}$$

Substituting Eq. (8.100) for T_{liq} and rearranging shows that the interface will be *stable* if

$$\frac{G}{V} \geq -\frac{m_\ell C_0 (1 - k_0)}{k_0 D_\ell} \tag{8.102}$$

Strictly speaking this is a phenomenological, rather than a rigorous, analysis, as it relies on an assumption about the relationship between temperature gradient and stability. However, this stability criterion has been verified experimentally in a large number of materials, and its predictions are reliable.

In processes in which it is essential to maintain a stable plane interface, such as growing crystals for semiconductors, Eq. (8.102) determines the maximum possible growth speed. As an example, a temperature gradient of $G = 10^4$ K/m is approximately as large as one can maintain in an industrial process. Using $D_\ell = 10^{-8}$ m²/s, $m_\ell = -10$ K/wt.%, $C_0 = 0.1$ wt.%, and $k_0 = 0.01$ gives a solidification speed of $V = 10^{-6}$ mm/s. At this speed it takes nearly 12 days to grow a crystal that is 1 m long. This explains why semiconductor crystals must be grown so slowly.

8.3.5 STABILITY OF THE PLANAR INTERFACE (ADVANCED)

The analysis in the previous section tells us when the interface becomes unstable, but it does not tell us what the instabilities look like. When an experiment of this type is performed, the interface develops small protrusions, and the constitutional undercooling condition tells us that the disturbances should grow. In practice the disturbances have a characteristic length, and if we wish to describe the evolution of microstructures we must understand where this length comes from. This determination of length scales and interface shapes is usually called *pattern selection*, and it is one of the most difficult and challenging problems in materials processing. We will approach the problem by performing a linear stability analysis, starting from plane front solidification. The reader should understand that predicting the full microstructural

8.3 SOLIDIFICATION MICROSTRUCTURE DEVELOPMENT

pattern of dendrite shapes and their primary and secondary arm spacings is not yet possible. However, the analysis presented in this section provides an important starting point, and is essential to understanding the current research in the subject.

Introduction to Linear Stability Analysis

We begin with a short introduction to linear stability analysis. The subject is introduced through a simple example, and we indicate what parts belong to the general method, and what parts are specific to the problem at hand.

The overall procedure starts by determining a basic state, which may or may not be stable. One then introduces a small perturbation to that state, linearizes the governing equation and boundary conditions about the base state, and seeks solutions that depend exponentially on time. The argument of the exponential function may have both real and imaginary parts, and we will conclude that the system is stable, neutrally stable, or unstable based on whether the real part of the exponent is negative, zero, or positive, respectively.

Consider a simple pendulum, Fig. 8.16, a problem we studied in Chapter 1. This time we immediately simplify the problem by assuming that the pendulum is rigid, but we retain the possibility that the pendulum might swing through large angles θ. The governing equation is then given by

$$\ddot{\theta} + \frac{g}{L}\sin\theta = 0 \tag{8.103}$$

To simplify the presentation, we scale the time as before,

$$t^* = \sqrt{\frac{g}{L}}\,t \tag{8.104}$$

which reduces the governing equation to

$$\frac{d^2\theta}{dt^{*2}} + \sin\theta = 0 \tag{8.105}$$

From this point onward we drop the superscript asterisk from t for clarity.

We would like to find *basic states* of the system in which there is no motion, that is, where $\dot{\theta} = 0$. (The overdots represent derivatives with respect to t^*.) We designate

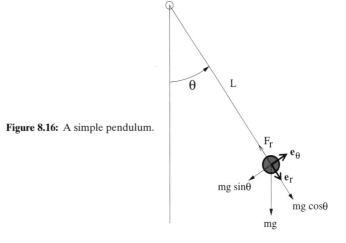

Figure 8.16: A simple pendulum.

these basic state solutions as θ_0. Substituting into Eq. (8.105) tells us that

$$\sin \theta_0 = 0 \tag{8.106}$$

which has two solutions, $\theta_0 = 0$ and $\theta_0 = \pi$. Experience tells us that $\theta_0 = 0$ is stable, whereas $\theta + 0 = \pi$ is unstable. Let us see how linear stability analysis gives the same result.

First, we must define what is meant by stable. We say that the basic state is *stable* if small perturbations decay over time, such that the system returns to the basic state. The basic state is *unstable* if the perturbations grow with time in an unbounded way. The system is said to be *neutrally stable* if a perturbation neither grows nor decays in time.

To examine the stability of the system, we write its behavior as

$$\theta(t) = \theta_0 + \theta_1(t) \tag{8.107}$$

where $\theta_1(t)$ represents a small, time-dependent perturbation from the basic state θ_0; θ_1 is assumed to be small, in a way we will characterize shortly. Substituting this into the governing equation, we have

$$\frac{d^2}{dt^2}(\theta_0 + \theta_1) + \sin(\theta_0 + \theta_1) = 0 \tag{8.108}$$

This is expanded to read

$$\ddot{\theta}_0 + \ddot{\theta}_1 + \sin \theta_0 \cos \theta_1 + \sin \theta_1 \cos \theta_0 = 0 \tag{8.109}$$

For both of the basic states, $\sin \theta_0 = 0$. For the basic state $\theta_0 = 0$ we have $\cos \theta_0 = 1$, whereas for the basic state $\theta_0 = \pi$ we have $\cos \theta_0 = -1$. From the definition of θ_0 we have $\ddot{\theta}_0 = 0$. This reduces the governing equation to

$$\ddot{\theta}_1 \pm \sin \theta_1 = 0 \tag{8.110}$$

where the plus sign corresponds to the basic state $\theta_0 = 0$, and the minus sign corresponds to $\theta_0 = \pi$.

In a linear stability analysis we assume that the perturbations are small, and we linearize the system of equations whenever necessary. To that end, we represent $\sin \theta_1$ by a truncated Taylor series, such that $\sin \theta_1 \approx \theta_1$. Equation (8.110) then becomes

$$\ddot{\theta}_1 \pm \theta_1 = 0 \tag{8.111}$$

To solve this system we require initial conditions on θ_1, which we choose as $\theta_1(0) = \theta_1^*$ and $\dot{\theta}_1(0) = 0$. In other words, we will move the pendulum away from its basic state by a small amount θ_1^* and then release it.

We write the solution as $\theta_1 = \theta_1^* e^{\sigma t}$. The exponent σ is a complex number, and what we are seeking is the sign of the real part of σ. Substituting this form into Eq. (8.111) yields the characteristic equation for σ,

$$\sigma^2 \pm 1 = 0 \tag{8.112}$$

which has the solutions $\sigma = i, 1$ (where $i = \sqrt{-1}$) for the basic states $\theta_0 = 0, \pi$, respectively. Finally, applying the initial condition, we find the perturbed solutions for

8.3 SOLIDIFICATION MICROSTRUCTURE DEVELOPMENT

the two cases to be

$$\theta_1 = \begin{cases} \theta_1^* e^{it}, & \theta_0 = 0 \\ \theta_1^* e^{t}, & \theta_0 = \pi \end{cases} \quad (8.113)$$

These solutions show that the basic state $\theta_0 = 0$ is neutrally stable, because the solution oscillates without growth or decay about the basic state. If we included damping in the model then the perturbations would decay, and the state $\theta_0 = 0$ would be stable. In contrast, the basic state $\theta_0 = \pi$ is unstable, because the perturbed solution grows without bound in time. These are the principal results of the linear stability analysis.

Note that the solution for $\theta_0 = \pi$ predicts that the disturbance eventually becomes infinite, which is physically incorrect. This is not a problem, however, because our solution is valid only for small perturbations, that is, where $\sin\theta_1 \approx \theta_1$, and the disturbance will grow exponentially when it is this small. The important conclusion is that *any* infinitesimal disturbance will move the solution far away from the basic state $\theta_0 = \pi$.

To recap, in linear stability analysis we first find a basic state, and then we introduce small perturbations on that basic state and examine their evolution. The time-dependent growth or decay of small perturbations shows whether the basic state is stable or unstable.

There is one important limitation to this analysis. Because we consider only infinitesimal disturbances, we learn nothing about the evolution of disturbances once they reach finite amplitude. If linear stability analysis shows that a basic state is stable to infinitesimal disturbances, we cannot be sure that it will be stable to larger disturbances. As an example, consider a marble sitting in a small depression at the top of a mound. If the marble is moved slightly and released, it returns to it original position, but if the marble is moved out of the depression and released, it moves far away. However, if linear stability analysis shows that a system is *unstable* with respect to small disturbances, then we can be certain that the system is unstable (or at best neutrally stable) to larger disturbances. Even with this limitation, linear stability analysis is a powerful tool for understanding the behavior of dynamic systems.

Linear Stability Analysis of a Binary Alloy in Planar Front Growth

With the tool of linear stability analysis in hand, we now revisit the problem of planar front growth. The analysis uses the same basic state that we found earlier, but we are now equipped to perform the stability analysis. The analysis presented here was first done by Mullins and Sekerka (1964).

In all of the solidification problems we considered thus far, the interface was assumed to be flat. On a flat interface the temperature of a pure material is its thermodynamic melting point, and for an alloy the temperature is related to the composition through the equilibrium phase diagram. When the interface is curved, however, there is an adjustment to the thermodynamic melting point caused by the energy of the curved interface. In such a case, the interface temperatures in the liquid T_ℓ^i and in the solid T_s^i are given by

$$T_\ell^i = T_m + m_\ell C_\ell^i - 2\hat{\Gamma}\kappa_m \quad (8.114)$$

$$T_s^i = T_m + m_s C_s^i - 2\hat{\Gamma}\kappa_m \quad (8.115)$$

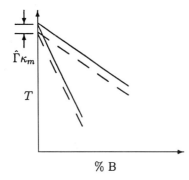

Figure 8.17: Correction of the equilibrium phase diagram (solid lines) for curvature (dashed lines).

Here T_m is the melting point of the pure material, m_ℓ and m_s are the slopes of the liquidus and solidus curves, $\hat{\Gamma}$ is a material parameter equal to the liquid–solid interfacial energy Γ divided by the entropy of fusion L_f, and κ_m is the local mean curvature of the interface (see Appendix A.6). Equations (8.114) and (8.115) are known as the *Gibbs–Thomson equations*. Although the value of $\hat{\Gamma}$ is typically small (of the order of 10^{-9} m), the correction that is due to curvature is crucial to the pattern selection problem. These equations can be applied to pure materials by setting the solid and liquid concentrations to zero.

A portion of the equilibrium phase diagram, corrected for curvature, is depicted in Fig. 8.17. Notice that the curvature effect shifts the liquidus and solidus curves to lower temperatures, but the separation between them at any given temperature is unchanged. Thus, we can still write the difference between the compositions of the liquid and solid at the interface in terms of C_ℓ^i and k_0, the segregation coefficient:

$$\Delta C_0 = C_\ell^i - C_s^i = (1 - k_0) C_\ell^i \tag{8.116}$$

Returning to the unidirectional solidification problem, we assume that the interface is moving at constant velocity in the z direction, and that the temperatures far away from the interface are fixed. For simplicity, all material properties are assumed to be constant and the thermal properties of the solid and liquid are taken to be equal. These assumptions simplify the math, but they can be relaxed if necessary. For the basic state we use a planar interface, and the temperature and concentration in the solid and liquid must then satisfy the following equations:

$$\frac{\partial T_s}{\partial t} - V \frac{\partial T_s}{\partial z} = \alpha \frac{\partial^2 T_s}{\partial z^2}, \quad z < 0 \tag{8.117}$$

$$\frac{\partial T_\ell}{\partial t} - V \frac{\partial T_\ell}{\partial z} = \alpha \frac{\partial^2 T_\ell}{\partial z^2}, \quad z > 0 \tag{8.118}$$

$$\frac{\partial C_\ell}{\partial t} - V \frac{\partial C_\ell}{\partial z} = D_\ell \frac{\partial^2 C_\ell}{\partial z^2}, \quad z < 0 \tag{8.119}$$

$$\frac{\partial C_s}{\partial t} - V \frac{\partial C_s}{\partial z} = D_s \frac{\partial^2 C_s}{\partial z^2}, \quad z > 0 \tag{8.120}$$

We now scale these equations. As in other problems involving semi-infinite domains, the boundary conditions impose no obvious length or time scale. We use a

8.3 SOLIDIFICATION MICROSTRUCTURE DEVELOPMENT

length scale L that is initially unknown, and we choose the time scale to be L/V; C_ℓ and C_s are already dimensionless (but not scaled), but we leave them as they are for the moment. Temperature is scaled by using a reference temperature T_{ref} and reference temperature difference ΔT. The proper choices for these temperature scales will become clear a little later. The scaled variables are now

$$t^* = \frac{V}{L}t, \quad z^* = \frac{z}{L}, \quad \theta = \frac{T - T_{\text{ref}}}{\Delta T} \tag{8.121}$$

and these lead to governing equations of

$$\frac{VL}{\alpha}\left(\frac{\partial \theta_s}{\partial t^*} - \frac{\partial \theta_s}{\partial z^*}\right) = \frac{\partial^2 \theta_s}{\partial z^{*2}} \tag{8.122}$$

$$\frac{VL}{\alpha}\left(\frac{\partial \theta_\ell}{\partial t^*} - \frac{\partial \theta_\ell}{\partial z^*}\right) = \frac{\partial^2 \theta_\ell}{\partial z^{*2}} \tag{8.123}$$

$$\frac{VL}{D_\ell}\left(\frac{\partial C_\ell}{\partial t^*} - \frac{\partial C_\ell}{\partial z^*}\right) = \frac{\partial^2 C_\ell}{\partial z^{*2}} \tag{8.124}$$

$$\frac{VL}{D_s}\left(\frac{\partial C_s}{\partial t^*} - \frac{\partial C_s}{\partial z^*}\right) = \frac{\partial^2 C_s}{\partial z^{*2}} \tag{8.125}$$

As we found before, there are several choices for L. Recognizing that, for metallic systems, $\alpha \gg D_\ell \gg D_s$, we choose to scale on diffusion in the liquid, and we choose $L = D_\ell/V = \delta_c$, where δ_c is a diffusion boundary layer thickness. Making this substitution in the governing equations yields

$$\frac{D_\ell}{\alpha}\left(\frac{\partial \theta_s}{\partial t^*} - \frac{\partial \theta_s}{\partial z^*}\right) = \frac{\partial^2 \theta_s}{\partial z^{*2}} \tag{8.126}$$

$$\frac{D_\ell}{\alpha}\left(\frac{\partial \theta_\ell}{\partial t^*} - \frac{\partial \theta_\ell}{\partial z^*}\right) = \frac{\partial^2 \theta_\ell}{\partial z^{*2}} \tag{8.127}$$

$$\frac{\partial C_\ell}{\partial t^*} - \frac{\partial C_\ell}{\partial z^*} = \frac{\partial^2 C_\ell}{\partial z^{*2}} \tag{8.128}$$

$$\frac{D_\ell}{D_s}\left(\frac{\partial C_s}{\partial t^*} - \frac{\partial C_s}{\partial z^*}\right) = \frac{\partial^2 C_s}{\partial z^{*2}} \tag{8.129}$$

Now, because $D_\ell/\alpha \ll 1$ and $D_\ell/D_s \gg 1$, we have

$$0 = \frac{\partial^2 \theta_s}{\partial z^{*2}} \tag{8.130}$$

$$0 = \frac{\partial^2 \theta_\ell}{\partial z^{*2}} \tag{8.131}$$

$$\frac{\partial C_\ell}{\partial t^*} - \frac{\partial C_\ell}{\partial z^*} = \frac{\partial^2 C_\ell}{\partial z^{*2}} \tag{8.132}$$

$$\frac{\partial C_s}{\partial t^*} - \frac{\partial C_s}{\partial z^*} = 0 \tag{8.133}$$

We conclude that the temperature field is "frozen" in space (i.e., independent of time), even though the concentration field is time dependent. This is sometimes called the *frozen temperature approximation*. Integrating the energy equation, we find the temperatures to be

$$\theta_s = c_1^s z^* + c_2^s$$
$$\theta_\ell = c_1^\ell z^* + c_2^\ell \qquad (8.134)$$

Temperature must be continuous at the interface, and we must also satisfy the Stefan condition there. The concentration solution will be needed to do this, so we tend to that next.

Recall that the coordinate system is chosen so that the position of the plane interface remains fixed at $z^* = 0$. For the basic state, all of the variables are invariant in time in this coordinate system. We designate these basic solutions with a superscript zero. Setting the time derivative of C_s^0 to zero reduces the diffusion equation to

$$\frac{\partial C_s^0}{\partial z^*} = 0 \qquad (8.135)$$

Clearly C_s^0 is constant. Setting the time derivative of C_ℓ^0 to zero, we find that

$$-\frac{\partial C_\ell^0}{\partial z^*} = \frac{\partial^2 C_\ell^0}{\partial z^{*2}} \qquad (8.136)$$

The boundary conditions on C_ℓ^0 are exactly those given in the previous section: $C_\ell^0 \to C_0$ as $z \to \infty$, and

$$^i C_\ell^0 (1 - k_0) = -\frac{D_s}{D_\ell}\frac{\partial C_s^0}{\partial z^*} - \frac{\partial C_\ell^0}{\partial z^*} \qquad (8.137)$$

at the interface. Here the prescript i refers to the interfacial value. Equation (8.135) tells us that the first term on the right-hand side is zero, leaving a boundary condition of

$$^i C_\ell^0 (1 - k_0) = \frac{\partial C_\ell^0}{\partial z^*} \qquad (8.138)$$

This is exactly the problem for the liquid phase concentration we solved before, so we simply copy the solution, Eq. (8.99):

$$C_\ell^0 = C_0 \left(1 - \frac{1 - k_0}{k_0} e^{-z^*}\right) \qquad (8.139)$$

Note that at the interface ($z^* = 0$), we have $C_\ell^0 = C_0/k_0$, and because we have $C_s = k_0 C_\ell$ at the interface, then $C_s^0 = C_0$.

Before going further, we rescale the concentration field so that it is truly of order one. The maximum value is attained at the interface, C_0/k_0, and the minimum appears at infinity, where $C_\ell^0 = C_0$. We therefore define

$$C_\ell^* = \frac{C_0/k_0 - C_\ell}{C_0/k_0 - C_0} = \frac{C_0/k_0 - C_\ell}{\Delta C_0} \qquad (8.140)$$

where ΔC_0 represents the difference in concentration between the liquid and solid at the solidus temperature. Note that in this scheme $0 \leq C_\ell^* \leq 1$. Substituting the

8.3 SOLIDIFICATION MICROSTRUCTURE DEVELOPMENT

dimensionless form into the basic solution for C_ℓ yields

$$C_\ell^{0*} = 1 - e^{-z^*} \tag{8.141}$$

We can now return to the temperature field and evaluate the constants of integration in Eqs. (8.134). We anticipate the temperature gradient will be the same in the solid and liquid, and we will demonstrate that this is a reasonable assumption. Therefore we discuss only one temperature T and its associated dimensionless form θ. We first redimensionalize the temperature, recognizing that the temperature varies at most linearly with position, and we write

$$T = T'_{\text{ref}} + G_T z \tag{8.142}$$

where T'_{ref} is a new reference temperature that we will evaluate in a moment, and G_T is the imposed temperature gradient. In practice one normally thinks of G_T as a process parameter that can be set independently. To evaluate the reference temperature, recall that the coupling condition required by local equilibrium at the interface. If the concentration in the liquid is C_0/k_0, then $C_s = C_0$ and the interface temperature must be the solidus temperature, $T_m + m_\ell C_0/k_0$. Thus,

$$T = T_m + \frac{m_\ell C_0}{k_0} + G_T z \tag{8.143}$$

Now, substituting $z^* = z/\delta_c$ and rearranging gives

$$\theta = \frac{T - T_m - m_\ell C_0/k_0}{G_T D_\ell/V} = z^* \tag{8.144}$$

The denominator of this expression, $G_T D_\ell/V$, is the characteristic temperature difference ΔT. Note that this is simply the temperature change over the length scale δ_c.

The temperature field in the presence of the moving interface cannot really have the same gradient in both phases, even when the thermal conductivity of the solid and liquid are equal, because this does not satisfy the Stefan condition at the interface:

$$\rho L_f V = k_s \frac{\partial T_s}{\partial z} - k_\ell \frac{\partial T_\ell}{\partial z} \tag{8.145}$$

Scaling this equation by using the dimensionless variable θ given in Eq. (8.144), and invoking the assumption that $k_s = k_\ell$, we find

$$\frac{\rho L_f V}{k_s G_T} = \frac{\partial \theta_s}{\partial z^*} - \frac{\partial \theta_\ell}{\partial z^*} \tag{8.146}$$

The left-hand side of this expression can be evaluated for any given set of experimental conditions. If we choose parameters for aluminum, solidified at a speed of 10^{-6} m/s in a temperature gradient of 10^4 K/m, values that are typical of crystal growth processes, we find that the left-hand side evaluates to 5×10^{-4}. Thus, the temperature gradients can indeed be considered equal. This approximation is not required to do the stability analysis, but it is accurate for many cases and makes the analysis easier to follow.

We are now ready to examine the stability of the basic state for planar front growth. We consider the possibility that the interface is perturbed by a small

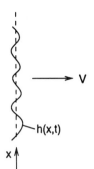

Figure 8.18: A schematic of the perturbed interface.

disturbance, as illustrated in Fig. 8.18. The perturbed interface position is given by the function $h(x, t)$, and we wish to determine the conditions under which such a perturbation will grow or decay. The solidification rate of the perturbed interface is given by

$$v_i = V + \frac{\partial h}{\partial t} \tag{8.147}$$

Even in our reference frame, where the average interface position is fixed and the material travels at velocity V, the interface position varies with time.

Recalling that the time scale was given as $t^* = Vt/L$ and using δ_c for the characteristic length L, we find that the scaled time variable is

$$t^* = \frac{V^2}{D_\ell} t \tag{8.148}$$

Defining h^* as h/δ_c, and $v_i^* = v_i/V$, we obtain

$$v_i^* = 1 + \frac{\partial h^*}{\partial t^*} \tag{8.149}$$

We will also need the unit normal vector to the interface in the following analysis. From Appendix A.6 this is

$$\mathbf{n} = \frac{\mathbf{e}_z - \frac{\partial h}{\partial x}\mathbf{e}_x}{\sqrt{1 + \left(\frac{\partial h}{\partial x}\right)^2}} \tag{8.150}$$

We now assume that the perturbation is small, such that $\frac{\partial h}{\partial x} \ll 1$, and set the denominator to one. In dimensionless form, we then have

$$\mathbf{n} = \mathbf{e}_z - \frac{\partial h^*}{\partial x^*}\mathbf{e}_x \tag{8.151}$$

with $x^* = x/\delta_c$.

The stability analysis begins by writing a formal expansion for the concentration field and interface position:

$$C_\ell^* = C_\ell^{0*} + C_\ell^{1*} \tag{8.152}$$

$$h^* = h^{0*} + h^{1*} \tag{8.153}$$

8.3 SOLIDIFICATION MICROSTRUCTURE DEVELOPMENT

Note that we have already found the basic state solution C_ℓ^{0*}, and $h^{0*} = 0$ because the interface is flat and stationary in the basic state. Also, because the scaling analysis eliminated the transient terms from the energy equation, the temperature field is unperturbed, and $\theta = z^*$. For convenience of notation we will use h^* for the scaled interface position in the perturbed state. The scaled equation for the liquid concentration is determined in the same way as for the basic state, but the equation is now two dimensional:

$$\frac{\partial C_\ell^*}{\partial t} - \frac{\partial C_\ell^*}{\partial z^*} = \nabla^2 C_\ell^* \tag{8.154}$$

Substituting Eq. (8.152) for C_ℓ^* yields

$$\left(\frac{\partial C_\ell^{0*}}{\partial t^*} + \frac{\partial C_\ell^{1*}}{\partial t^*} \right) - \left(\frac{\partial C_\ell^{0*}}{\partial z^*} + \frac{\partial C_\ell^{1*}}{\partial z^*} \right) = \nabla^2 C_\ell^{0*} + \nabla^2 C_\ell^{1*} \tag{8.155}$$

To simplify this expression, note that the terms involving C_ℓ^{0*} equate to zero because the basic state (also called the zero-order solution) satisfies

$$\frac{\partial C_\ell^{0*}}{\partial t^*} - \frac{\partial C_\ell^{0*}}{\partial z^*} = \nabla^2 C_\ell^{0*} \tag{8.156}$$

This leaves

$$\frac{\partial C_\ell^{1*}}{\partial t^*} - \frac{\partial C_\ell^{1*}}{\partial z^*} = \nabla^2 C_\ell^{1*} \tag{8.157}$$

We will also need boundary conditions for C_ℓ^{1*}. As $z^* \to \infty$, $C_\ell^* \to 1$. However, the zero-order solution also gives $C_\ell^{0*} \to 1$ as $z^* \to \infty$, so we require

$$C_\ell^{1*}(\infty) = 0 \tag{8.158}$$

Boundary conditions on the interface are more complicated to derive. We intend to linearize the boundary condition, and to this end we write a Taylor series for the concentration on $z^* = h^*$, truncating at the first term:

$$C_\ell^*(h^*) \approx C_\ell^*(0) + h^* \left. \frac{\partial C_\ell^*}{\partial z^*} \right|_{z^*=0} \tag{8.159}$$

Here the gradient of C_ℓ^* is evaluated at the unperturbed interface, $z^* = 0$. Substituting Eq. (8.152) for C_ℓ^* gives

$$C_\ell^*(h^*) = C_\ell^{0*}(0) + C_\ell^{1*}(0) + h^* \left(\left. \frac{\partial C_\ell^{0*}}{\partial z^*} \right|_{z^*=0} + \left. \frac{\partial C_\ell^{1*}}{\partial z^*} \right|_{z^*=0} \right) \tag{8.160}$$

Now, by our choice of scales, $C_\ell^{0*} = 0$. We also linearize the boundary condition by assuming that products of small perturbations (such as $h^* \partial C_\ell^{1*}/\partial z^*$) are negligible. This leaves

$$C_\ell^*(h^*) = C_\ell^{1*}(0) + h^* \tag{8.161}$$

The important thing about this expression is that C_ℓ^{1*} is evaluated at $z^* = 0$, rather than $z^* = h^*$, which makes the solution of the eventual differential equation tractable. This procedure is sometimes called "transferring the boundary condition" to the undisturbed interface, but this transfer is really a consequence of linearizing the boundary condition.

We now obtain the linearized form of the Gibbs–Thomson condition. In dimensional form we have, on $z = h$,

$$T^i = T_0 + m_\ell C_\ell(h) - 2\hat{\Gamma}\kappa_m \tag{8.162}$$

where κ_m is the mean curvature of the interface. Introducing the dimensionless variables, we find that this equation becomes

$$\theta^i = -\frac{m_\ell C_0(1-k_0)V}{G_T D_\ell k_0} C_\ell^*(h^*) - \frac{2\hat{\Gamma} V^2}{G_T D_\ell^2}\kappa_m^* \tag{8.163}$$

where $\kappa_m^* = \delta_c \kappa_m$. We recognize the first dimensionless group from our earlier analysis of morphological stability. We call this group the *morphological number*, M:

$$M = -\frac{m_\ell C_0(1-k_0)V}{G_T D_\ell k_0} \tag{8.164}$$

Note that $M \geq 0$ for any system. If $k_0 < 1$ then $m_\ell < 0$, and if $k_0 > 1$ then $m_\ell > 0$, so M is always positive. The second dimensionless group will be carried as is a little while longer. To simplify Eq. (8.163), first recall that the temperature follows the frozen temperature approximation, so that $\theta^i = z^* = h^*$. We also substitute $C_\ell^*(h^*)$ from Eq. (8.161). Finally, for small disturbances in two dimensions,

$$2\kappa_m^* = -\frac{d^2 h^*}{dx^{*2}} \tag{8.165}$$

Substituting all of these expressions into Eq. (8.163) yields

$$h^* = Mh^* + MC_\ell^{1*} + \frac{\hat{\Gamma} V^2}{G_T D_\ell^2}\frac{d^2 h^*}{dx^{*2}} \tag{8.166}$$

After some rearrangement of terms, we have

$$C_\ell^{1*} = \left(\frac{1}{M} - 1\right)h^* - \frac{\hat{\Gamma} V^2}{M G_T D_\ell^2}\frac{d^2 h^*}{dx^{*2}} \tag{8.167}$$

The last dimensionless group is given the symbol S, because it is a measure of the importance of surface tension. We have

$$S = \frac{\hat{\Gamma} V^2}{M G_T D_\ell^2} = -\frac{\hat{\Gamma} V k_0}{D_\ell m_\ell C_0(1-k_0)} \tag{8.168}$$

and we write the perturbed version of the Gibbs–Thomson condition as

$$C_\ell^{1*}(0) = \left(\frac{1}{M} - 1 - S\frac{d^2}{dx^{*2}}\right)h^* \tag{8.169}$$

The linearization of the flux condition proceeds in exactly the same manner as the Gibbs–Thomson equation. We therefore leave it as an exercise for the reader to

8.3 SOLIDIFICATION MICROSTRUCTURE DEVELOPMENT

show that

$$\frac{\partial C_\ell^{1*}(0)}{\partial z^*} = (k_0 - 1)C_\ell^{1*}(0) + k_0 h^* + \frac{\partial h^*}{\partial t^*} \tag{8.170}$$

Notice that the evaluation of C_ℓ^{1*} is again at $z^* = 0$.

With the differential equation and boundary conditions in hand, we may now find the solution for the perturbed concentration field and interface position. We seek normal mode solutions of the form

$$C_\ell^{1*} = A(z^*) \exp\{\sigma t^* + inx\} \tag{8.171}$$

$$h^* = H \exp\{\sigma t^* + inx\} \tag{8.172}$$

where n, the wave number, governs the spacing of the perturbations. We will be most interested in determining the conditions under which σ takes on positive and negative values, and we will associate these cases with instability and stability, respectively.

Substituting the normal mode forms into the differential equation for C_ℓ^{1*}, Eq. (8.157), yields an ordinary differential equation for $A(z^*)$:

$$\frac{d^2 A}{dz^{*2}} + \frac{dA}{dz^*} - (\sigma + n^2) = 0 \tag{8.173}$$

This has the solution

$$C_\ell^{1*} = \left\{ B_1 \exp\left[\left(-\frac{1}{2} + \frac{1}{2}\sqrt{1 + 4(\sigma + n^2)}\right) z^*\right] \right.$$
$$\left. + B_2 \exp\left[\left(-\frac{1}{2} - \frac{1}{2}\sqrt{1 + 4(\sigma + n^2)}\right) z^*\right] + B_3 \right\} \exp\{\sigma t * + inx\} \tag{8.174}$$

The constants $B_1 - B_3$ are evaluated by using the boundary conditions. As $z^* \to \infty$, we have $C_\ell^{1*} = 0$. Clearly the first term can grow unboundedly for some values of n, regardless of the sign of σ, so we require that $B_1 = 0$. The second term vanishes at ∞ for all values of n, so we do not learn anything about B_2. Finally, B_3 must be zero in order to satisfy the condition $C_\ell^{1*} = 0$. Thus, we are left with the following solution for the perturbed concentration:

$$C_\ell^{1*} = B_2 \exp\left[\left(-\frac{1}{2} - \frac{1}{2}\sqrt{1 + 4(\sigma + n^2)}\right) z^*\right] \exp\{\sigma t * + inx\} \tag{8.175}$$

We now consider the two boundary conditions on the interface. Note that the linearization of the boundary conditions "transferred" them to $z^* = 0$, where the first spatial exponential term is one. Substituting the perturbed concentration field from Eq. (8.175) into the Gibbs–Thomson condition of Eq. (8.169) gives

$$B_2 - \left(\frac{1}{M} - 1 + Sn^2\right) H = 0 \tag{8.176}$$

Similarly, substituting Eq. (8.175) into the flux condition Eq. (8.170) yields

$$\left[\frac{1}{2}\left(1 - \sqrt{1 + 4(\sigma + n^2)}\right) - k_0\right] B_2 - (k_0 + \sigma)H = 0 \tag{8.177}$$

Equations (8.176) and (8.177) are two simultaneous, linear, homogeneous equations for B_2 and H. Therefore, a nontrivial solution exists if and only if the determinant of the coefficient matrix vanishes. This leads to the following expression:

$$\sqrt{1+4(\sigma+n^2)} = 1 - 2k_0 - \frac{2(k_0+\sigma)}{\frac{1}{M}-1+Sn^2} \tag{8.178}$$

Equation (8.178) contains the information we are looking for, but it is not very clear. To improve the situation, we consider the state of marginal stability, where $\sigma = 0$. In that case, Eq. (8.178) reduces, after some rearrangement of terms, to

$$\frac{1}{M} = 1 - Sn^2 + \frac{2k_0}{1 - 2k_0 - \sqrt{1+4n^2}} \tag{8.179}$$

Figure 8.19 shows a graph of $1/M$ versus n for various values of S, with $k_0 = 0.1$. It is easy to confirm that stability ($\sigma < 0$) corresponds to the region above the neutral stability curve, and that instability corresponds to combinations of $1/M$ and n that fall below the curve. For $S = 0$, the limit of vanishing surface tension, we see that there are possible unstable states for all wave numbers, so that to ensure stability of the interface, we require $1/M > 1$. This is precisely the result we obtained earlier by using the constitutional supercooling argument.

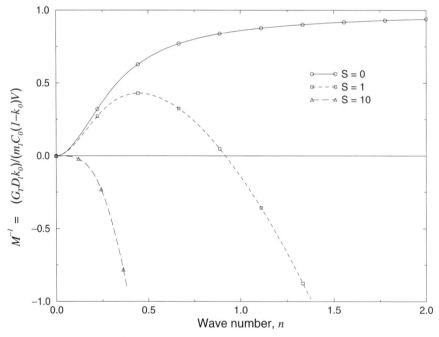

Figure 8.19: Neutral stability curves for plane front growth at different values of S, with $k_0 = 0.1$. Values of $1/M$ above the curve are stable; values below are unstable. Negative values of $1/M$ are physically unattainable.

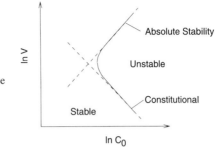

Figure 8.20: Neutral stability curve in V–C_0 space (Merchant and Davis, 1990).

For an intermediate value of S, say $S = 1$, there is a range of unstable wave numbers between zero and a short wavelength limit, n_s. The most dangerous mode is at the maximum, n_c. This wave number is unstable at the largest value of $1/M$ and is therefore the most likely wave number to be observed. Thus, the presence of surface tension adds a length scale selection to the stability problem and tells us the spacing between the cells as cellular growth begins. Finally, we see from the figure that at sufficiently large values of S, say $S = 10$, there are no wave numbers for which an instability can occur. This condition is called *absolute stability*, and it is observed at very high growth speeds. (Recall that S is proportional to the growth velocity V.) This latter condition is observed only at extremely high solidification rates, as found in laser processing and rapid quenching of small droplets. This phenomenon also cannot be predicted without surface tension in the model.

M and S are convenient parameters for the stability analysis, but they are not entirely satisfactory on a physical basis. It would be more useful to have a stability diagram in terms of processing parameters, such as V and C_0. Merchant and Davis (1990) have performed such an analysis by rescaling the system using capillary length scales, with the result shown schematically in Fig. 8.20. Here we see that the interface is always stable for low enough concentration C_0. Above this limit there is a range of unstable velocities for each composition. The lower stability curve represents long wavelength disturbances, and it corresponds to the constitutional supercooling limit; the upper curve corresponds to absolute stability, or stability imposed by surface tension on short wavelength disturbances.

As a final note, the linear stability analysis shown here is useful for demonstrating the importance of surface tension in pattern selection, but there is much more to the analysis of microstructure. For example, the linear analysis is not valid for finite-size disturbances, and a more complete description would require a full nonlinear analysis. Interested readers are referred to the treatise by Davis (1999).

8.4 SUMMARY

This chapter has examined problems related to species transport in materials processing. For dilute systems, the governing equations for mass transport have the same form as the governing equations for energy transport, and we can make an analogy between diffusion problems and heat conduction problems. Many familiar solutions for transient heat conduction, such as the error function solution for a semi-infinite body, can be applied directly to the diffusion of a solute in a solid. When two species are interdiffusing the governing equations have the same general form, but the observed diffusion coefficient for either species is no longer equal to its intrinsic diffusion coefficient.

An important difference between heat conduction and molecular diffusion arises from the disparity in length and time scales. For metals, thermal diffusivity in either the liquid or the solid is typically 10^4 times greater than molecular diffusivity in the liquid, which in turn is approximately 10^4 times greater than diffusivity in the solid. This leads to a difference in length and time scales that is important to the microstructure of metals. Diffusion is often incomplete in the liquid and negligible in the solid.

The equilibrium phase diagram is also important in determining solidified microstructures. At any given temperature, liquid at one composition is in equilibrium with solid at another composition. This causes solute (the minor component) to be rejected at the solid–liquid interface, at a rate proportional to the solidification rate. Using information from the phase diagram and some assumptions, one can develop simple models that relate the mass fraction of solid in a sample to its temperature. These relationships, the inverse lever rule and the Scheil equation, represent two extreme cases that bound most real situations.

More detailed analyses of solidification combine information from the phase diagram with models of solute diffusion near the solid–liquid interface. These calculations give a maximum solidification speed at which a planar interface will be stable, called the *constitutional supercooling limit*. This limit determines the maximum rate at which semiconductor crystals can be grown. At higher solidification speeds the interface first forms cells or fingers, and then dendrites, causing the solute to be distributed nonuniformly in the solid, an effect known as *segregation*. A linear stability analysis of the plane interface can give the spacing of the cells, but the prediction of dendrite shapes and spacing is a challenging problem that is still the subject of much current research.

BIBLIOGRAPHY

T. F. Bower, H. D. Brody, and M. C. Flemings. Measurement of solute redistribution in dendritic solidification. *Trans. AIME* 236:624, 1966.

W. D. Callister. *Materials Science and Engineering*. Wiley, New York, 4th edition, 1997.

S. H. Davis. *Theory of Solidification*. Cambridge University Press, Cambridge, 1999.

M. C. Flemings. *Solidification Processing*. McGraw-Hill, New York, 1974.

M. E. Glicksman. *Diffusion in Solids*. Wiley Interscience, New York, 2000.

W. Kurz and D. J. Fisher. *Fundamentals of Solidification*. Trans Tech Publications, Aedermannsdorf, Switzerland, 3rd edition, 1989.

G. J. Merchant and S. H. Davis. Morphological instability in rapid directional solidification. *Acta Metall. Mater.* 38:2683, 1990.

W. W. Mullins and R. F. Sekerka. Stability of a planar interface during solidification of a binary alloy. *J. Appl. Phys.* 35:444–451, 1964.

D. R. Poirier and G. H. Geiger. *Transport Phenomena in Materials Processing*. TMS–AIME Warrendale, PA, 1994.

H. F. Weinberger. *A First Course in Partial Differential Equations*. Xerox, Lexington, MA, 1965.

EXERCISES

1. **Self-diffusion coefficient.** Self-diffusion coefficients are often measured by using radioactive tracers. In this technique, a thin layer of radioactive isotope is plated onto one end of a long cylindrical rod. The original thickness of the layer is $2\delta_0$. A second rod is then placed in contact with the surface of the plating, forming a geometry like that shown in Fig. 8.2. After heat treatment at a prescribed temperature for a prescribed time, the couple is sectioned and the concentration of the radioactive isotopes is measured. What should the functional form for the concentration be? How could you determine the diffusion coefficient from this distribution?

2. **Concentrations in a weldment.** Two pieces of 1020 steel (0.2 wt. %C) are butt-welded together by using a steel filler rod containing 0.8 wt. %C. The weld is 10 mm wide. After welding, the assembly is heated to 600°C for 8 h to promote diffusion of the two different compositions. Assume that the initial carbon distribution has the form shown in Fig. 8.2, and that the diffusion coefficient satisfies $D_0 = 6.2 \times 10^{-7}$ m²/s and $E_D = 80$ kJ/mol K. Compute and plot the carbon content in the welded slab after the heat treatment. What can you conclude about the effectiveness of such a heat treatment?

3. **Carburization of a gear surface.** We wish to carburize a gear made of 1020 steel to a carbon content of 0.4% at a depth of 0.5 mm below the surface. We will place the gear in a furnace at 900°C where the atmosphere is controlled to make the surface concentration stay at 0.9% C. The diffusion coefficient parameters are $D_0 = 2 \times 10^{-5}$ m²/s and $E_D = 142$ kJ/mol K.

    (a) How long will the process take at 900°C?

    (b) Determine a temperature at which the process would be completed in half of the time.

4. **Continuous carburization.** A plain carbon steel plate is passed through a carburizing furnace to produce a hardenable layer on its surface, as shown below. The initial concentration of carbon in the plate is uniform at 0.3% C; that is, 1030 steel is used. The furnace temperature is 1200°C; and the furnace atmosphere is controlled such that the surface of the plate is maintained at 1.2% C. The plate moves through the furnace at constant velocity V.

    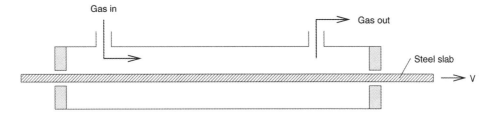

    (a) Write the differential equation and boundary conditions that govern the concentration distribution within the steel plate. Reduce them by making appropriate assumptions. Write the boundary conditions at the upper and lower surfaces of the plate.

(b) Scale the equations, assuming that the plate has thickness $2H$ and that the furnace has length L. What dimensionless group appears?

(c) Analyze the case in which the plate is 0.02 m thick and travels at a constant velocity of $V = 0.015$ m/s, and the furnace is 2 m long. Show that the carburization problem can be treated as a pseudotime problem under these conditions, that is, that axial diffusion can be neglected (see Section 7.1.1). Define a pseudotime variable and solve for the concentration through the thickness of the slab along the length of the furnace. Plot the concentration through the thickness at the furnace exit.

5. **Solidification microstructures in steel.** A series of experiments was performed where the cooling rate through the freezing range of an Fe–0.1% C alloy was systematically varied. The secondary dendrite arm spacing λ_2 was measured in each experiment, and the following empirical relationship between λ_2 and the local solidification time t_f was determined by using data taken over over a range of solidification times from 3 s to 1000 s:

$$\lambda_2 = 37.6 t_f^{0.384}$$

Here λ_2 is in micrometers and t_f is in seconds. The relevant portion of the Fe–C phase diagram is shown below.

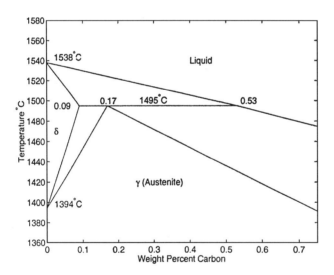

(a) Determine the values of the dimensionless groups in Eqs. (8.84)–(8.86) for this alloy, using the following values:

$$\alpha = 6 \times 10^{-6}\,\text{m}^2/\text{s}, \quad D_L = 2 \times 10^{-8}\,\text{m}^2/\text{s}, \quad D_s = 6 \times 10^{-9}\,\text{m}^2/\text{s}$$

(b) From these calculations, would you expect Fe–C alloys to follow the lever rule or the Scheil equation under normal conditions? Plot the appropriate f_s–T curve (use $k_0 = 0.19$).

6. **Solidification microstructures in Al–Si alloys.** The Al-rich corner of the Al–Si phase diagram is shown below. Compute and plot the f_s–T curve for an alloy

of 1.4% Si, using both the Scheil equation and lever rule. How much eutectic is predicted for each case?

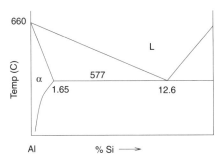

7 **Homogenization.** In a dendritic structure, such as that shown in Fig. 8.10, the result of nonequilibrium solidification is to produce an inhomogeneous concentration distribution across the dendrite arm. The concentration distribution may be represented by the Gulliver–Scheil equation, but in this problem we simplify this form and assume that the concentration distribution at the end of solidification is given by

$$C = C_0 - \Delta C_0 \cos \frac{2\pi x}{\lambda_2}$$

where C_0 is the mean alloy concentration and ΔC_0 is the initial difference between the maximum and minimum concentrations. It is often desirable to homogenize the material after solidification by raising the temperature and allowing diffusion to smooth out the concentration distribution.

(a) Write the differential equations and boundary conditions that govern the homogenization process.

(b) Assume that the structure is two dimensional and that $\lambda_2/L_d \ll 1$. Scale and reduce the governing equations appropriately. This should leave you with a one-dimensional diffusion problem with periodic boundary conditions.

(c) Solve the differential equation. Hint: Look for a separation of variables solution.

(d) The *residual segregation index δ* is defined as

$$\delta = \frac{C_{max} - C_{min}}{\Delta C_0}$$

Compute the residual segregation index from your solution. At what time will the segregation index equal 0.01, that is, the total segregation is 1% of its original value?

(e) How would you solve this problem if you wanted to use the Gulliver–Scheil equation as a measure of the original concentration profile?

8 **Solidification with diffusion and convection.** Our solution for the concentration profile ahead of a plane solidification front assumed that there was no convection in the liquid. However, in some cases the effects of convection

cannot be ignored. One rather practical approach to dealing with this situation is described in this problem.

One assumes that the steady-state mass transport equation still governs the solute distribution ahead of the interface, so we still have

$$-V\frac{\partial C_\ell}{\partial z} = D_\ell \frac{\partial^2 C_\ell}{\partial z^2}$$

As before, the solution to this equation is

$$C_\ell = c_1 + c_2 \exp\left(-\frac{Vz}{D_\ell}\right)$$

To account for convection, the solutal boundary layer is assumed to have a thickness δ, where δ is a known quantity. Specifically, one uses the boundary condition $C_\ell = C_0$ at $x = \delta$. Here C_0 is the liquid composition far away from the interface. This particular assumption is a bit questionable, because it requires C_ℓ to fall below C_0 for $z \gg \delta$. Perhaps a better way to regard this assumption is to say that it gives a realistic result close to the interface (closer than δ) by contracting the boundary layer in a way that is consistent with the diffusion solution. Once one has made this assumption, the rest of the development follows logically.

(a) Using the solution for C_ℓ and the boundary condition given above, plus the boundary condition for solute rejection at the interface, determine the constants c_1 and c_2.

(b) Find the liquid and solid concentrations at the interface.

(c) Show that the ratio between the solid concentration and C_0 is given by

$$k' = \frac{k_0}{k_0 + (1 - k_0)\exp(-V\delta/D_\ell)}$$

The quantity k' is called the *nonequilibrium partition coefficient*. It is sometimes used to model solute distributions when both convection and diffusion are important in the liquid.

9 *Use of the nonequilibrium partition coefficient.* One may treat plane front solidification problems that have both diffusion and convection by adapting the analysis we used to derive the Scheil equation. It is necessary to neglect any transient effects; then the result is identical to the Scheil equation except for the substitution of k' for k_0. k' is the nonequilibrium partition coefficient, which is defined in the previous problem.

(a) Use this rule to calculate solute concentrations as a function of fraction solid for the cases of $\delta = 0$, $\delta = 1.7(D_\ell/V)$, and $\delta = \infty$. Assume $C_0 = 2$ wt.%. Graph all three solutions on one graph. Take the maximum value of C_s to be 10 wt.% and use $k_0 = 0.3$.

(b) Evaluate k' for the two limiting cases of $\delta = \infty$ and $\delta = 0$. Show that $\delta = \infty$ is identical to the problem solved in the text for diffusion with no convection, at least as far as the concentration of the solid that forms at steady state. Similarly, show that $\delta = 0$ is identical to the Scheil equation, which assumes a perfectly mixed liquid.

10 **Derivation of linearized flux condition (advanced).** Linearize the flux condition to show that Eq. (8.170) is correct. First scale the flux condition,

$$C_\ell(1 - k_0)\mathbf{v} \cdot \hat{\mathbf{n}} = -D_\ell \nabla C_\ell \cdot \hat{\mathbf{n}}$$

using $z* = z/\delta_c$. Then introduce the perturbation expansion for C_ℓ^* to show that the linearized form of the boundary condition becomes

$$\frac{\partial C_\ell^{1*}}{\partial z^*} = \frac{\partial h^*}{\partial t^*} + k_0 h^* - (1 - k_0)C_\ell^{1*}$$

evaluated at $z^* = 0$.

To solve this problem, you should follow the procedure given in the chapter where the linearized Gibbs–Thomson condition is derived. That is, expand each expression (C_ℓ^*, $\partial C_\ell^*/\partial z^*$, etc.) in a Taylor series around $z^* = 0$, and neglect terms containing products of the perturbations.

APPENDIX A

Mathematical Background

This appendix summarizes some important mathematical background. We first cover vector and tensor algebra and the relationship between different styles of notation. Other topics include the divergence theorem, the curvature of curves and surfaces, and the Gaussian error function. The material is intended primarily for reference purposes, so no derivations or proofs are provided. Readers desiring more background can consult the bibliography at the end of this appendix.

In this appendix the symbols **u**, **v**, and **w** represent arbitrary vectors, and σ and τ represent arbitrary tensors.

A.1 SCALARS, VECTORS, AND TENSORS: DEFINITIONS AND NOTATION

A.1.1 SCALARS AND VECTORS

A *scalar* is a quantity that has magnitude but no direction. Physical quantities that are scalars include temperature, pressure, and density. Scalar quantities are denoted by lightface italic symbols: T, p, ρ.

A *vector* is a quantity having both magnitude and direction. We frequently visualize a vector as an arrow in space: the direction of the arrow gives the direction of the vector, and the length of the arrow gives the vector's magnitude. Vector quantities include velocity, surface traction, and heat flux.

This text uses several different notations for vectors. In one of these, called *Gibbs notation*, the vector is denoted by a boldface letter: **v**, **t**, **q**. The bold letter represents the entire vector, and this notation is not tied to any particular choice of coordinate system. A convenient notation for the blackboard or handwritten work is to underline letters that represent vectors.

A *unit vector* is a vector of unit length. In any orthogonal coordinate system, the coordinate directions at any point in space are defined by three mutually perpendicular unit vectors. For Cartesian, or rectangular, coordinates we will call the unit vectors $\mathbf{e}_x, \mathbf{e}_y, \mathbf{e}_z$ or, alternately, $\mathbf{e}_1, \mathbf{e}_2, \mathbf{e}_3$. These are shown in Fig. A.1. The coordinates are either (x, y, z) or (x_1, x_2, x_3). Both notations refer to the same coordinate system, and we may choose whichever is convenient for the problem at hand.

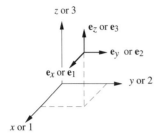

Figure A.1: Unit vectors in Cartesian (rectangular) coordinates.

In any given coordinate system, one can express a vector in terms of its components:

$$\mathbf{v} = v_x \mathbf{e}_x + v_y \mathbf{e}_y + v_z \mathbf{e}_z$$
$$= v_1 \mathbf{e}_1 + v_2 \mathbf{e}_2 + v_3 \mathbf{e}_3 \qquad (A.1)$$

The components, either (v_x, v_y, v_z) or (v_1, v_2, v_3), are the projections of the vector $\mathbf{v}$ along each unit vector. Each component of a vector is a scalar. If a different coordinate system is chosen then the components of the vector will be different in the new coordinate system (see Section A.1.4).

A.1.2 TENSORS

A vector associates a scalar component with each coordinate direction. A *tensor* associates a scalar component with each ordered pair of coordinate directions. (Strictly speaking, these are second-order tensors. Tensors of higher orders exist as well.) In addition, a quantity must obey the coordinate transformation rule discussed in Section A.1.4 to be a tensor. Physical quantities that are tensors include stress, rate of deformation, and thermal conductivity.

Tensors will also be denoted by boldface letters: $\boldsymbol{\sigma}$, $\mathbf{D}$, $\mathbf{k}$. For the blackboard or handwritten work, symbols representing tensors can be doubly underlined. It is customary to write a tensor as the matrix of its components:

$$\boldsymbol{\tau} = \begin{bmatrix} \tau_{11} & \tau_{12} & \tau_{13} \\ \tau_{21} & \tau_{22} & \tau_{23} \\ \tau_{31} & \tau_{32} & \tau_{33} \end{bmatrix} \qquad (A.2)$$

The *unit dyads* play the same role for tensors that the unit vectors play for vectors. A unit dyad is an ordered pair of unit vectors, such as $\mathbf{e}_1\mathbf{e}_1$, $\mathbf{e}_1\mathbf{e}_2$, etc. [Actually these are dyadic products; see Eq. (A.22)]. The order of the vectors in a unit dyad is significant: $\mathbf{e}_1\mathbf{e}_2$ is different from $\mathbf{e}_2\mathbf{e}_1$. Writing out the tensor in terms of its unit dyads gives

$$\begin{aligned}\boldsymbol{\tau} = &\ \tau_{11}\mathbf{e}_1\mathbf{e}_1 + \tau_{12}\mathbf{e}_1\mathbf{e}_2 + \tau_{13}\mathbf{e}_1\mathbf{e}_3 \\ &+ \tau_{21}\mathbf{e}_2\mathbf{e}_1 + \tau_{22}\mathbf{e}_2\mathbf{e}_2 + \tau_{23}\mathbf{e}_2\mathbf{e}_3 \\ &+ \tau_{31}\mathbf{e}_3\mathbf{e}_1 + \tau_{32}\mathbf{e}_3\mathbf{e}_2 + \tau_{33}\mathbf{e}_3\mathbf{e}_3\end{aligned} \qquad (A.3)$$

A.1.3 INDICIAL NOTATION

Another style of notation for vectors and tensors is *indicial notation*, also called Cartesian tensor notation. Indicial notation allows expressions and equations to be

A.1 SCALARS, VECTORS, AND TENSORS: DEFINITIONS AND NOTATION

written very compactly, provided one is using a Cartesian coordinate system. Any subscript on a vector or tensor component may be replaced by a lower-case letter, such as i, j, or k. These letters stand for the indices 1, 2, or 3, and they are interpreted according to the following rules.

1. If an index is repeated in an expression, then the expression is summed over all possible values of the index. For example, $u_i v_i$ is indicial notation for $\sum_{i=1}^{3} u_i v_i$, and τ_{kk} means $\sum_{k=1}^{3} \tau_{kk}$. This rule is called the *summation convention*.
2. If a letter subscript in an equation is not repeated, then it is a *dummy index*, and one of two interpretations is possible. One may regard the equation as a set of scalar equations, each equation being generated by assigning a particular value to the dummy index. Thus, $u_i = v_i$ implies the three equations $u_1 = v_1$, $u_2 = v_2$, and $u_3 = v_3$. (The summation convention does not apply in this case because the two i's are on opposite sides of the equation.)

    Alternately, one may create a vector or tensor equation by multiplying both sides of the equation by the corresponding unit vectors and summing over the dummy indices. That is, $u_i = v_i$ means $\mathbf{u} = \mathbf{v}$. With this in mind, we can write Eqs. (A.1) and (A.3) as

$$\mathbf{v} = v_i$$
$$\boldsymbol{\tau} = \tau_{ij} \tag{A.4}$$

These equations establish the correspondence between indicial notation and Gibbs notation.

A third rule of indicial notation, used in some texts, is that a subscript preceded by a comma denotes partial differentiation with respect to the corresponding spatial coordinate. In this style, $v_{i,j}$ means $\partial v_i / \partial x_j$. Also, $v_{i,t}$ indicates a partial time derivative, $\partial v_i / \partial t$. We do not use this feature of indicial notation in this text.

A.1.4 COORDINATE TRANSFORMATION

Although we commonly deal with vectors and tensors in terms of their components in some coordinate system, vectors and tensors do not rely on the coordinate system for their existence. Consequently, one can describe the same vector or tensor by its components in some other coordinate system. The task of computing the components in one coordinate system from the components in another system is called *coordinate transformation*. Here we consider only transformation between two Cartesian coordinate systems. Transformation between general coordinate systems is discussed in a variety of texts (e.g., Aris, 1962).

Consider the two coordinate systems shown in Fig. A.2. We know the components v_i of the vector $\mathbf{v}$ in the 1, 2, 3 system, and we wish to compute its components v'_k in the $1', 2', 3'$ system. First we define the *transformation tensor* $\mathbf{Q}$, whose component Q_{ik} is the cosine of the angle between the unit vectors $\mathbf{e}'_i$ and $\mathbf{e}_k$. Because of the properties of the dot product we could define $\mathbf{Q}$ by using

$$Q_{ik} = \mathbf{e}'_i \cdot \mathbf{e}_k \tag{A.5}$$

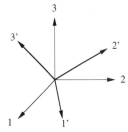

Figure A.2: Two Cartesian coordinate systems.

Here i takes on the values $1', 2', 3'$, while k takes on the values $1, 2, 3$. The transformation rule for vectors is

$$v'_i = Q_{ik} v_k \tag{A.6}$$

where the summation convention has been used. The transformation rule for tensors is similar:

$$\tau'_{ij} = Q_{ik} \tau_{kl} Q_{jl} \tag{A.7}$$

The transformation tensor **Q** is *orthogonal*, which is to say that its inverse equals its transpose; see Eq. (A.42). Some authors (e.g., Aris, 1962) use these transformation rules as the starting point to define vectors and tensors.

A.1.5 THE UNIT TENSOR AND THE KRONECKER DELTA

The tensor whose matrix of components has ones on the diagonal and zeros elsewhere is called the *unit tensor* **I**, or the *identity tensor*:

$$\mathbf{I} = \begin{bmatrix} 1 & 0 & 0 \\ 0 & 1 & 0 \\ 0 & 0 & 1 \end{bmatrix} \tag{A.8}$$

In indicial notation the unit tensor is written as δ_{ij}. This is consistent with the use of δ_{ij} to denote the Kronecker delta:

$$\delta_{ij} = \begin{cases} 1 & \text{if } i = j \\ 0 & \text{if } i \neq j \end{cases} \tag{A.9}$$

The components of the unit tensor are the same in any orthogonal coordinate system; thus we say that the unit tensor is *isotropic*.

A.1.6 TRANSPOSES, SYMMETRY, AND ANTISYMMETRY

The *transpose* of a tensor has the same meaning as a matrix transpose of the tensor components. We will denote a transpose by a superscript T. We can write out the transpose of τ as

$$\tau^T = \begin{bmatrix} \tau_{11} & \tau_{21} & \tau_{31} \\ \tau_{12} & \tau_{22} & \tau_{32} \\ \tau_{13} & \tau_{23} & \tau_{33} \end{bmatrix} \tag{A.10}$$

Compare this to Eq. (A.3). In indicial notation this is simply

$$(\tau_{ij})^T = \tau_{ji} \tag{A.11}$$

If a tensor is equal to its transpose, that is if

$$\tau_{ij} = \tau_{ji} \tag{A.12}$$

then the tensor is *symmetric*. A tensor that equals the negative of its transpose,

$$\tau_{ij} = -\tau_{ji} \tag{A.13}$$

is *antisymmetric* or *skew symmetric*. The diagonal components of an antisymmetric tensor always equal zero.

A.2 VECTOR AND TENSOR ALGEBRA

The major operations in vector and tensor algebra are given here. Each operation is "defined" by writing Gibbs notation on the left-hand side of the equation and indicial notation on the right. Remember that the summation convention applies in indicial notation.

A.2.1 ADDITION AND SUBTRACTION

Vectors and tensors are added (or subtracted) by adding (or subtracting) corresponding pairs of components:

$$\mathbf{u} + \mathbf{v} = u_i + v_i \tag{A.14}$$
$$\boldsymbol{\tau} + \boldsymbol{\sigma} = \tau_{ij} + \sigma_{ij} \tag{A.15}$$

All components must refer to the same coordinate system.

A.2.2 MULTIPLICATION

There are many different ways to multiply vectors and tensors together, each with a different physical interpretation. To multiply a vector or tensor by a scalar s, simply multiply each component by s:

$$s\mathbf{u} = su_i \tag{A.16}$$
$$s\boldsymbol{\tau} = s\tau_{ij} \tag{A.17}$$

Scalar multiplication changes the magnitude of a vector or tensor without affecting its directionality.

The *dot product* of two vectors is given by

$$\mathbf{u} \cdot \mathbf{v} = u_i v_i \tag{A.18}$$

The dot product is a scalar quantity, equal to the product of magnitude of $\mathbf{u}$, the magnitude of $\mathbf{v}$, and the cosine of the angle between them. If $\mathbf{u}$ is a unit vector then

($\mathbf{v} \cdot \mathbf{u}$) is the component of $\mathbf{v}$ in the direction of $\mathbf{u}$. If two vectors are perpendicular to one another, their dot product equals zero.

The *cross product* of two vectors is found from

$$\mathbf{u} \times \mathbf{v} = u_i v_j \epsilon_{ijk} \tag{A.19}$$

This equation uses the *permutation symbol*, which is defined such that

$$\epsilon_{ijk} = \begin{cases} 1 & \text{if } ijk = 123, 231, 312 \\ -1 & \text{if } ijk = 321, 213, 132 \\ 0 & \text{otherwise} \end{cases} \tag{A.20}$$

The cross product of two vectors is a third vector, perpendicular to both of the original vectors, whose length equals the area of the parallelogram defined by the original vectors.

The scalar triple product of three vectors,

$$\mathbf{u} \cdot [\mathbf{v} \times \mathbf{w}] = u_i v_j w_k \epsilon_{ijk} \tag{A.21}$$

is the volume of a parallelepiped defined by the vectors $\mathbf{u}$, $\mathbf{v}$, and $\mathbf{w}$.

The *dyadic product* of two vectors is a second-order tensor:

$$\mathbf{uv} = u_i v_j \tag{A.22}$$

The *double dot product* of two tensors is a scalar:

$$\boldsymbol{\tau} : \boldsymbol{\sigma} = \tau_{ij} \sigma_{ji} \tag{A.23}$$

The *dot product* (or single-dot product) of two tensors is another tensor.

$$\boldsymbol{\tau} \cdot \boldsymbol{\sigma} = \tau_{ik} \sigma_{kj} \tag{A.24}$$

Note in Eqs. (A.23) and (A.24) that each dot collects one index from each operand for summation. The single-dot product is carried out exactly like the matrix multiplication of the two tensors. It is customary to abbreviate $(\boldsymbol{\tau} \cdot \boldsymbol{\tau})$ as $\boldsymbol{\tau}^2$, $(\boldsymbol{\tau}^2 \cdot \boldsymbol{\tau})$ as $\boldsymbol{\tau}^3$, and so forth.

Taking the dot product of any tensor with the unit tensor gives the original tensor:

$$\boldsymbol{\tau} \cdot \mathbf{I} = \mathbf{I} \cdot \boldsymbol{\tau} = \boldsymbol{\tau} \tag{A.25}$$

There are two ways to make a *vector–tensor dot product*:

$$\begin{aligned} \boldsymbol{\tau} \cdot \mathbf{v} &= \tau_{ij} v_j \\ \mathbf{v} \cdot \boldsymbol{\tau} &= v_i \tau_{ij} \end{aligned} \tag{A.26}$$

Both products are vectors that differ in magnitude and direction from $\mathbf{v}$. These operations are carried out in the same way as matrix–vector multiplication. The dot product of any vector with the unit tensor is the original vector:

$$\begin{aligned} \mathbf{I} \cdot \mathbf{v} &= \mathbf{v} \\ \mathbf{v} \cdot \mathbf{I} &= \mathbf{v} \end{aligned} \tag{A.27}$$

A.2.3 THE TRACE OF A TENSOR

The *trace* of a tensor is the scalar sum of its diagonal components:

$$\text{tr}\,\tau = \tau_{ii} \tag{A.28}$$

A tensor whose trace equals zero is said to be *deviatoric*.

A.2.4 SCALAR INVARIANTS AND MAGNITUDES

An *invariant* is a scalar quantity that can be computed from the components of a vector or tensor, and it has the same value no matter what coordinate system the components refer to. That is, the quantity is invariant under coordinate transformation. A vector has one independent invariant, its magnitude:

$$|\mathbf{v}| = \sqrt{\mathbf{v} \cdot \mathbf{v}} = \sqrt{v_i v_i} \tag{A.29}$$

A second-order tensor has three independent invariants. One way to choose these is (e.g., Bird et al., 1987, p. 568)

$$\begin{aligned} I &= \text{tr}(\tau) = \tau_{ii} \\ II &= \text{tr}(\tau^2) = \tau : \tau = \tau_{ij}\tau_{ji} \\ III &= \text{tr}(\tau^3) = \tau_{ij}\tau_{jk}\tau_{ki} \end{aligned} \tag{A.30}$$

A subscript may be appended to indicate which tensor the invariants refer to, as I_τ. An alternate set of invariants containing exactly the same information (e.g., Malvern, 1969, p. 89) is

$$\begin{aligned} I_1 &= \text{tr}(\tau) = \tau_{ii} \\ I_2 &= \frac{1}{2}[(\text{tr}\,\tau)^2 - \text{tr}(\tau^2)] = \frac{1}{2}[(\tau_{ii})^2 - \tau_{ij}\tau_{ji}] \\ I_3 &= \det(\tau) = \frac{1}{6}\epsilon_{ijk}\epsilon_{pqr}\tau_{ip}\tau_{jq}\tau_{kr} \end{aligned} \tag{A.31}$$

Regardless of which set of invariants is used, these quantities are called the first, second, and third invariants, respectively.

The *scalar magnitude*, or simply the magnitude, of a tensor is defined as

$$\begin{aligned} |\tau| = \tau &= \sqrt{\frac{1}{2}\tau : \tau^T} \\ &= \sqrt{\frac{1}{2}\tau_{ij}\tau_{ij}} \end{aligned} \tag{A.32}$$

The notation is that τ (lightface italic) is the scalar magnitude of the tensor $\boldsymbol{\tau}$ (boldface). If $\boldsymbol{\tau}$ is symmetric, then

$$|\tau| = \sqrt{\frac{1}{2}II_\tau} \tag{A.33}$$

A.2.5 EIGENVALUES AND EIGENVECTORS

For any tensor τ there exist special vectors $\mathbf{e}_{(i)}$ called *eigenvectors*, which, when multiplied by τ, change length but not direction. That is,

$$\tau \cdot \mathbf{e}_{(i)} = \lambda_{(i)} \mathbf{e}_{(i)} \tag{A.34}$$

The subscripts in parentheses serve only to label the eigenvalues and eigenvectors, and there is no sum on (i). The scalar factor $\lambda_{(i)}$ by which the eigenvector changes its length is called its *eigenvalue*. Without loss of generality we can require the eigenvectors to have unit length. Equation (A.34) can be manipulated to read

$$\left(\lambda_{(i)} \mathbf{I} - \tau\right) \cdot \mathbf{e}_{(i)} = 0 \tag{A.35}$$

This equation can have a nonzero solution for $\mathbf{e}_{(i)}$ only if

$$\det(\lambda \mathbf{I} - \tau) = 0 \tag{A.36}$$

where λ represents any of the $\lambda_{(i)}$'s. Expanding the determinant gives a cubic equation for the eigenvalues, called the *characteristic equation* of the tensor:

$$\lambda^3 - I_1 \lambda^2 + I_2 \lambda - I_3 = 0 \tag{A.37}$$

This equation has three roots and, if the tensor τ is symmetric, the roots will all be real, though not necessarily distinct. One can write the tensor in terms of it eigenvalues and eigenvectors as

$$\tau = \lambda_{(1)} \mathbf{e}_{(1)} \mathbf{e}_{(1)} + \lambda_{(2)} \mathbf{e}_{(2)} \mathbf{e}_{(2)} + \lambda_{(3)} \mathbf{e}_{(3)} \mathbf{e}_{(3)} \tag{A.38}$$

The three eigenvectors of a symmetric tensor are also mutually perpendicular, and they can form the basis for a coordinate system. If τ is transformed to this coordinate system, it becomes diagonal:

$$\tau = \begin{bmatrix} \lambda_{(1)} & 0 & 0 \\ 0 & \lambda_{(2)} & 0 \\ 0 & 0 & \lambda_{(3)} \end{bmatrix} \tag{A.39}$$

The coordinate axes defined by the eigenvalues are called the *principal axes* of the tensor, and the eigenvalues are its *principal values*.

The eigenvalues of a tensor are also invariant under coordinate transformation, so the set of three eigenvalues $[\lambda_{(1)}, \lambda_{(2)}, \lambda_{(3)}]$ is yet another alternative to Eqs. (A.30) and (A.31) as the three invariants of a tensor.

An important related result is the *Cayley–Hamilton theorem*, which states that a tensor satisfies its own characteristic equation:

$$\tau^3 - I_1\tau^2 + I_2\tau - I_3\mathbf{I} = 0 \qquad (A.40)$$

This means that τ^3 can always be replaced by an expression containing $\tau^2, \tau, \mathbf{I}$, and the invariants.

A.2.6 THE INVERSE OF A TENSOR

The inverse of a tensor is defined such that the dot product of the tensor with its inverse equals the identity tensor:

$$\tau \cdot \tau^{-1} = \mathbf{I} \qquad (A.41)$$

The inverse of a tensor may be computed by taking the inverse of its component matrix. The inverse of any particular tensor may or may not exist, depending on the value of the tensor's determinant.

A tensor whose inverse equals its transpose,

$$\tau^{-1} = \tau^T \qquad (A.42)$$

is said to be *orthogonal*.

A.3 DIFFERENTIAL OPERATIONS IN RECTANGULAR COORDINATES

A.3.1 THE GRADIENT OPERATOR

The gradient operator is symbolized by ∇, called *del* or *nabla*. This quantity is a vector operator. That is, it functions much like a vector, but instead its "components" are differential operators. In Cartesian coordinates this operator is

$$\nabla \equiv \mathbf{e}_x \frac{\partial}{\partial x} + \mathbf{e}_y \frac{\partial}{\partial y} + \mathbf{e}_z \frac{\partial}{\partial z} \qquad (A.43)$$

or, using indicial notation,

$$\nabla = \frac{\partial}{\partial x_i} \qquad (A.44)$$

A.3.2 GRADIENTS OF SCALAR AND VECTORS FIELDS

Suppose that $s(\mathbf{x})$ is a scalar function of position, that is, a scalar field. Then ∇ operating on s produces a vector called the *gradient* of s:

$$\nabla s = \frac{\partial s}{\partial x_i} \qquad (A.45)$$

(As usual, we define operations by equating their Gibbs notation to their indicial notation.) The gradient vector points in the direction in which s increases most rapidly, and its magnitude is equal to the derivative of s in that direction.

One may also compute the gradient of a vector field:

$$\nabla \mathbf{v} = \frac{\partial v_j}{\partial x_i} \tag{A.46}$$

Note that this is a dyadic product of ∇ and $\mathbf{v}$, so the gradient of a vector is a second-order tensor.

According to the rules of dyadic products, Eq. (A.22), the first index comes from ∇ and the second index from $\mathbf{v}$. Thus, $\partial v_j / \partial x_i$ is the ij component of $\nabla \mathbf{v}$. This seems counterintuitive, and many authors swap the indices and define the ij component of $\nabla \mathbf{v}$ as $\partial v_i / \partial x_j$. However, in this text we follow the dyadic product convention. For convenience we define a velocity gradient tensor $L_{ij} = \partial v_i / \partial x_j$, so that $\mathbf{L} = (\nabla \mathbf{v})^T$.

A.3.3 DIVERGENCE OF VECTOR AND TENSOR FIELDS

The dot product of ∇ with a vector field $\mathbf{v}(\mathbf{x})$ is called the *divergence* of $\mathbf{v}$, and it is a scalar field:

$$\nabla \cdot \mathbf{v} = \frac{\partial v_i}{\partial x_i} \tag{A.47}$$

Note the summation on the right-hand side. If the vector field is generated by taking the gradient of a scalar function, then we get the *Laplacian* of the scalar field:

$$\nabla^2 s \equiv \nabla \cdot (\nabla s) = \sum_{i=1}^{3} \frac{\partial^2 s}{\partial x_i^2} \tag{A.48}$$

∇^2 is called the Laplacian operator.

One may also take the divergence of a tensor field,

$$\nabla \cdot \boldsymbol{\tau} = \frac{\partial \tau_{ij}}{\partial x_i} \tag{A.49}$$

which is an operation that produces a vector.

A.3.4 ALGEBRAIC PROPERTIES OF ∇

Many vector and tensor identities can be proved by using the basic results given so far. We particularly want to point out that, although certain vector and tensor dot products are associative, for example,

$$\mathbf{u} \cdot (s\mathbf{v}) = s(\mathbf{u} \cdot \mathbf{v})$$
$$\mathbf{u} \cdot (\boldsymbol{\tau} \cdot \mathbf{v}) = (\mathbf{u} \cdot \boldsymbol{\tau}) \cdot \mathbf{v} \tag{A.50}$$

the corresponding operations involving ∇ are not. In these examples the correct relations are

$$\nabla \cdot (s\mathbf{v}) = (\nabla s) \cdot \mathbf{v} + s(\nabla \cdot \mathbf{v})$$
$$\nabla \cdot (\boldsymbol{\tau} \cdot \mathbf{v}) = (\nabla \cdot \boldsymbol{\tau}) \cdot \mathbf{v} + \boldsymbol{\tau} : (\nabla \mathbf{v}) \tag{A.51}$$

A.4 VECTORS AND TENSORS IN CYLINDRICAL AND SPHERICAL COORDINATES

A convenient approach to proving these and other similar propositions is to convert the first expression into indicial notation, manipulate it by using the usual rules of scalar algebra and calculus, and then return the result to Gibbs notation.

A.4 VECTORS AND TENSORS IN CYLINDRICAL AND SPHERICAL COORDINATES

A.4.1 COORDINATE CURVES AND UNIT VECTORS

In a cylindrical coordinate system (Fig. A.3), each point in space possesses a unique set of values for the coordinates (r, θ, z). Any curve generated by fixing two of these coordinates and varying the third is called a *coordinate curve*. In cylindrical coordinates the r coordinate curves are lines radiating out from the z axis and perpendicular to it; the θ curves are circles centered on the z axis and lying in planes normal to it; and the z curves are lines parallel to the z axis. One coordinate curve of each type emanates from any point, and the curves are mutually perpendicular at that point. Thus, cylindrical coordinates are one type of *orthogonal curvilinear* coordinate system.

In any three-dimensional coordinate system we can define three unit vectors at any point, each one tangent to a coordinate curve and pointing in the direction in which the coordinate increases. The unit vectors in cylindrical coordinates are $\mathbf{e}_r$, $\mathbf{e}_\theta$, and $\mathbf{e}_z$. These are also shown in Fig. A.3. Note that the direction of the unit vectors will change if the θ coordinate of the point changes.

Spherical coordinates are defined similarly, with the coordinates (r, θ, ϕ) determining the coordinate curves and the unit vectors $\mathbf{e}_r$, $\mathbf{e}_\theta$, and $\mathbf{e}_\phi$. These are shown in Fig. A.4. Note that the unit vectors $\mathbf{e}_r$ and $\mathbf{e}_\theta$ in spherical coordinates are *not* the same as $\mathbf{e}_r$ and $\mathbf{e}_\theta$ in cylindrical coordinates. Spherical coordinates are also orthogonal and curvilinear.

A.4.2 VECTOR AND TENSOR COMPONENTS

Vectors and tensors in cylindrical and spherical coordinates are defined in terms of the unit vectors, just as in Cartesian coordinates. Thus, Eqs. (A.1) and (A.3) can be written as Cylindrical coordinates:

$$\mathbf{v} = v_r \mathbf{e}_r + v_\theta \mathbf{e}_\theta + v_z \mathbf{e}_z \tag{A.52}$$

$$\begin{aligned}\boldsymbol{\tau} = {} & \tau_{rr}\mathbf{e}_r\mathbf{e}_r + \tau_{r\theta}\mathbf{e}_r\mathbf{e}_\theta + \tau_{rz}\mathbf{e}_r\mathbf{e}_z \\ & + \tau_{\theta r}\mathbf{e}_\theta\mathbf{e}_r + \tau_{\theta\theta}\mathbf{e}_\theta\mathbf{e}_\theta + \tau_{\theta z}\mathbf{e}_\theta\mathbf{e}_z \\ & + \tau_{zr}\mathbf{e}_z\mathbf{e}_r + \tau_{z\theta}\mathbf{e}_z\mathbf{e}_\theta + \tau_{zz}\mathbf{e}_z\mathbf{e}_z \end{aligned} \tag{A.53}$$

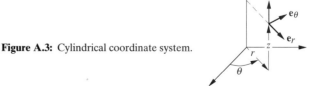

Figure A.3: Cylindrical coordinate system.

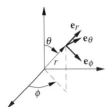

Figure A.4: Spherical coordinate system.

Spherical coordinates:

$$\mathbf{v} = v_r \mathbf{e}_r + v_\theta \mathbf{e}_\theta + v_\phi \mathbf{e}_\phi$$

$$\boldsymbol{\tau} = \tau_{rr}\mathbf{e}_r\mathbf{e}_r + \tau_{r\theta}\mathbf{e}_r\mathbf{e}_\theta + \tau_{r\phi}\mathbf{e}_r\mathbf{e}_\phi$$
$$+ \tau_{\theta r}\mathbf{e}_\theta\mathbf{e}_r + \tau_{\theta\theta}\mathbf{e}_\theta\mathbf{e}_\theta + \tau_{\theta\phi}\mathbf{e}_\theta\mathbf{e}_\phi \quad (A.54)$$
$$+ \tau_{\phi r}\mathbf{e}_\phi\mathbf{e}_r + \tau_{\phi\theta}\mathbf{e}_\phi\mathbf{e}_\theta + \tau_{\phi\phi}\mathbf{e}_\phi\mathbf{e}_\phi \quad (A.55)$$

A.4.3 ALGEBRAIC OPERATIONS

When performing any of the vector or tensor algebraic operations defined in Section A.2, one only deals with vectors and tensors defined at a single point in space. Then all the vector and tensor components are based on the same set of unit vectors, and all of the operations proceed just as they do in rectangular coordinates. One can adapt the indicial notation by allowing repeated indices to cycle through (r, θ, z) or (r, θ, ϕ). Here are a few examples in spherical coordinates:

$$\mathbf{u} \cdot \mathbf{v} = u_r v_r + u_\theta v_\theta + u_\phi v_\phi$$
$$\mathrm{tr}(\boldsymbol{\tau}) = \tau_{rr} + \tau_{\theta\theta} + \tau_{\phi\phi} \quad (A.56)$$

A.4.4 DIFFERENTIAL OPERATIONS

Whereas the algebraic operations are unchanged from Cartesian coordinates, the differential operations are quite different. This occurs for two reasons. First, *scale factors* are required to account for the change in position created by a change in coordinate. For instance, in the cylindrical system a coordinate change $d\theta$ produces a position change of $r\,d\theta$. Because the ∇ operator is the derivative with respect to position, it must contain these scale factors. This leads to the following expressions for the gradient of a *scalar* function:

$$\nabla s = \mathbf{e}_r \frac{\partial s}{\partial r} + \mathbf{e}_\theta \frac{1}{r} \frac{\partial s}{\partial \theta} + \mathbf{e}_z \frac{\partial s}{\partial z} \quad (A.57)$$

$$\nabla s = \mathbf{e}_r \frac{\partial s}{\partial r} + \mathbf{e}_\theta \frac{1}{r} \frac{\partial s}{\partial \theta} + \mathbf{e}_\phi \frac{1}{r \sin\theta} \frac{\partial s}{\partial \phi} \quad (A.58)$$

for cylindrical and spherical coordinates, respectively.

Second, the directions of the unit vectors themselves change with position. For example, in cylindrical coordinates we have $(\partial \mathbf{e}_r / \partial \theta) = \mathbf{e}_\theta$. (You might make a sketch to verify this, and also figure out $\partial \mathbf{e}_\theta / \partial \theta$.) So, when ∇ operates on a vector or tensor, we get terms from the derivatives of the unit vectors, as well from the derivatives of the scalar components. A nice explanation of this is provided by Bird et al. (1987), and

A.6 CURVATURE OF CURVES AND SURFACES

we will not pursue the details here. The reader may note that Appendix B contains expressions for $\nabla \cdot \mathbf{q}$, $\nabla \cdot \boldsymbol{\tau}$, and $\nabla^2 T$ embedded in the various equations. This brief discussion should help explain why these expressions have the forms they do.

A.5 THE DIVERGENCE THEOREM

A theorem of vector calculus called the *divergence theorem* relates certain integrals over a volume V to surface integrals over the surface S that encloses V. Note that S must be a *closed* surface. The most common form applies to the divergence of a vector field:

$$\int_V \nabla \cdot \mathbf{v} \, dV = \int_S \mathbf{n} \cdot \mathbf{v} \, dS \tag{A.59}$$

Here $\mathbf{n}$ is the unit vector pointing outward from the surface at dS (see Fig. 2.1). There are also analogous theorems for scalar and vector fields:

$$\int_V \nabla s \, dV = \int_S \mathbf{n} s \, dS \tag{A.60}$$

$$\int_V \nabla \cdot \boldsymbol{\tau} \, dV = \int_S \mathbf{n} \cdot \boldsymbol{\tau} \, dS \tag{A.61}$$

A.6 CURVATURE OF CURVES AND SURFACES

A classical treatment of this subject is provided by Struik (1961). Here we summarize some important results without derivation.

A.6.1 CURVES

The geometrical interpretation of curvature is probably familiar. Consider a curve in space C. As Fig. A.5 suggests, in the neighborhood of any point P on this curve, the

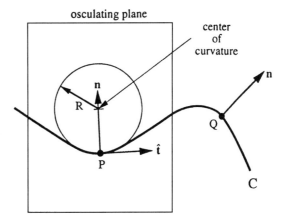

Figure A.5: A curve in space, showing the tangent vector $\hat{\mathbf{t}}$, the normal vector $\mathbf{n}$, the osculating plane, and the radius of curvature R.

curve can be locally approximated by part of an arc of a circle. The radius of this circle R is called the *radius of curvature*, and $\kappa = 1/R$ is called the *curvature* at point P.

More formally, suppose that the curve is defined by giving its position $\mathbf{x}$ as a function of arc length s measured along the curve:

$$\mathbf{x} = \mathbf{x}(s) \tag{A.62}$$

The vector $\hat{\mathbf{t}}$ that is tangent to the curve at any point is equal to the derivative of this function:

$$\hat{\mathbf{t}} = \frac{d\mathbf{x}}{ds} \tag{A.63}$$

Note that $\hat{\mathbf{t}}$ is a unit vector. [We use the caret to distinguish $\hat{\mathbf{t}}$ from the surface traction vector $\mathbf{t}$ defined in Eq. (2.49)]. For a straight line, $\hat{\mathbf{t}}$ does not change as we move along the curve. Thus, the rate of change in $\hat{\mathbf{t}}$ measures how much the curve changes direction, that is, its curvature. The derivative of $\hat{\mathbf{t}}$ is called the curvature vector, $\mathbf{k}$:

$$\mathbf{k} = \frac{d\hat{\mathbf{t}}}{ds} \tag{A.64}$$

Because $\hat{\mathbf{t}}$ is a unit vector it can only change direction, not magnitude, so the infinitesimal change $d\hat{\mathbf{t}}$ must be perpendicular to $\hat{\mathbf{t}}$ itself. Consequently, $\mathbf{k}$ must also be normal to $\hat{\mathbf{t}}$.

We factor $\mathbf{k}$ into two parts: a unit vector $\mathbf{n}$ called the *principal normal* to the curve, and a scalar magnitude κ, the *curvature*:

$$\mathbf{k} = \kappa \mathbf{n}, \quad \mathbf{n} \cdot \mathbf{n} = 1 \tag{A.65}$$

The vectors $\hat{\mathbf{t}}$ and $\mathbf{n}$ define the *osculating plane* at the point P. The center of curvature (the center of the circle that approximates the curve locally) lies in this plane a distance $|R|$ from the curve, along the axis defined by the normal vector. The radius of curvature is

$$R = 1/\kappa \tag{A.66}$$

We must also pay attention to the sign convention for curvature; $\mathbf{k}$ always points toward the center of curvature, that is, to the inside of the curve. This direction of $\mathbf{k}$ is uniquely determined by the shape of the curve; if we measure arc length in the opposite direction, both ds and $d\hat{\mathbf{t}}$ change sign, and $\mathbf{k}$ is unchanged. However, we can choose the direction of $\mathbf{n}$ arbitrarily. One convention is to choose a normal vector $\mathbf{n}$ that always points to the same side of the curve. Then the curvature κ will be positive if $\mathbf{n}$ points toward the inside of the curve (like point P in Fig. A.5), and negative if $\mathbf{n}$ points toward the outside (as at point Q). We will use Eq. (A.66) as the definition of R, so R may also be positive or negative, following the sign of κ.

The preceding definitions are general and apply to any curve in three-dimensional space. If the curve lies in a plane, then this plane will contain both $\hat{\mathbf{t}}$ and $\mathbf{n}$, as well as the center of curvature.

An important special case is a curve that lies in a plane, say $z = 0$, and is described by a function $y = f(x)$. In this case the change in coordinate dx for an arc length

A.6 CURVATURE OF CURVES AND SURFACES

movement ds is

$$\frac{dx}{ds} = \frac{1}{\sqrt{1+f'^2}} \tag{A.67}$$

where $f' = df/dx$. Using this in the formulas above, we find that the components of the tangent and normal vectors are

$$t_x = n_y = \frac{1}{\sqrt{1+f'^2}}, \quad t_y = -n_x = \frac{f'}{\sqrt{1+f'^2}} \tag{A.68}$$

and the curvature is

$$\kappa = \frac{f''}{\sqrt{1+f'^2}} \tag{A.69}$$

If the slope of the curve is small, $f' \ll 1$, then Eq. (A.69) reduces to the familiar form

$$\kappa = f'' = \frac{d^2 f}{dx^2} \tag{A.70}$$

The signs of the curvatures given by Eqs. (A.69) and (A.70) are consistent with **n** pointing upward; that is, $n_y > 0$.

A.6.2 SURFACES

To extend this idea of curvature to surfaces, consider a point P lying on a surface S, as shown in Fig. A.6. At this point construct the unit vector **n** normal to the *surface*. Then choose a plane containing **n**; this is called a *normal plane*. The intersection of the normal plane and the surface define a curve, called a *normal curve* of the surface. The curvature κ of this curve, as defined in the previous section, is called the *normal curvature*. This is a property of both the surface shape, and of the choice of the normal plane. We can choose an infinite number of normal planes by twisting the plane using **n** as an axis, and each normal plane will give its own value of the normal curvature. However, two particular normal planes will give the largest and smallest

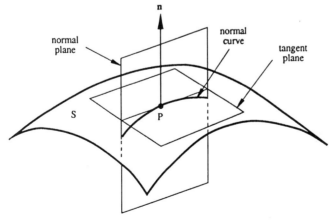

Figure A.6: A normal curve at point P on surface S, defined by the intersection between the surface and a normal plane; **n** is the unit vector normal to the surface.

values of the normal curvature, and these two planes will be mutually perpendicular. The two curvatures associated with these planes are called the *principal curvatures*, and we will denote them by κ_1 and κ_2. Corresponding to these are the principal radii of curvature:

$$R_1 = \frac{1}{\kappa_1}, \quad R_2 = \frac{1}{\kappa_2} \tag{A.71}$$

The surface normal **n** is also perpendicular to every normal curve, so κ_1 and κ_2 will be positive if their surface curves are concave toward **n**, and negative if their surface curves are convex towards **n**. The centers of curvature for R_1 and R_2 will both lie on the axis defined by **n**, but in general they will be at different distances from the surface and may be on opposite sides of the surface.

Two important scalar measures of surface curvature are the *mean curvature*,

$$\kappa_m = \frac{1}{2}(\kappa_1 + \kappa_2) \tag{A.72}$$

and the *Gaussian* or *total curvature*,

$$\kappa_G = \kappa_1 \kappa_2 \tag{A.73}$$

Common symbols are H for mean curvature and K for Gaussian curvature. For a plane, both principal curvatures are zero, as are the mean and Gaussian curvature. A surface shaped like a saddle has one positive and one negative principal curvature. Such a surface will have negative Gaussian curvature, and could possibly have zero mean curvature. A cylinder has one principal curvature equal to zero, so its Gaussian curvature is zero. Any surface that has zero Gaussian curvature everywhere is called a *developable surface*, and it can be created by rolling up a flat sheet of material. A cone is developable; a sphere is not.

The principal curvatures κ_1 and κ_2, as well as the mean and Gaussian curvature, are *invariant* properties of the surface. That is, for any point on a given surface their values are independent of the choice of coordinate system, and independent of the mathematical description of the surface. In fact κ_1 and κ_2 are eigenvalues of a particular tensor, so this is just another example of the concept of tensor invariants, though the details necessary to show this are beyond the scope of this discussion.

Computing the curvature of a general surface is fairly involved (see Aris, 1962, for a treatment in generalized coordinate systems). However, two special cases can be treated simply. The first is a prismatic surface, which is generated by choosing a curve in a plane and then sweeping the curve in the direction normal to the plane. This is the type of surface that arises when two-dimensional problems are analyzed. If the initial curve lies in the x–y plane and is swept in the z direction, then κ_1 is the curvature of the generating curve as given by Eq. (A.69), and κ_2 equals zero.

A second special case is a surface of revolution. Here we choose a cylindrical coordinate system. The surface is generated by choosing a curve $r = g(z)$ in the r–z plane, and rotating the curve about the z axis. Figure A.7 shows an example. This type of surface arises, for instance, in the extrusion of axisymmetric shapes, and in the analysis of blown film extrusion (e.g., Agassant et al., 1991, p. 252). For this surface, the curves defining the principal curvatures at any point are the meridians (curves of

A.7 THE GAUSSIAN ERROR FUNCTION

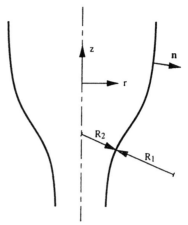

Figure A.7: A surface of revolution $r = g(z)$, showing the centers of curvature and the principal radii of curvature R_1 and R_2.

constant θ, including the curves shown in Fig. A.7), and circles of constant z. For the curvature of the meridian we can apply Eq. (A.69) in the r–z plane, while the other center of curvature will lie on the intersection of the surface normal and the z axis, as shown in Fig. A.7. Thus, the two principal radii of curvature are

$$R_1 = \frac{(1+g'^2)^{3/2}}{g''}, \quad R_2 = -g\sqrt{1+g'^2} \tag{A.74}$$

with $g' = dg/dz$ and $g'' = d^2g/dz^2$. These formulas assume that the normal vector is directed radially outward, as shown in Fig. A.7. Then R_2 will always be negative, whereas R_1 can be either positive or negative. If the normal vector is directed inward then both radii will change sign.

A.7 THE GAUSSIAN ERROR FUNCTION

A function that appears frequently in heat and mass transfer problems is the *Gaussian error function*, or simply the error function. The error function is defined as

$$\text{erf}(\zeta) \equiv \frac{2}{\sqrt{\pi}} \int_0^\zeta \exp(-u^2)\, du \tag{A.75}$$

The error function has the following properties:

$$\text{erf}(0) = 0 \tag{A.76}$$

$$\text{erf}(\infty) = 1 \tag{A.77}$$

$$\text{erf}(-u) = -\text{erf}(u) \tag{A.78}$$

$$\frac{d}{dx}(\text{erf}(u)) = \frac{2}{\sqrt{\pi}} \exp(-u^2) \frac{du}{dx} \tag{A.79}$$

The quantity $(1 - \text{erf}(u))$ is the *complementary error function*:

$$\text{erfc}(u) \equiv 1 - \text{erf}(u) \tag{A.80}$$

Numerical values for the error function appear in Table A.1. It is sometimes convenient to be able to evaluate the error function approximately, without referring to

Table A.1 The Error Function and $xe^{x^2}\text{erf}(x)$

x	$\text{erf}(x)$	$xe^{x^2}\text{erf}(x)$	x	$\text{erf}(x)$	$xe^{x^2}\text{erf}(x)$	x	$\text{erf}(x)$	$xe^{x^2}\text{erf}(x)$
0.00	0.0000	0.0000	0.50	0.5205	0.3342	1.00	0.8427	2.2907
0.02	0.0226	0.0005	0.52	0.5379	0.3666	1.04	0.8586	2.6338
0.04	0.0451	0.0018	0.54	0.5549	0.4011	1.08	0.8733	3.0280
0.06	0.0676	0.0041	0.56	0.5716	0.4380	1.12	0.8868	3.4819
0.08	0.0901	0.0073	0.58	0.5879	0.4774	1.16	0.8991	4.0054
0.10	0.1125	0.0114	0.60	0.6039	0.5193	1.20	0.9103	4.6106
0.12	0.1348	0.0164	0.62	0.6194	0.5640	1.24	0.9205	5.3115
0.14	0.1569	0.0224	0.64	0.6346	0.6117	1.28	0.9297	6.1252
0.16	0.1790	0.0294	0.66	0.6494	0.6626	1.32	0.9381	7.0717
0.18	0.2009	0.0374	0.68	0.6638	0.7167	1.36	0.9456	8.1752
0.20	0.2227	0.0464	0.70	0.6778	0.7745	1.40	0.9523	9.4648
0.22	0.2443	0.0564	0.72	0.6914	0.8360	1.44	0.9583	10.9753
0.24	0.2657	0.0675	0.74	0.7047	0.9017	1.48	0.9637	12.7486
0.26	0.2869	0.0798	0.76	0.7175	0.9716	1.52	0.9684	14.8354
0.28	0.3079	0.0932	0.78	0.7300	1.0463	1.56	0.9726	17.2970
0.30	0.3286	0.1079	0.80	0.7421	1.1259	1.60	0.9763	20.2078
0.32	0.3491	0.1238	0.82	0.7538	1.2109	1.64	0.9796	23.6581
0.34	0.3694	0.1410	0.84	0.7651	1.3015	1.68	0.9825	27.7582
0.36	0.3893	0.1596	0.86	0.7761	1.3984	1.72	0.9850	32.6424
0.38	0.4090	0.1796	0.88	0.7867	1.5018	1.76	0.9872	38.4755
0.40	0.4284	0.2011	0.90	0.7969	1.6122	1.80	0.9891	45.4593
0.42	0.4475	0.2242	0.92	0.8068	1.7303	1.84	0.9907	53.8422
0.44	0.4662	0.2490	0.94	0.8163	1.8565	1.88	0.9922	63.9305
0.46	0.4847	0.2755	0.96	0.8254	1.9916	1.92	0.9934	76.1025
0.48	0.5027	0.3038	0.98	0.8342	2.1360	1.96	0.9944	90.8269

tables. Some useful engineering approximations are (Kurz and Fisher, 1989, p. 167)

$$\text{erf}(\zeta) \simeq \left[1 - \exp\left(-\frac{4\zeta^2}{\pi}\right)\right]^{1/2} \tag{A.81}$$

$$\text{erf}(\zeta) \simeq 1 \quad \text{for } \zeta > 2 \tag{A.82}$$

$$\text{erf}(\zeta) \simeq \frac{2\zeta}{\sqrt{\pi}} \quad \text{for } \zeta < 0.2 \tag{A.83}$$

These formulas have truncation errors less than 0.01. For more precise work, the approximation

$$\text{erf}(\zeta) \simeq 1 - (0.3480242 s - 0.0958798 s^2 + 0.7478556 s^3) \exp(-\zeta^2)$$
$$s = \frac{1}{1 + 0.47047\zeta} \tag{A.84}$$

EXERCISES 345

has a truncation error less than $\pm 2.5 \times 10^{-5}$ (Abramowitz and Stegun, 1965, Art. 7.1.25). This approximation can easily be incorporated into a computer program or spreadsheet, and it is just as accurate as Table A.1.

BIBLIOGRAPHY

M. Abramowitz and I. A. Stegun, *Handbook of Mathematical Functions.* Dover, New York, 1965.

J.-F. Agassant, P. Avenas, J.-Ph. Sergent, and P. J. Carreau, *Polymer Processing: Principles and Modeling.* Hanser, Munich, 1991.

R. Aris. *Vectors, Tensors, and the Basic Equations of Fluid Mechanics.* Prentice-Hall, Englewood Cliffs, NJ, 1962.

R. B. Bird, R. C. Armstrong, and O. Hassager. *Dynamics of Polymeric Liquids. Volume 1: Fluid Mechanics.* Wiley, New York, 2nd edition, 1987.

W. Kurz and D. J. Fisher. *Fundamentals of Solidification.* Trans Tech Publications, Aedermannsdorf, Switzerland, 3rd edition, 1989.

L. E. Malvern, *Introduction to the Mechanics of a Continuous Medium.* Prentice-Hall, Englewood Cliffs, NJ, 1969.

D. J. Struik. *Lectures on Classical Differential Geometry.* Addison-Wesley, Reading, MA, 2nd edition, 1961.

EXERCISES

1   *Vector and tensor notation.* To refresh your understanding of different notations for vector and tensor algebra, perform the following translations.

   (a) Write out these expressions in terms of Cartesian (rectangular) coordinates (x, y, z) and components $(v_x, v_y,$ etc.). Do not use indicial notation.

   ∇P

   $\text{tr}(\mathbf{D})$

   $\mathbf{u} \cdot \mathbf{v}$

   (b) Write these expressions in Gibbs (boldface) notation:

   $$\frac{\partial v_x}{\partial x} + \frac{\partial v_y}{\partial y} + \frac{\partial v_z}{\partial z}$$

   $$\frac{\partial T}{\partial t} + v_x \frac{\partial T}{\partial x} + v_y \frac{\partial T}{\partial y} + v_z \frac{\partial T}{\partial z}$$

   (c) Write out an expression for ∇T in cylindrical coordinates.

   (d) Rewrite the following equation in Gibbs (boldface) notation:

   $$\frac{D A_{ij}}{Dt} = (W_{ik} A_{kj} - A_{ik} W_{kj}) + (D_{ik} A_{kj} + A_{ik} D_{kj})$$
   $$- 2 \left(\frac{A_{ij} A_{kl} D_{lk}}{A_{mm}} \right) + 2 D_r (\delta_{ij} - 3 A_{ij})$$

   Note: D_r is a scalar.

   (e) Rewrite the following equations in indicial notation:

   $\mathbf{q} = -\mathbf{k} \cdot \nabla T$

   $\boldsymbol{\tau} = 2\eta \mathbf{D} : \mathbf{pppp}$

(f) Consider the tensor given by

$$\mathbf{k} = (k_a - k_t)\mathbf{pp} + k_t\mathbf{I}$$

Write out the components of $\mathbf{k}$ in matrix form, when $\mathbf{p}$ is a unit vector that points in the x_1 direction.

2 *The double dot product.* Show that

$$\mathbf{A} : \mathbf{B} = \mathrm{tr}(\mathbf{A} \cdot \mathbf{B})$$

for any tensors $\mathbf{A}$ and $\mathbf{B}$. This is easy to do by using indicial notation.

3 *A tensor identity.*

(a) Show that the double dot product of any symmetric tensor (say τ) with any antisymmetric tensor (say $\mathbf{W}$) equals zero. Suggestion: write out the product in terms of the components, and then use the properties of symmetric and antisymmetric tensors.

(b) Using your result from part (a), show that

$$\tau : \mathbf{L} = \tau : \mathbf{D}$$

4 *Coordinate transformation.* Consider two Cartesian coordinate systems, x, y, z and x', y', z'. The z and z' axes are identical, and the x' axis is rotated 30° counterclockwise from the x axis.

(a) Write out the components of the transformation matrix $\mathbf{Q}$ that transforms vectors and tensors from the x, y, z system to the x', y', z' system.

(b) In the unprimed axes, a vector $\mathbf{v}$ has components $\{1.732, 1.0, 1.0\}^T$. Find the components of $\mathbf{v}'$, the same vector expressed in the primed system.

(c) Calculate the magnitude of $\mathbf{v}$ by using its components in the unprimed axes. Also calculate the magnitude by using the primed components. Compare the values.

(d) A tensor has components

$$\mathbf{A} = \begin{bmatrix} 1.5 & 0 & 0 \\ 0 & 2.0 & 0 \\ 0 & 0 & 1.0 \end{bmatrix}$$

Find its components in the primed system.

(e) Calculate the three invariants of $\mathbf{A}$ by using its components in the unprimed axes. Repeat by using the components in the primed axes. Compare the two sets of values.

5 *The transformation matrix.* Show that the transformation matrix $\mathbf{Q}$ has the following properties:

$$\mathbf{e}'_i = Q_{ij}\mathbf{e}_j$$
$$\mathbf{e}_j = Q_{ij}\mathbf{e}'_i$$
$$\mathbf{Q} \cdot \mathbf{Q}^T = \mathbf{I}$$

EXERCISES

Note that summation is implied on the right-hand side of the first two equations. Hint: Use Eq. (A.5).

6   *The unit tensor.* Prove that the unit tensor is isotropic. That is, **I** has the same components in any orthogonal coordinate system. Do this by transforming **I** to an arbitrary coordinate system, using the transformation matrix **Q**.

7   *Tensor invariants.* Following the pattern of the proof in Section 6.4.2:

  (a) Prove that tr **A** is invariant under coordinate transformation between two orthogonal coordinate systems, for any tensor **A**.

  (b) Prove that tr $\mathbf{A}^2$ and tr $\mathbf{A}^3$ are invariant under coordinate transformation between two orthogonal coordinate systems.

8   *Curvature of a paraboloid.* The tip shape of a dendrite is sometimes approximated as a paraboloid of revolution. That is, the surface near the tip is given by $r = cz^2$ where c is a constant. At the very tip ($z = 0$), find the principal radii of curvature, the mean curvature, and the Gaussian curvature.

APPENDIX B

Balance and Kinematic Equations

The following tables are adapted from Bird, Stewart, and Lightfoot (1960); Bird, Armstrong, and Hassager (1987); and Tucker (1989).

B.1 CONTINUITY EQUATION: GENERAL FORM

Rectangular coordinates:

$$\frac{\partial \rho}{\partial t} + \frac{\partial}{\partial x}(\rho v_x) + \frac{\partial}{\partial y}(\rho v_y) + \frac{\partial}{\partial z}(\rho v_z) = 0 \tag{B.1}$$

Cylindrical coordinates:

$$\frac{\partial \rho}{\partial t} + \frac{1}{r}\frac{\partial}{\partial r}(\rho r v_r) + \frac{1}{r}\frac{\partial}{\partial \theta}(\rho v_\theta) + \frac{\partial}{\partial z}(\rho v_z) = 0 \tag{B.2}$$

Spherical coordinates:

$$\frac{\partial \rho}{\partial t} + \frac{1}{r^2}\frac{\partial}{\partial r}(\rho r^2 v_r) + \frac{1}{r \sin \theta}\frac{\partial}{\partial \theta}(\rho v_\theta \sin \theta) + \frac{1}{r \sin \theta}\frac{\partial}{\partial \phi}(\rho v_\phi) = 0 \tag{B.3}$$

B.2 CONTINUITY EQUATION: CONSTANT ρ

Rectangular coordinates:

$$\frac{\partial v_x}{\partial x} + \frac{\partial v_y}{\partial y} + \frac{\partial v_z}{\partial z} = 0 \tag{B.4}$$

Cylindrical coordinates:

$$\frac{1}{r}\frac{\partial}{\partial r}(r v_r) + \frac{1}{r}\frac{\partial v_\theta}{\partial \theta} + \frac{\partial v_z}{\partial z} = 0 \tag{B.5}$$

Spherical coordinates:

$$\frac{1}{r^2}\frac{\partial}{\partial r}(r^2 v_r) + \frac{1}{r \sin \theta}\frac{\partial}{\partial \theta}(v_\theta \sin \theta) + \frac{1}{r \sin \theta}\frac{\partial v_\phi}{\partial \phi} = 0 \tag{B.6}$$

B.3 RATE-OF-DEFORMATION TENSOR

Rectangular coordinates:

$$D_{xx} = \frac{\partial v_x}{\partial x} \tag{B.7}$$

$$D_{yy} = \frac{\partial v_y}{\partial y} \tag{B.8}$$

$$D_{zz} = \frac{\partial v_z}{\partial z} \tag{B.9}$$

$$D_{xy} = D_{yx} = \frac{1}{2}\left[\frac{\partial v_x}{\partial y} + \frac{\partial v_y}{\partial x}\right] \tag{B.10}$$

$$D_{yz} = D_{zy} = \frac{1}{2}\left[\frac{\partial v_y}{\partial z} + \frac{\partial v_z}{\partial y}\right] \tag{B.11}$$

$$D_{zx} = D_{xz} = \frac{1}{2}\left[\frac{\partial v_z}{\partial x} + \frac{\partial v_x}{\partial z}\right] \tag{B.12}$$

Cylindrical coordinates:

$$D_{rr} = \frac{\partial v_r}{\partial r} \tag{B.13}$$

$$D_{\theta\theta} = \frac{1}{r}\frac{\partial v_\theta}{\partial \theta} + \frac{v_r}{r} \tag{B.14}$$

$$D_{zz} = \frac{\partial v_z}{\partial z} \tag{B.15}$$

$$D_{r\theta} = D_{\theta r} = \frac{1}{2}\left[r\frac{\partial}{\partial r}\left(\frac{v_\theta}{r}\right) + \frac{1}{r}\frac{\partial v_r}{\partial \theta}\right] \tag{B.16}$$

$$D_{\theta z} = D_{z\theta} = \frac{1}{2}\left[\frac{\partial v_\theta}{\partial z} + \frac{1}{r}\frac{\partial v_z}{\partial \theta}\right] \tag{B.17}$$

$$D_{zr} = D_{rz} = \frac{1}{2}\left[\frac{\partial v_z}{\partial r} + \frac{\partial v_r}{\partial z}\right] \tag{B.18}$$

Spherical coordinates:

$$D_{rr} = \frac{\partial v_r}{\partial r} \tag{B.19}$$

$$D_{\theta\theta} = \frac{1}{r}\frac{\partial v_\theta}{\partial \theta} + \frac{v_r}{r} \tag{B.20}$$

$$D_{\phi\phi} = \frac{1}{r\sin\theta}\frac{\partial v_\phi}{\partial \phi} + \frac{v_r}{r} + \frac{v_\theta \cot\theta}{r} \tag{B.21}$$

$$D_{r\theta} = D_{\theta r} = \frac{1}{2}\left[r\frac{\partial}{\partial r}\left(\frac{v_\theta}{r}\right) + \frac{1}{r}\frac{\partial v_r}{\partial \theta}\right] \tag{B.22}$$

$$D_{\theta\phi} = D_{\phi\theta} = \frac{1}{2}\left[\frac{\sin\theta}{r}\frac{\partial}{\partial \theta}\left(\frac{v_\phi}{\sin\theta}\right) + \frac{1}{r\sin\theta}\frac{\partial v_\theta}{\partial \phi}\right] \tag{B.23}$$

$$D_{r\phi} = D_{\phi r} = \frac{1}{2}\left[\frac{1}{r\sin\theta}\frac{\partial v_r}{\partial \phi} + r\frac{\partial}{\partial r}\left(\frac{v_\phi}{r}\right)\right] \tag{B.24}$$

B.4 VORTICITY TENSOR

Rectangular coordinates:

$$W_{xy} = -W_{yx} = \frac{1}{2}\left[\frac{\partial v_x}{\partial y} - \frac{\partial v_y}{\partial x}\right] \tag{B.25}$$

$$W_{yz} = -W_{zy} = \frac{1}{2}\left[\frac{\partial v_y}{\partial z} - \frac{\partial v_z}{\partial y}\right] \tag{B.26}$$

$$W_{zx} = -W_{xz} = \frac{1}{2}\left[\frac{\partial v_z}{\partial x} - \frac{\partial v_x}{\partial z}\right] \tag{B.27}$$

Cylindrical coordinates:

$$W_{r\theta} = -W_{\theta r} = \frac{1}{2}\left[\frac{1}{r}\frac{\partial v_r}{\partial \theta} - \frac{1}{r}\frac{\partial}{\partial r}(rv_\theta)\right] \tag{B.28}$$

$$W_{\theta z} = -W_{z\theta} = \frac{1}{2}\left[\frac{\partial v_\theta}{\partial z} - \frac{1}{r}\frac{\partial v_z}{\partial \theta}\right] \tag{B.29}$$

$$W_{zr} = -W_{rz} = \frac{1}{2}\left[\frac{\partial v_z}{\partial r} - \frac{\partial v_r}{\partial z}\right] \tag{B.30}$$

Spherical coordinates:

$$W_{r\theta} = -W_{\theta r} = \frac{1}{2}\left[\frac{1}{r}\frac{\partial v_r}{\partial \theta} - \frac{1}{r}\frac{\partial}{\partial r}(rv_\theta)\right] \tag{B.31}$$

$$W_{\theta\phi} = -W_{\phi\theta} = \frac{1}{2}\left[\frac{1}{r\sin\theta}\frac{\partial v_\theta}{\partial \phi} - \frac{1}{r\sin\theta}\frac{\partial}{\partial \theta}(v_\phi \sin\theta)\right] \tag{B.32}$$

$$W_{\phi r} = -W_{r\phi} = \frac{1}{2}\left[\frac{1}{r}\frac{\partial}{\partial r}(rv_\phi) - \frac{1}{r\sin\theta}\frac{\partial v_r}{\partial \phi}\right] \tag{B.33}$$

B.5 GENERAL EQUATION OF MOTION

Rectangular coordinates:

$$\rho\left(\frac{\partial v_x}{\partial t} + v_x\frac{\partial v_x}{\partial x} + v_y\frac{\partial v_x}{\partial y} + v_z\frac{\partial v_x}{\partial z}\right) = -\frac{\partial p}{\partial x} + \left(\frac{\partial \tau_{xx}}{\partial x} + \frac{\partial \tau_{yx}}{\partial y} + \frac{\partial \tau_{zx}}{\partial z}\right) + \rho b_x \tag{B.34}$$

$$\rho\left(\frac{\partial v_y}{\partial t} + v_x\frac{\partial v_y}{\partial x} + v_y\frac{\partial v_y}{\partial y} + v_z\frac{\partial v_y}{\partial z}\right) = -\frac{\partial p}{\partial y} + \left(\frac{\partial \tau_{xy}}{\partial x} + \frac{\partial \tau_{yy}}{\partial y} + \frac{\partial \tau_{zy}}{\partial z}\right) + \rho b_y \tag{B.35}$$

$$\rho\left(\frac{\partial v_z}{\partial t} + v_x\frac{\partial v_z}{\partial x} + v_y\frac{\partial v_z}{\partial y} + v_z\frac{\partial v_z}{\partial z}\right) = -\frac{\partial p}{\partial z} + \left(\frac{\partial \tau_{xz}}{\partial x} + \frac{\partial \tau_{yz}}{\partial y} + \frac{\partial \tau_{zz}}{\partial z}\right) + \rho b_z \tag{B.36}$$

B.5 GENERAL EQUATION OF MOTION

Cylindrical coordinates:

$$\rho \left(\frac{\partial v_r}{\partial t} + v_r \frac{\partial v_r}{\partial r} + \frac{v_\theta}{r} \frac{\partial v_r}{\partial \theta} - \frac{v_\theta^2}{r} + v_z \frac{\partial v_r}{\partial z} \right)$$

$$= -\frac{\partial p}{\partial r} + \left(\frac{1}{r} \frac{\partial}{\partial r}(r\tau_{rr}) + \frac{1}{r} \frac{\partial \tau_{r\theta}}{\partial \theta} - \frac{\tau_{\theta\theta}}{r} + \frac{\partial \tau_{rz}}{\partial z} \right) + \rho b_r \quad (B.37)$$

$$\rho \left(\frac{\partial v_\theta}{\partial t} + v_r \frac{\partial v_\theta}{\partial r} + \frac{v_\theta}{r} \frac{\partial v_\theta}{\partial \theta} + \frac{v_r v_\theta}{r} + v_z \frac{\partial v_\theta}{\partial z} \right)$$

$$= -\frac{1}{r}\frac{\partial p}{\partial \theta} + \left(\frac{1}{r^2} \frac{\partial}{\partial r}(r^2 \tau_{r\theta}) + \frac{1}{r} \frac{\partial \tau_{\theta\theta}}{\partial \theta} + \frac{\partial \tau_{\theta z}}{\partial z} \right) + \rho b_\theta \quad (B.38)$$

$$\rho \left(\frac{\partial v_z}{\partial t} + v_r \frac{\partial v_z}{\partial r} + \frac{v_\theta}{r} \frac{\partial v_z}{\partial \theta} + v_z \frac{\partial v_z}{\partial z} \right)$$

$$= -\frac{\partial p}{\partial z} + \left(\frac{1}{r} \frac{\partial}{\partial r}(r\tau_{rz}) + \frac{1}{r} \frac{\partial \tau_{\theta z}}{\partial \theta} + \frac{\partial \tau_{zz}}{\partial z} \right) + \rho b_z \quad (B.39)$$

Spherical coordinates:

$$\rho \left(\frac{\partial v_r}{\partial t} + v_r \frac{\partial v_r}{\partial r} + \frac{v_\theta}{r} \frac{\partial v_r}{\partial \theta} + \frac{v_\phi}{r \sin \theta} \frac{\partial v_r}{\partial \phi} - \frac{v_\theta^2 + v_\phi^2}{r} \right)$$

$$= -\frac{\partial p}{\partial r} + \left(\frac{1}{r^2} \frac{\partial}{\partial r}(r^2 \tau_{rr}) + \frac{1}{r \sin \theta} \frac{\partial}{\partial \theta}(\tau_{r\theta} \sin \theta) \right.$$

$$\left. + \frac{1}{r \sin \theta} \frac{\partial \tau_{r\phi}}{\partial \phi} - \frac{\tau_{\theta\theta} + \tau_{\phi\phi}}{r} \right) + \rho b_r \quad (B.40)$$

$$\rho \left(\frac{\partial v_\theta}{\partial t} + v_r \frac{\partial v_\theta}{\partial r} + \frac{v_\theta}{r} \frac{\partial v_\theta}{\partial \theta} + \frac{v_\phi}{r \sin \theta} \frac{\partial v_\theta}{\partial \phi} + \frac{v_r v_\theta}{r} - \frac{v_\phi^2 \cot \theta}{r} \right)$$

$$= -\frac{1}{r}\frac{\partial p}{\partial \theta} + \left(\frac{1}{r^2} \frac{\partial}{\partial r}(r^2 \tau_{r\theta}) + \frac{1}{r \sin \theta} \frac{\partial}{\partial \theta}(\tau_{\theta\theta} \sin \theta) \right.$$

$$\left. + \frac{1}{r \sin \theta} \frac{\partial \tau_{\theta\phi}}{\partial \phi} + \frac{\tau_{r\theta}}{r} - \frac{\cot \theta}{r} \tau_{\phi\phi} \right) + \rho b_\theta \quad (B.41)$$

$$\rho \left(\frac{\partial v_\phi}{\partial t} + v_r \frac{\partial v_\phi}{\partial r} + \frac{v_\theta}{r} \frac{\partial v_\phi}{\partial \theta} + \frac{v_\phi}{r \sin \theta} \frac{\partial v_\phi}{\partial \phi} + \frac{v_\phi v_r}{r} + \frac{v_\theta v_\phi}{r} \cot \theta \right)$$

$$= -\frac{1}{r \sin \theta}\frac{\partial p}{\partial \phi} + \left(\frac{1}{r^2} \frac{\partial}{\partial r}(r^2 \tau_{r\phi}) + \frac{1}{r} \frac{\partial \tau_{\theta\phi}}{\partial \theta} \right.$$

$$\left. + \frac{1}{r \sin \theta} \frac{\partial \tau_{\phi\phi}}{\partial \phi} + \frac{\tau_{r\phi}}{r} + \frac{2 \cot \theta}{r} \tau_{\theta\phi} \right) + \rho b_\phi \quad (B.42)$$

B.6 NAVIER–STOKES EQUATION: CONSTANT ρ AND μ

Rectangular coordinates:

$$\rho\left(\frac{\partial v_x}{\partial t} + v_x\frac{\partial v_x}{\partial x} + v_y\frac{\partial v_x}{\partial y} + v_z\frac{\partial v_x}{\partial z}\right) = -\frac{\partial p}{\partial x} + \mu\left(\frac{\partial^2 v_x}{\partial x^2} + \frac{\partial^2 v_x}{\partial y^2} + \frac{\partial^2 v_x}{\partial z^2}\right) + \rho b_x$$
(B.43)

$$\rho\left(\frac{\partial v_y}{\partial t} + v_x\frac{\partial v_y}{\partial x} + v_y\frac{\partial v_y}{\partial y} + v_z\frac{\partial v_y}{\partial z}\right) = -\frac{\partial p}{\partial y} + \mu\left(\frac{\partial^2 v_y}{\partial x^2} + \frac{\partial^2 v_y}{\partial y^2} + \frac{\partial^2 v_y}{\partial z^2}\right) + \rho b_y$$
(B.44)

$$\rho\left(\frac{\partial v_z}{\partial t} + v_x\frac{\partial v_z}{\partial x} + v_y\frac{\partial v_z}{\partial y} + v_z\frac{\partial v_z}{\partial z}\right) = -\frac{\partial p}{\partial z} + \mu\left(\frac{\partial^2 v_z}{\partial x^2} + \frac{\partial^2 v_z}{\partial y^2} + \frac{\partial^2 v_z}{\partial z^2}\right) + \rho b_z$$
(B.45)

Cylindrical coordinates:

$$\rho\left(\frac{\partial v_r}{\partial t} + v_r\frac{\partial v_r}{\partial r} + \frac{v_\theta}{r}\frac{\partial v_r}{\partial \theta} - \frac{v_\theta^2}{r} + v_z\frac{\partial v_r}{\partial z}\right)$$
$$= -\frac{\partial p}{\partial r} + \mu\left(\frac{\partial}{\partial r}\left(\frac{1}{r}\frac{\partial}{\partial r}(rv_r)\right) + \frac{1}{r^2}\frac{\partial^2 v_r}{\partial \theta^2} - \frac{2}{r^2}\frac{\partial v_\theta}{\partial \theta} + \frac{\partial^2 v_r}{\partial z^2}\right) + \rho b_r$$
(B.46)

$$\rho\left(\frac{\partial v_\theta}{\partial t} + v_r\frac{\partial v_\theta}{\partial r} + \frac{v_\theta}{r}\frac{\partial v_\theta}{\partial \theta} + \frac{v_r v_\theta}{r} + v_z\frac{\partial v_\theta}{\partial z}\right)$$
$$= -\frac{1}{r}\frac{\partial p}{\partial \theta} + \mu\left(\frac{\partial}{\partial r}\left(\frac{1}{r}\frac{\partial}{\partial r}(rv_\theta)\right) + \frac{1}{r^2}\frac{\partial^2 v_\theta}{\partial \theta^2} + \frac{2}{r^2}\frac{\partial v_r}{\partial \theta} + \frac{\partial^2 v_\theta}{\partial z^2}\right) + \rho b_\theta$$
(B.47)

$$\rho\left(\frac{\partial v_z}{\partial t} + v_r\frac{\partial v_z}{\partial r} + \frac{v_\theta}{r}\frac{\partial v_z}{\partial \theta} + v_z\frac{\partial v_z}{\partial z}\right)$$
$$= -\frac{\partial p}{\partial z} + \mu\left(\frac{1}{r}\frac{\partial}{\partial r}\left(r\frac{\partial v_z}{\partial r}\right) + \frac{1}{r^2}\frac{\partial^2 v_z}{\partial \theta^2} + \frac{\partial^2 v_z}{\partial z^2}\right) + \rho b_z$$
(B.48)

Spherical coordinates:

$$\rho\left(\frac{\partial v_r}{\partial t} + v_r\frac{\partial v_r}{\partial r} + \frac{v_\theta}{r}\frac{\partial v_r}{\partial \theta} + \frac{v_\phi}{r\sin\theta}\frac{\partial v_r}{\partial \phi} - \frac{v_\theta^2 + v_\phi^2}{r}\right)$$
$$= -\frac{\partial p}{\partial r} + \mu\left(\frac{1}{r^2}\frac{\partial^2}{\partial r^2}(r^2 v_r) + \frac{1}{r^2\sin\theta}\frac{\partial}{\partial \theta}\left(\sin\theta\frac{\partial v_r}{\partial \theta}\right) + \frac{1}{r^2\sin^2\theta}\frac{\partial^2 v_r}{\partial \phi^2}\right) + \rho b_r$$
(B.49)

$$\rho\left(\frac{\partial v_\theta}{\partial t} + v_r\frac{\partial v_\theta}{\partial r} + \frac{v_\theta}{r}\frac{\partial v_\theta}{\partial \theta} + \frac{v_\phi}{r\sin\theta}\frac{\partial v_\theta}{\partial \phi} + \frac{v_r v_\theta}{r} - \frac{v_\phi^2 \cot\theta}{r}\right)$$

$$= -\frac{1}{r}\frac{\partial p}{\partial \theta} + \mu\left(\frac{1}{r^2}\frac{\partial}{\partial r}\left(r^2\frac{\partial v_\theta}{\partial r}\right) + \frac{1}{r^2}\frac{\partial}{\partial \theta}\left(\frac{1}{\sin\theta}\frac{\partial}{\partial \theta}(v_\theta \sin\theta)\right)\right.$$

$$\left.+ \frac{1}{r^2 \sin^2\theta}\frac{\partial^2 v_\theta}{\partial \phi^2} + \frac{2}{r^2}\frac{\partial v_r}{\partial \theta} - \frac{2\cos\theta}{r^2 \sin^2\theta}\frac{\partial v_\phi}{\partial \phi}\right) + \rho b_\theta \qquad (B.50)$$

$$\rho\left(\frac{\partial v_\phi}{\partial t} + v_r\frac{\partial v_\phi}{\partial r} + \frac{v_\theta}{r}\frac{\partial v_\phi}{\partial \theta} + \frac{v_\phi}{r\sin\theta}\frac{\partial v_\phi}{\partial \phi} + \frac{v_\phi v_r}{r} + \frac{v_\theta v_\phi}{r}\cot\theta\right)$$

$$= -\frac{1}{r\sin\theta}\frac{\partial p}{\partial \phi} + \mu\left(\frac{1}{r^2}\frac{\partial}{\partial r}\left(r^2\frac{\partial v_\phi}{\partial r}\right) + \frac{1}{r^2}\frac{\partial}{\partial \theta}\left(\frac{1}{\sin\theta}\frac{\partial}{\partial \theta}(v_\phi \sin\theta)\right)\right.$$

$$\left.+ \frac{1}{r^2 \sin^2\theta}\frac{\partial^2 v_\phi}{\partial \phi^2} + \frac{2}{r^2 \sin\theta}\frac{\partial v_r}{\partial \phi} + \frac{2\cos\theta}{r^2 \sin^2\theta}\frac{\partial v_\theta}{\partial \phi}\right) + \rho b_\phi \qquad (B.51)$$

B.7 HEAT FLUX VECTOR: ISOTROPIC MATERIAL

Rectangular coordinates:

$$q_x = -k\frac{\partial T}{\partial x} \qquad (B.52)$$

$$q_y = -k\frac{\partial T}{\partial y} \qquad (B.53)$$

$$q_z = -k\frac{\partial T}{\partial z} \qquad (B.54)$$

Cylindrical coordinates:

$$q_r = -k\frac{\partial T}{\partial r} \qquad (B.55)$$

$$q_\theta = -k\frac{1}{r}\frac{\partial T}{\partial \theta} \qquad (B.56)$$

$$q_z = -k\frac{\partial T}{\partial z} \qquad (B.57)$$

Spherical coordinates:

$$q_r = -k\frac{\partial T}{\partial r} \qquad (B.58)$$

$$q_\theta = -k\frac{1}{r}\frac{\partial T}{\partial \theta} \qquad (B.59)$$

$$q_\phi = -k\frac{1}{r\sin\theta}\frac{\partial T}{\partial \phi} \qquad (B.60)$$

B.8 ENERGY BALANCE: GENERAL FORM

Rectangular coordinates:

$$\rho c_p \left(\frac{\partial T}{\partial t} + v_x \frac{\partial T}{\partial x} + v_y \frac{\partial T}{\partial y} + v_z \frac{\partial T}{\partial z} \right) + \rho L_p \frac{Dp}{Dt}$$

$$= -\left[\frac{\partial q_x}{\partial x} + \frac{\partial q_y}{\partial y} + \frac{\partial q_z}{\partial z} \right] + \left\{ \tau_{xx} \frac{\partial v_x}{\partial x} + \tau_{yy} \frac{\partial v_y}{\partial y} + \tau_{zz} \frac{\partial v_z}{\partial z} \right\}$$

$$+ \left\{ \tau_{xy} \left(\frac{\partial v_x}{\partial y} + \frac{\partial v_y}{\partial x} \right) + \tau_{yz} \left(\frac{\partial v_y}{\partial z} + \frac{\partial v_z}{\partial y} \right) \right.$$

$$\left. + \tau_{zx} \left(\frac{\partial v_z}{\partial x} + \frac{\partial v_x}{\partial z} \right) \right\} + \rho \dot{R} \quad \text{(B.61)}$$

Cylindrical coordinates:

$$\rho c_p \left(\frac{\partial T}{\partial t} + v_r \frac{\partial T}{\partial r} + \frac{v_\theta}{r} \frac{\partial T}{\partial \theta} + v_z \frac{\partial T}{\partial z} \right) + \rho L_p \frac{Dp}{Dt}$$

$$= -\left[\frac{1}{r} \frac{\partial}{\partial r}(r q_r) + \frac{1}{r} \frac{\partial q_\theta}{\partial \theta} + \frac{\partial q_z}{\partial z} \right] + \left\{ \tau_{rr} \frac{\partial v_r}{\partial r} + \tau_{\theta\theta} \frac{1}{r} \left(\frac{\partial v_\theta}{\partial \theta} + v_r \right) + \tau_{zz} \frac{\partial v_z}{\partial z} \right\}$$

$$+ \left\{ \tau_{r\theta} \left[r \frac{\partial}{\partial r} \left(\frac{v_\theta}{r} \right) + \frac{1}{r} \frac{\partial v_r}{\partial \theta} \right] + \tau_{\theta z} \left(\frac{1}{r} \frac{\partial v_z}{\partial \theta} + \frac{\partial v_\theta}{\partial z} \right) \right.$$

$$\left. + \tau_{zr} \left(\frac{\partial v_z}{\partial r} + \frac{\partial v_r}{\partial z} \right) \right\} + \rho \dot{R} \quad \text{(B.62)}$$

Spherical coordinates:

$$\rho c_p \left(\frac{\partial T}{\partial t} + v_r \frac{\partial T}{\partial r} + \frac{v_\theta}{r} \frac{\partial T}{\partial \theta} + \frac{v_\phi}{r \sin \theta} \frac{\partial T}{\partial \phi} \right) + \rho L_p \frac{Dp}{Dt}$$

$$= -\left[\frac{1}{r^2} \frac{\partial}{\partial r}(r^2 q_r) + \frac{1}{r \sin \theta} \frac{\partial}{\partial \theta}(q_\theta \sin \theta) + \frac{1}{r \sin \theta} \frac{\partial q_\phi}{\partial \phi} \right]$$

$$+ \left\{ \tau_{rr} \frac{\partial v_r}{\partial r} + \tau_{\theta\theta} \left(\frac{1}{r} \frac{\partial v_\theta}{\partial \theta} + \frac{v_r}{r} \right) + \tau_{\phi\phi} \left(\frac{1}{r \sin \theta} \frac{\partial v_\phi}{\partial \phi} + \frac{v_r}{r} + \frac{v_\theta \cot \theta}{r} \right) \right\}$$

$$+ \left\{ \tau_{r\theta} \left(\frac{\partial v_\theta}{\partial r} + \frac{1}{r} \frac{\partial v_r}{\partial \theta} - \frac{v_\theta}{r} \right) + \tau_{\theta\phi} \left(\frac{1}{r} \frac{\partial v_\phi}{\partial \theta} + \frac{1}{r \sin \theta} \frac{\partial v_\theta}{\partial \phi} - \frac{\cot \theta}{r} v_\phi \right) \right.$$

$$\left. + \tau_{\phi r} \left(\frac{\partial v_\phi}{\partial r} + \frac{1}{r \sin \theta} \frac{\partial v_r}{\partial \phi} - \frac{v_\phi}{r} \right) \right\} + \rho \dot{R} \quad \text{(B.63)}$$

B.9 ENERGY BALANCE: CONSTANT ρ, K, AND μ

Rectangular coordinates:

$$\rho c_p \left(\frac{\partial T}{\partial t} + v_x \frac{\partial T}{\partial x} + v_y \frac{\partial T}{\partial y} + v_z \frac{\partial T}{\partial z} \right)$$

$$= k \left[\frac{\partial^2 T}{\partial x^2} + \frac{\partial^2 T}{\partial y^2} + \frac{\partial^2 T}{\partial z^2} \right] + 2\mu \left\{ \left(\frac{\partial v_x}{\partial x} \right)^2 + \left(\frac{\partial v_y}{\partial y} \right)^2 + \left(\frac{\partial v_z}{\partial z} \right)^2 \right\}$$

$$+ \mu \left\{ \left(\frac{\partial v_x}{\partial y} + \frac{\partial v_y}{\partial x} \right)^2 + \left(\frac{\partial v_y}{\partial z} + \frac{\partial v_z}{\partial y} \right)^2 \right.$$

$$\left. + \left(\frac{\partial v_z}{\partial x} + \frac{\partial v_x}{\partial z} \right)^2 \right\} + \rho \dot{R} \qquad (B.64)$$

Cylindrical coordinates:

$$\rho c_p \left(\frac{\partial T}{\partial t} + v_r \frac{\partial T}{\partial r} + \frac{v_\theta}{r} \frac{\partial T}{\partial \theta} + v_z \frac{\partial T}{\partial z} \right) = k \left[\frac{1}{r} \frac{\partial}{\partial r} \left(r \frac{\partial T}{\partial r} \right) + \frac{1}{r^2} \frac{\partial^2 T}{\partial \theta^2} + \frac{\partial^2 T}{\partial z^2} \right]$$

$$+ 2\mu \left\{ \left(\frac{\partial v_r}{\partial r} \right)^2 + \left[\frac{1}{r} \left(\frac{\partial v_\theta}{\partial \theta} + v_r \right) \right]^2 + \left(\frac{\partial v_z}{\partial z} \right)^2 \right\}$$

$$+ \mu \left\{ \left[\frac{1}{r} \frac{\partial v_r}{\partial \theta} + r \frac{\partial}{\partial r} \left(\frac{v_\theta}{r} \right) \right]^2 + \left(\frac{\partial v_\theta}{\partial z} + \frac{1}{r} \frac{\partial v_z}{\partial \theta} \right)^2 \right.$$

$$\left. + \left(\frac{\partial v_z}{\partial r} + \frac{\partial v_r}{\partial z} \right)^2 \right\} + \rho \dot{R} \qquad (B.65)$$

Spherical coordinates:

$$\rho c_p \left(\frac{\partial T}{\partial t} + v_r \frac{\partial T}{\partial r} + \frac{v_\theta}{r} \frac{\partial T}{\partial \theta} + \frac{v_\phi}{r \sin \theta} \frac{\partial T}{\partial \phi} \right)$$

$$= k \left[\frac{1}{r^2} \frac{\partial}{\partial r} \left(r^2 \frac{\partial T}{\partial r} \right) + \frac{1}{r^2 \sin \theta} \frac{\partial}{\partial \theta} \left(\sin \theta \frac{\partial T}{\partial \theta} \right) + \frac{1}{r^2 \sin^2 \theta} \frac{\partial^2 T}{\partial \phi^2} \right]$$

$$+ 2\mu \left\{ \left(\frac{\partial v_r}{\partial r} \right)^2 + \left(\frac{1}{r} \frac{\partial v_\theta}{\partial \theta} + \frac{v_r}{r} \right)^2 + \left(\frac{1}{r \sin \theta} \frac{\partial v_\phi}{\partial \phi} + \frac{v_r}{r} + \frac{v_\theta \cot \theta}{r} \right)^2 \right\}$$

$$+ \mu \left\{ \left[r \frac{\partial}{\partial r} \left(\frac{v_\theta}{r} \right) + \frac{1}{r} \frac{\partial v_r}{\partial \theta} \right]^2 + \left[\frac{1}{r \sin \theta} \frac{\partial v_r}{\partial \phi} + r \frac{\partial}{\partial r} \left(\frac{v_\phi}{r} \right) \right]^2 \right.$$

$$\left. + \left[\frac{\sin \theta}{r} \frac{\partial}{\partial \theta} \left(\frac{v_\phi}{\sin \theta} \right) + \frac{1}{r \sin \theta} \frac{\partial v_\theta}{\partial \phi} \right]^2 \right\} + \rho \dot{R} \qquad (B.66)$$

BIBLIOGRAPHY

R. B. Bird, R. C. Armstrong, and O. Hassager. *Dynamics of Polymeric Liquids. Volume 1: Fluid Mechanics.* Wiley, New York, 2nd edition, 1987.

R. B. Bird, W. E. Stewart, and E. N. Lightfoot. *Transport Phenomena.* Wiley, New York, 1960.

C. L. Tucker III, editor. *Fundamentals of Computer Modeling for Polymer Processing.* Carl Hanser Verlag, Munich, 1989.

Index

absolute stability, 319
activation energy
   diffusion, 284
   viscosity, 213–4
advection of heat, 70, 239–50
aluminum, properties, 119
amplification factor, 17
angular momentum, conservation of, 34
Arrhenius model, 213–4, 284
asymptotic expansions, matched, 73
atomic fraction. *See* concentration

balance equations. *See* continuity *and* equation of motion *and* energy balance Bernoulli's equation, 44–45, 162–3, 166
Bessel function, 65
biaxial elongation, 230
Bingham fluid, 210–1
   tube flow, 211–3
Biot number, 77
   in advection, 244
   in conduction, 99
body forces, 34
boundary layers, 69–76
   buoyancy-driven flow, 272–3
   concentration, 304
   momentum, 169–70, 171
   numerical solution of, 73–76
   scaling, 72–73
   thermal, 243
Boussinesq approximation, 261, 264–5
Boussinesq number, 77, 264
Brinkman number, 77, 256, 257
buoyancy-driven flow, 259–74
   energy balance, 264
   in a square cavity, 271–4
   in a thin cavity, 266–71, 280–1
   inertial and body force balance, 262
   mass balance, 263–4
   momentum balance, 261–3
   viscous and body force balance, 280–1

capacity, traffic flow, 9
capillary number, 77, 155–6
capillary rheometer. *See* rheometer
carburization, 285–287
Carreau model, 209, 213.
Cartesian tensor notation. *See* indicial notation
case hardening. *See* carburization
Cauchy stress, 34
Cauchy stress principle, 34
Cayley–Hamilton theorem, 335
center of curvature, 339–40
CFL (Courant–Friedrichs–Lewy) condition, 18
chain rule, 4, 27, 49, 64
characteristic values, 61–2
choke, 164
Chvorinov's rule, 113, 128
co-injection molding, 162
complementary error function. *See* error function
compression molding, 55, 236–7
concentration
   atomic fraction, 283
   mass, 282
   mass fraction, 282
concentric cylinders
   axial drag flow, 145
      power law fluid, 201
   axial pressure flow, 145
   tangential drag flow, 145
      power law fluid, 201
conduction. *See* Fourier's law *and* heat conduction
configuration, 28
constitutive equation
   heat conduction, 47–8
   Newtonian fluid, 37
   traffic flow, 8–9
continuity equation
   conservative form, 30
   constant density, 31
   differential form, 30
   integral form, 31
   integrated, 226
   tables, 52, 348

continuity equation (*Cont.*)
   traffic flow, 8
continuous casting, 244–5
contraction, flow in, 66, 80
control volume, 7–8, 31
convected derivative, 216
coordinate curves, 337
coordinates
   Cartesian, 327
   cylindrical, 337
   material, 24, 53
   orthogonal curvilinear, 337
   spatial, 24, 53
   spherical, 337–8
coordinate transformation, 203–5, 329–30
copper, properties, 120
Couette flow, 140, 145
   See also concentric cylinders, tangential drag flow
Courant number, 17
creeping flow. *See* Stokes flow
Cross model, 209, 213
curvature
   of a curve, 339–40
      plane curve, 341
      vector, 340
   of a surface
      mean 108, 152, 342
      normal, 341
      principal radii, 342
      surface of revolution, 343
      total, 342
cylinder, heat conduction, 63
Czochralski crystal growth, 303

D/Dt. *See* material derivative
Deborah number, 214
deformation gradient tensor, 29, 58
del. *See* gradient operator
delta function, 248
   See also Kronecker delta
dendrite, 192, 295, 299
   arm spacing, 299
density
   constant, 31
   reference value, 260
   traffic flow, 7
developing flow, 168–70
die swell, 215
diffusion
   analogy with heat conduction, 284–285
   couples, 293
   encapsulated layer, 287–289
   equation, 284
   length and time scales, 285
   numerical, 75
   similarity solution for, 286
   with surface reaction, 289–293
   See also interdiffusion

diffusivity, mass
   effect of temperature, 284
   intrinsic, 283, 284–295
   temperature dependence, 284
dilatant. *See* shear thickening
dimensionless groups, table, 77
dimensionless variables. *See* scaled variables
Dirac delta function. *See* delta function
directional solidification. *See* solidification
Dirichlet boundary condition, 89
disks
   radial pressure flow, 146
      power law fluid, 201–2
   torsional drag flow, 146
displacement functions, 25, 53
divergence, vector and tensor fields, 336
divergence theorem, 339
DNS (direct numerical simulation), turbulence, 173–4
dummy index, 329
dyadic product, 332

eigenvalues, 334
eigenvectors, 334
elastic recovery, 215
elongational viscosity. *See* viscosity, elongational
elongational flows, 230
end effects, heat conduction, 65–66
energy
   balance equation, 45–50
      names for terms, 88
      tables, 52, 354–5
   internal, 46
   kinetic, 45
enthalpy, 49
enthalpy–temperature curve, 298
entrance length, 168, 170
entrance loss, 147
equation of motion
   general form, 35–6
   integral form, 36
   Newtonian fluid, 37
   tables, 350–1
equation of state, thermodynamic, 49
equilibrium phase diagram. *See* phase diagram
error function, 96, 343
   approximations, 344
   complementary, 96, 343
   properties, 343
   table, 344
Euler angles, 203–4
Euler–Cauchy stress principle, 34
Euler equation, 43, 44
Eulerian representation, 26
eutectic, 107, 296, 302–3
exponential model, temperature-dependent viscosity, 214
extension thickening, 214
extra stress. *See* stress, extra

extrusion, 140, 199
    die flow, 147–9
    pumping model, 140–3

fiber spinning, 32–33, 54, 183
Fick's law, 282
finite difference derivatives, 12, 73
first law of thermodynamics, 45
flow conductance, 218
    power law fluid, 226–7
flow instabilities
    inertial, 171–3
    elastic, 215
flow rate
    fluid, 139, 199
    traffic, 7, 9
fountain flow, 160–1, 220
Fourier number, 77
Fourier series, 103
Fourier's law
    isotropic materials, 48, 353
    anisotropic materials, 122
Fourier stability analysis, 16
frame indifference. See material objectivity
free boundary problems, 107, 150
    fluid boundary condition, 152, 153
Froude number, 77
frozen temperature approximation, 311–312
FTCS scheme, 12
fully developed flow, 138, 168

gas constant, 213, 284
gate. See ingate
Gaussian curvature. See curvature, total
Gaussian error function. See error function
generalized Hele-Shaw approximation. See
    Hele–Shaw approximation
generalized Newtonian fluid, 191, 205–6, 209–10
Gibbs, J. W., 48
Gibbs notation, 327
Gibbs–Thomson equation, 309–310
    linearized, 316
Giesekus model, 216
Gordon-Schowalter derivative. See convected
    derivative
gradient
    cylindrical and spherical coordinates, 338
    pressure, 135, 227
    scalar field, 335
    surface, 152
    temperature, 48
    vector field, 336
    velocity, 37–8, 336
gradient operator, 335
    algebraic properties, 336
Graetz number, 77, 241
Grashof number, 77, 263
Gulliver. See Scheil

hardenability, 125, 277–8
heat conduction, 46
heat conduction
    equation, 88
    boundary conditions, 89
heat diffusivity, 119, 124
heat flux vector, 46, 48
    tables, 353
heat generation, internal, 46
heat of fusion. See latent heat of fusion
heat source, moving, 247–50
heat transfer coefficient, typical values, 89
Hele–Show approximation
    compression molding, 236–7
    governing equations, 218–9
    power law fluid, 226–7
    radial flow, 236
    nonisothermal, 227, 278–9
history-dependent fluid, 192
homogereous deformation, 230
homogenization, 323

identity tensor. See unit tensor
incompressible fluid, 31, 260
indicial notation, 328–9
ingate, 164
injection molding. See mold filling
insulation, critical radius, 121–2
interdiffusion, 293–295
interfacial tension. See surface tension
invariants, 203
    surface curvature, 342
    tensor, 205
    vector, 204
inverse lever rule. See lever rule
inviscid flow, 43, 44, 162
irrotational flow, 43–4

Jacobian, of deformation, 29, 58–59
Jominy test, 124–5, 277–8

$\bar{k} - \varepsilon$ model, 176
Kirkendall effect, 293
Kolmogorov microscale, 174
Kronecker delta, 330

ladle, flow from, 162–4
Lagrange's theorem, 44
Lagrangian representation, 26
Laplace's equation, 44, 90, 219
Laplacian operator, 336
latent heat
    of fusion, 106
        enhanced for superheat, 118
        incorporated in enthalpy, 298
    of pressure change, 49
    of volume change, 49
Lax's Method, 12, 23
LES (large eddy simulation), turbulence, 177

lever rule, 297
Lewis number, 77
linearization, 308
liquidus temperature, 296, 310
   local, 305–6
local thermodynamic equilibrium, 301, 304
   curved interface, 309
lubrication approximation, 138

Marangoni flow, 152, 187
mass balance equation. *See* continuity equation
mass conservation, 30
mass flux,
   advective, 30
   diffusive, 283–4
mass fraction. *See* concentration
material derivative, 27–28
material description, 24–6
material objectivity, 202
Maxwell model, 216
melt fracture, 182, 215
melting temperature, 106
   curved interface, 309–10
melting, temperature distribution, 106, 276
mixed boundary condition, 89
mixing length, 176
modeling, steps for, 1–2
modulus of geometrical section. *See* surface area to volume ratio
mold filling, 138, 157–160, 246, 280
   flow front, 160–1
   metal casting, 164–6
   numerical simulation, 220, 227–8
   surface tension effects, 154–7
momentum balance equation. *See* equation of motion
moving boundary problems, 107, 150, 220
   fluid boundary condition, 152
multimode models, viscoelastic fluid, 216

nabla. *See* gradient operator
Nahme–Griffith number, 77, 258–9, 280
natural convection. *See* buoyancy-driven flow
Navier-Stokes equation, 39, 56
   filtered, 177
   Reynolds-averaged, 174–6
   tables, 352–3
Neumann boundary condition, 89
Newtonian fluid, 37–9
Newton's law, 3, 35
normal curve, 341
normal mode solutions, 317
normal plane, 341
normal, principal, 340
normal stress, in shear flow, 214–5
numerical diffusion. *See* diffusion, numerical
Nusselt number, 77

objectivity. *See* material objectivity
orthogonal matrix. *See* tensor, orthogonal
oscillations, numerical, 74, 76
osculating plane, 339–40
oxidation, 289–293

paint sag, 237
parabolic scaling constant, 293
parallel plates
   combined pressure and drag flow, 134–6, 139
   drag flow, 39–41
      combined viscous and thermal effects, 257–9
      heated plates, 50–52
      multidirectional, 206–8
      power law fluid, 193–6
      startup, 84
      temperature-dependent viscosity, 250–4
      temperature, with dissipation, 254–7
   pressure flow, 144
      combined viscous and thermal effects, 259, 280
      power law fluid, 200
      temperature-dependent viscosity, 254
parameters of a problem, 61
partition coefficient
   equilibrium, 297, 324
   nonequilibrium, 324
pattern selection, 306
Pearson number, 77, 253–4
Péclet number, 70, 77, 241
   buoyancy-driven flow, 264–5
   continuous casting, 245
   grid, 74
pendulum
   period, 5
   richer models, 20–21
   simple model, 2–5
   stability, 307–309
permutation symbol, 332
phase change, 106
phase diagram, 296–298
   aluminum–copper, 298
   aluminum–silicon, 323
   curvature correction, 310
   iron-carbon, 322
planar elongation, 230
planar front growth, concentration, 304
Poiseuille flow. *See* tube, pressure flow
potential flow, 43–4
power law fluid, 193, 206, 209
power law index, 193
Prandtl mixing length. *See* mixing length
Prandtl number, 77, 245–7, 265
precipitation hardening, 125
pressure
   definition, 34–35
   hydrostatic distribution, 42–3
   modified, 41–2, 261

product solutions, 126
pseudoplastic. *See* shear thinning
pseudotime, 241–2
PTT (Phan-Thien and Tanner) model, 216
purely viscous fluid, 190, 192

radiation boundary condition, 89
radius of curvature. *See* curvature
RANS. *See* Navier-Stokes equation, Reynolds-averaged
rate-of-deformation tensor, 38
    tables, 349
Rayleigh number, 77, 265, 170, 171, 173
reference configuration, 29
relaxation time, viscoelastic, 192, 214
Reynolds number, 68, 77, 134–5
    buoyancy-driven flow, 262, 265
Reynolds stress, 175–6
Reynolds transport theorem, 32, 35, 45, 283
rheology, 190
rheometer
    capillary, 234–6
    Couette, 41, 140, 201
rod climbing, 215
Rosenthal solution, 248–50
runners, 164
    balancing, 185–6, 234

sand, properties, 119
scalar, 327
scaled equation, 62
scaled variable, 4, 60–1
scaling
    for numerical solution, 78–9
    isotropic, 78–79
    parameter variation, 80–1
    rules, 62
Scheil
    equation, 302
    solidification model, 301–303
Schmidt number, 77
screw extruder. *See* extrusion
secondary flows, 171–3
segregation coefficient. *See* partition coefficient
semi-infinite body
    concentration solution, 285–6
    temperature solution, 94–8
    velocity solution, 167–8
    with surface heat transfer coefficient, 98
separated flow, 170–1
separation of variables, 105
sharkskin, 215
shear rate, 139, 191
    See also strain rate, scalar
shear thickening, 191, 193, 199
shear thinning, 191, 193, 199, 214
Sherwood number, 77
Silly Putty, 192

similarity variable, 94–5
simple shear flow, 139, 191, 206, 230
skew-symmetric tensor. *See* tensor, anti-symmetric
slab
    moving, temperature distribution, 240–4
    steady heat conduction, 90–2
        composite wall, 92–3
    transient heat conduction
        arbitrary Biot number, 101–3
        computer program, 130–1
        different surface temperatures, 105–6
        large Biot number, 100–1, 103–5
        small Biot number, 99–100
        temperature graphs, 104
solidification, 295, 303–305
    directional
        concentration boundary layer, 304
        marginal stability condition, 306
        phenomenological stability, 305–306
        stability analysis, 309–316
    interface temperature, 117
    microstructure, 295–319
    nonequilibrium, 298–303
    temperature solution
        general case, 115–8
    interface control, 128
    mold control, 111–3
    solid control, 113–5
solidus temperature, 296, 310
solute, 282
solute balance, at liquid-solid interface, 302
solvent, 282
spatial description, 24–6
specific heat
    at constant pressure, 49
    at constant volume, 49
specific volume, 32
spin coating, 84, 185
spinneret, 183
sprue, 164–6
squeeze flow, lubricated, 55
stability analysis, 173
    amplification factor, 17–18
    basic states, 307
    CFL condition, 18
    directional solidification, 309–316
    linear, 307–309
stability
    definition, 308
    marginal, 318
    neutral, 308
state variables, thermodynamic, 48
steel, properties, 120
Stefan condition, 108–9, 110, 114, 116, 313
Stefan number, 77, 111
Stokes flow, 161
Stokes hypothesis, 39
strain rate, scalar, 50, 193, 195, 205–6, 226–7

stress
  Cauchy, 34
  deviatoric, 35
  extra, 35
    scalar magnitude, 210
  hydrostatic, 34
  subgrid, 177
  symmetry of, 34
  total, 34
summation convention, 28, 329
supercooling, constitutional, 306, 318
superheat, 109, 118
surface area to volume ratio, 100–1, 113
surface curvature. *See* curvature
surface, developable, 342
surface energy. *See* surface tension
surface forces, 33–34
surface gradient. *See* gradient
surface tension, 108, 151
  conversion factor, 156
  scaling, 154–7
surface traction, 34

Tammann scaling constant, 292
tangent vector, 339–40
Taylor microscale, 174
Taylor vortex, 172
tensor, 328
  addition, 331
  anti-symmetric, 331
  characteristic equation, 334
  cylindrical coordinates, 337
  deviatoric, 333
  dot product, 332
  double dot product, 332
  invariants, 333, 334
  inverse, 335
  isotropic, 330
  orthogonal, 204, 330, 335
  principal axes, 334
  principal values, 334
  scalar magnitude, 333
  scalar multiplication, 331
  spherical coordinates, 338
  square of, 332
  symmetric, 331
  trace, 333
  transformation. *See* transformation matrix
  transpose, 330–1
thermal conductivity
  anisotropic materials, 122
  isotropic material, 48
thermal diffusivity, 65
thermal expansion coefficient, 260
thermal resistance
  cylinder, 92
  heat transfer coefficient, 92
  wall, 91

thermocouple, 126–7
thixotropic fluid, 192
traffic flow, 6–11
  numerical solution, 11–19
  program, 13
  stability, 15–18
transformation diagram, 125–6
transformation matrix, 204, 329–30, 346
tube, pressure flow
  Bingham fluid, 211–3
  Newtonian fluid, 144
  power law fluid, 196–200
turbulence, 173–7

uniaxial elongation, 230
unit dyad, 328
unit tensor, 330
unit vectors
  Cartesian coordinates, 327
  cylindrical coordinates, 337
  spherical coordinates, 338
unstable, definition, 308
upwinding, 75

vector, 327
  cross product, 332
  cylindrical coordinates, 337
  dot product, 331
  scalar multiplication, 331
  spherical coordinates, 338
  triple product, 332
velocity, 27
  drift, 294
  fluctuation, 175
  gapwise average, 219, 224–5
  spatial average, 177
  time average, 174
velocity gradient tensor, 37–8, 336
velocity potential, 43
viscoelastic fluid, 192, 214–7
viscometer. *See* rheometer
viscoplastic material, 192
viscosity
  elongational, 230
  kinematic, 168
  Newtonian
    dilatational, 38
    shear, 37
  non-Newtonian, 191
  pressure dependence, 238
  temperature dependence, 213–4
    velocity distributions, 250–4, 257–9
  turbulent, 176, 177
  zero shear, 209
viscous dissipation, 88
  Newtonian fluid, 50, 57, 254–8
  power law fluid, 279
von Mises yield criterion, 211

INDEX

von Neumann stability analysis, 16
vorticity
   tensor, 38
      tables, 350
   vector, 43

wall. *See* slab
wave equation, traffic flow, 10–11, 22
Weber number, 187
welding
   hardness after, 277
   temperature distribution, 247–50
wire coating, 145, 182
WLF (Williams, Landel and Ferry) equation, 209, 213–4
work, rate of, 47

yield stress, 191, 210
   Papanastasiou model, 210
   velocity profiles, 211–3

zero shear viscosity, 209